CISM COURSES AND LECTURES

Series Editors:

The Rectors of CISM
Sandor Kaliszky - Budapest
Mahir Sayir - Zurich
Wilhelm Schneider - Wien

The Secretary General of CISM
Giovanni Bianchi - Milan

Executive Editor
Carlo Tasso - Udine

The series presents lecture notes, monographs, edited works and proceedings in the field of Mechanics, Engineering, Computer Science and Applied Mathematics.
Purpose of the series is to make known in the international scientific and technical community results obtained in some of the activities organized by CISM, the International Centre for Mechanical Sciences.

INTERNATIONAL CENTRE FOR MECHANICAL SCIENCES

COURSES AND LECTURES - No. 420

SEISMIC RESISTANT STEEL STRUCTURES

EDITED BY

FEDERICO M. MAZZOLANI
UNIVERSITY "FEDERICO II" OF NAPLES

VICTOR GIONCU
POLITEHNICA UNIVERSITY OF TIMISOARA

 Springer-Verlag Wien GmbH

This volume contains 300 illustrations

In order to make this volume available as economically and as
rapidly as possible the authors' typescripts have been
reproduced in their original forms. This method unfortunately
has its typographical limitations but it is hoped that they in no
way distract the reader.

ISBN 978-3-211-83329-2   ISBN 978-3-7091-2480-2 (eBook)
DOI 10.1007/978-3-7091-2480-2

PREFACE

In the design practice it is generally accepted that steel is an excellent material for seismic-resistant structures, due to its performances in terms of strength and ductility, as it is capable of withstanding substantial inelastic deformations. This is generally true, because in percentage the number of failures of steel structures has been always very small as compared to other constructional materials. But in the last decades, specialists have recognized that the so-called good ductility of steel structures under particular conditions may be only a dogma, which is denied by the reality. In fact, the recent earthquakes of Mexico City (1985), Loma Prieta (1989), Northridge (1994) and Kobe (1995) have seriously compromised this idyllic image as a perfect material for seismic areas. The performance of steel joints and members in some cases was very bad and the same type of damage was produced in different events, clearly showing that there are some lacks in the current design practice. Thus, it seems to be the right moment to analyze the progress recently achieved in conception, design and construction, considering the lessons learned from the last dramatic events.

The main conclusions raised from the up-to-dated research activity in this field are looking for the utilization of the multiple level design concept, the differentiation of earthquake types, the analysis of all factors influencing the steel structure behaviour during a strong ground motion, etc.

The international scientific community is also aware of the urgent need to investigate new topics (such as the influence of the strain rate on the cyclic behaviour of beam-to-column joints) and to improve the current seismic provisions consequently. In addition, the whole background of the modern seismic codes deserves to be completely reviewed in order to grasp the design rules which failed during the last earthquakes. This revision is aimed at the up-dating of seismic codes and in particular in Europe, at the improvement of Eurocode 8, whose application will be widespread in the next years during the so-called conversion phase.

The challenge for the future is the transfer of these achievements to practice, in order to fill the gap existing between the accumulated knowledge and design codes.

*In this perspective the European research project dealing with the "**Reliability of moment resistant connections of steel building frames in seismic areas**" (RECOS) has been recently worked out under the sponsorship of the European Community within the INCO-Copernicus joint research projects of the 4th Framework Program. The aim of this project has been to examine the behaviour of joints which -as it is well known- play a paramount influence on the seismic behaviour of steel frames.*

This project has been developed with the partnership of eight European Countries (Belgium, Bulgaria, France, Greece, Italy, Portugal, Romania, Slovenia), on the base of the knowledge and the experience of many outstanding specialists, some of them being involved in the preparation of this Volume.

Coming from the above mentioned scenario, the latest important results in the field of seismic resistant steel structures have been illustrated in the advanced professional training course held at the International Centre for Mechanical Sciences (CISM) in Udine (Italy) on October 12-22, 1999. It has been coordinated by the undersigned and organized in the way to progressively cover all the concerned aspects from phenomena, design criteria, experiment, application to more advanced steel structure typologies. The lectures have been addressed to structural engineers involved in seismic design as well as academic people interested in research in this field and also post-graduate students.

The present Volume is the output of this course, whose contents are organized through 7 Chapters covering the material illustrated during 34 lectures.

Chapter 1 (F. Mazzolani) is the introduction, giving general information on the steel structures in seismic zones, with reference to the main typologies and lessons to be learned from the recent earthquakes.

Chapter 2 (V. Gioncu) illustrates and comments the design criteria according to the up-to-dated philosophy, giving the relationship between local and global ductilities and considering also the different influence of the earthquake types (near-field and far-field sources).

Chapter 3 (H. Akiyama) is devoted to the methodology based on energy criteria, with reference to the energy input in different systems (from SDOF to MDOF), with the evaluation of the corresponding damage.

Chapter 4 (F. Mazzolani) deals with the design of moment resisting frames, starting from the code provisions, with the evaluation of q-factor and the ductile design methodologies, and passing through the analysis of the influence factors, like configuration, semi-rigidity, structural regularity, P-(effect, randomness of material and stress-skin effect due to claddings.

Chapter 5 (I. Vayas) goes into details on the design of braced frames both concentric and eccentric, not only of classical type but also considering special damping devices.

Chapter 6 (A. Plumier) is devoted to the important typology of composite structures, which have been analyzed in all details by means of the most advanced methodologies.

Chapter 7 (L. Calado) analyses the behaviour of connections both from the experimental point of view and by means of numerical models, giving a low cycle fatigue assessment.

It do not seem the case to underline the high quality level of these Chapters, because the Authors are very well known as outstanding experts in the specific fields.

The coordinators of the CISM Course, who are playing here in this Volume the role of both Authors and Editors are very happy to declare that they are extremely satisfied of the development of the Course which was really a success and they express their gratitude to the Colleagues who accepted to transfer their lectures into the Chapters of this Volume, with so high quality of experience and evidence.

Federico M. Mazzolani
Victor Gioncu

CONTENTS

Page

Preface

CHAPTER 1

Steel Structures in Seismic Zones
by F.M. Mazzolani 1

CHAPTER 2

Design Criteria for Seismic Resistant Steel Structures
by V. Gioncu 19

CHAPTER 3

Method Based on Energy Criteria
by H. Akiyama 101

CHAPTER 4

Design of Moment Resisting Frames
by F.M. Mazzolani 159

CHAPTER 5

Design of Braced Frames
by I. Vayas 241

CHAPTER 6

Seismic Resistant Composite Structures
by A. Plumier 289

CHAPTER 7

Design of Connections
by L. Calado 349

CHAPTER 1

STEEL STRUCTURES IN SEISMIC ZONES

F.M. Mazzolani

University "Federico II" of Naples, Naples, Italy

1 Introduction

Steel structures have been always considered as a suitable solution for constructions in high seismicity areas, due to the very good strength and ductility exhibited by the structural material, the high quality assurance guaranteed by the industrial production of steel shapes and plates and the reliability of connections built up both in workshop and in field (Mazzolani and Piluso, 1996). In spite of these natural advantages, researchers are concerned about the necessity that, in order to ensure ductile structural behaviour, special care must be paid mainly in conceiving dissipative zones, which have to be properly detailed, assuring stable hysteresis loops, able to dissipate the earthquake input energy with high efficiency (Bruneau et al., 1998). As a confirmation, during the recent seismic events of Northridge (Los Angeles, 17 January 1994) and Hyogoken-Nanbu (Kobe, 17 January 1995), even if the cases of collapse of steel buildings have been extremely rare, steel moment frame buildings, considered as highly ductile systems, exhibited an unexpected fragile behaviour. They presented many failures located at the beam-to-column connections, challenging the assumption of high ductility and demonstrating that the knowledge on steel moment frames is not yet complete. Hence, in order to improve constructional details and to propose new design solutions for achieving a correspondence between the design requirements and the actual structural response, the scientific community began to deepen the reasons of this poor behaviour: does it depends on the material quality, on the design concept, on the structural scheme, on the constructional details, on the code provision, or on the seismic input occurred (Mazzolani, 1998)? Most of these questions are still being analysed, but much more has been understood on the seismic behaviour of steel structures. Consequently, during the last years most of the recent knowledge has been already or is going to be introduced into the structural design provisions for seismic resistant design in all the earthquake prone Countries, giving rise to a new generation of seismic codes.

2 Seismic Design Methodologies

The earthquake-resistant design of steel structures is based on the same principles which all civil engineering constructions in seismic zones are based upon.

The traditional philosophy (Mazzolani, 1998) basically identifies three performance levels, which should be achieved for increasing intensity of earthquake actions: structural and non structural damage in frequent minor ground motions has to be prevented; the minimisation of both structural and non structural damage during occasional moderate seismic events has to be attained; finally collapse or serious damage in rare major earthquakes has to be avoided (Mazzolani and Piluso, 1994, and Mazzolani et al. 1995a). In some codes, the two first levels are concentrated in one, corresponding to a damage limitation level, which can be called serviceability limit state.

In this context, seismic resistant dissipative structures are usually designed to withstand severe earthquakes by asking for a proper combination of strength and energy dissipation capacity. This design goal is pursued in current seismic codes by means of provisions aiming at assuring a minimum level of strength and by means of design and detail rules for obtaining the required energy dissipation capacity (Mazzolani, 1991a). The minimum required level of strength is strictly dependent on the required energy dissipation capacity and, for buildings, it is commonly expressed by a coefficient, namely q-factor in Eurocode 8 or reduction factor R in the American codes UBC and AISC, which reduces the elastic design spectrum. Different values of the reduction factor are given by the seismic codes for each structural typology (EC8, 1994, UBC 1997, and AISC 1997).

At the same time a minimum level of lateral stiffness is required in order to reduce the discomfort occurring during frequent minor earthquakes.

It can be pointed out that the traditional philosophy leads to a great simplification in the assessment of the design method, because it allows to perform in a single shot both the check against the serviceability limit state and the collapse limit state. But as it often happens, simplifications induce some limitations. In fact certain structural typologies, dimensioned with current design procedures, present lesser amount of energy dissipation capacity than that required in order to prevent the collapse under the most severe design earthquake (Mazzolani and Piluso, 1995). Furthermore even in modern seismic codes the need to check the actual seismic inelastic behaviour of the structure is not specified, because it requires very cumbersome calculations. But the necessity to assess general provisions, which are always on the safe side, is unquestionable even if not yet completely fulfilled in the codes, specially for steel structures.

According to a more recent philosophy, a comprehensive design procedure should correlate the resistance of a structure at various limit states to the probability that the earthquake action can reach the intensity required to induce the corresponding failure modes (Mazzolani, 1998). In this way, the minimisation of the cost/benefit ratio, which takes into account the construction cost and the expected losses, can be attained for all the limit states occurring during the service life of the structure (Mazzolani and Piluso, 1996).

The complete achievement of the design objectives needs the use of multi-level design criteria, such as the so-called "performance based earthquake resistant design". It can be stated that also the traditional design approach qualitatively and partially agrees with this concept (Bertero et al. 1996, and Bertero,1996). A more detailed definition of the performance based earthquake resistant design has been supplied within the activities of SEAOC Vision 2000 Committee (SEAOC Vision 2000, 1995).

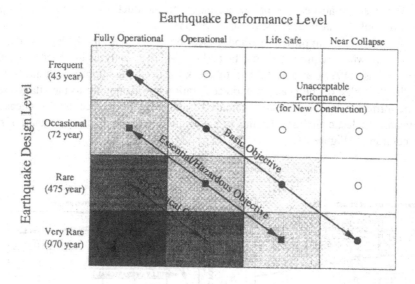

Figure 1. Seismic performance design objective matrix

The aim of this new approach is to provide the designers with the criteria for selecting the appropriate structural system and its layout and for proportioning and detailing both structural and non structural components, so that for specified levels of earthquake intensity the structural damage will be constrained within given limits. The coupling of a performance level with a specific level of ground motion provides a performance design objective. In this perspective, four performance levels have been proposed: *fully operational, operational, life safe* and *near collapse.* The correspondent requirements for different civil engineering facilities and constructional materials are also suggested (Figure 1).

In addition, four earthquake design levels are specified: *frequent, occasional, rare* and *very rare,* which are characterised by a return period equal to 43, 72, 475 and 970 years, respectively.

According to the framework of Figure 1, it can be recognised that the complete knowledge of the seismic performances of a structure should need sophisticated numerical procedure, because the quantitative evaluation of the structural damage requires non-linear dynamic analyses. As such analyses cannot be compulsory requested in common design practice, simplified procedures have to be suggested, leading to the evaluation of the seismic inelastic performance at different design criteria.

With reference to the European situation, the present phase of conversion from ENV to EN of Eurocode 8 could be a good opportunity to improve the provisions in such direction (Mazzolani, 1998). In particular the traditional design methods could be integrated by simple methods, which allow for verifying the design assumptions with regard to the energy dissipation capacity of the structure and therefore the fulfilment of the design performance objectives. At this scope an iterative procedure based on the ductility and energy dissipation capacity control could be introduced (Mazzolani, 1995a).

In this perspective a simplified method based on a trilinear modelling of the load-displacement curve which requires only elastic and rigid-plastic analyses has been proposed by Mazzolani and Piluso (1997), as a useful tool to check the design assumptions. It has been also used to predict the value of the peak ground acceleration corresponding to the attainement of the above performance levels. The approximate results have been successfully checked by means of dynamic inelastic analyses, leading to the evaluation of the ductility demand of multi-story MR steel frames (Figure 2).

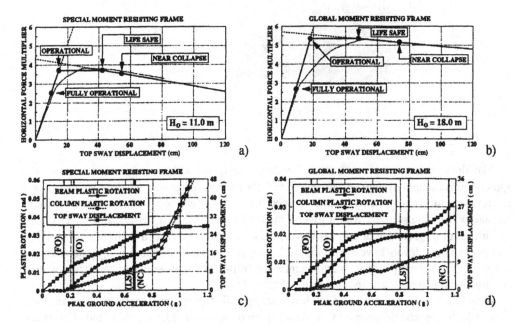

Figure 2. Results of a simplified method based on a trilinear modelling of the equivalent SDOF system (a and b) and its validation by means of inelastic dynamic analyses of special (c) and global (d) MR steel multistory frames

3   The European Seismic Codification Chronicle

The principles of design for the use of steel in seismic resistant structures represent a quite new subject, because steel is not the most popular structural material in European Countries (Mazzolani, 1999a). The lack of direct experience has been compensated by several research activities developed all over the World in the last decade (as reported in STESSA '94, STESSA '97 and the next STESSA 2000 Conferences).

In this direction, the Committee TC13 on "Seismic Design" of the European Convention for Constructional Steelworks (ECCS) plays a leader role of coordination, harmonisation and codification (Mazzolani, 1988). In 1988 the ECCS Committee published its first proposal of codification, the European Recommendations for Steel Structures in Seismic Zones (ECCS, 1988). This important issue represented in Europe the first international meeting point between the well consolidated "steel culture" and the more recent "seismic culture" (Mazzolani, 1991b).

In the special climate characterised by the early developments of the Eurocodes during the eighties, the ECCS-Committee TC13 devoted the main part of its activity as a consulting body of the drafting panel of EC8. The pressing deadlines for the issue of EC8 forced the ECCS-Committee TC13 to concentrate in a very short time the preparation of its Recommendations, which were immediately incorporated as the "Steel section" (Chapter 3 of Part 1.3) of the Eurocode 8, edition 1988, which is practically the same in the last edition of 1994 (Eurocode 8, 1994).

For these reasons, this first edition was not perfect and has to be considered as a first honest attempt to clarify the right way to proceed, in view that the results of the current research activity should help to fill some of the existing gaps (Mazzolani, 1992).

After many years, due to the big developments in the field of seismic resistant steel structures on international level, together with the lessons learned from the last terrible earthquakes, the scenario is completely changed and new areas are not yet explored (Mazzolani, 1995b).

One recent issue of ECCS-Committee TC13 has been the ECCS Manual on Design of Steel Structures in Seismic Zones (Mazzolani and Piluso, 1994). It intends to provide designers with the basic principles which the ECCS provisions, and therefore the EC8 ones, are based upon and, at the same time, to give some new results, which must be used in the improvement of the present codification. Therefore, this volume can be helpful during the conversion phase from ENV to EV of EC 8, together with the output of the STESSA Conferences and the results of the more recent research projects (see Section 7).

4   The Basic Structural Typologies

The main typologies of seismic-resistant dissipative structures can be classified according to the type of dissipative elements (Mazzolani, 1999a). Three different typologies can be recognised (Mazzolani and Piluso, 1994):

- Moment resisting frames (MRF);
- Concentrically braced frames (CBF);
- Eccentrically braced frames (EBF).

Moment resisting frames are rectilinear assemblages of beams and columns, with the beams rigidly connected to the columns. Their primary source of lateral stiffness and strength are the bending rigidity and strength of the frame members. Dissipative zones form in a large number and they are concentrated in discrete regions at the ends of the members, the so-called plastic hinges, which dissipate energy through a quite stable bending cyclic behaviour. In order to maximise the energy dissipation capacity, plastic hinges have to develop in beams and at the column base. The corresponding failure mode is called "global collapse mechanism". Moreover moment frames are also preferred for their architectural versatility: there are no bracing elements which block wall openings and the maximum flexibility for space utilisation is provided. This advantage is accompanied by a poor lateral stiffness of the whole structure, so that the member sizes are larger than those required for strength, due to the necessity to contain sway deflections within the drift limits, imposed by the codes (Mazzolani and Piluso, 1996, and Bruneau et al. 1998).

Concentrically braced frames resist lateral loads primarily by developing high axial forces in diagonal members. Only in some cases, the bending actions in moment resisting connections (when appropriate) can resist a small percentage of lateral loads. In general, the dissipative zones are represented by the tensile diagonals, because of the assumption usually made that the compression ones buckle. The inelastic cyclic performance of concentric braces is affected by an energy dissipation capacity degradation, because of the repeated buckling of diagonal bars. For this reasons the q reduction factor is assumed to be lower than the one corresponding to MRF. Besides, high elastic stiffness is achieved by the presence of diagonal bracing members (Mazzolani and Piluso, 1996, and Bruneau et al. 1998). The most typical configurations are shown in Figure 3:

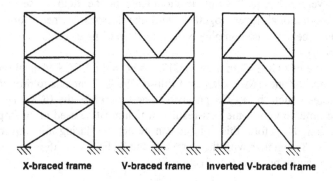

X-braced frame   V-braced frame   Inverted V-braced frame

Figure 3. Common configurations of concentrically braced frames (CBF)

Eccentrically braced frames are hybrid lateral force-resisting systems respect to the above mentioned ones. In fact they combine the individual advantages of moment-resisting frames and concentrically braced frames, assuring a high elastic stiffness, together with stable inelastic response under cyclic lateral loading and good ductility and energy dissipation capacity. They are characterised by diagonals eccentrically located in moment-resisting frames, producing a stiffening effect. In such a way the beams are divided in two or three parts, being the shortest one the dissipative element of the eccentrically braced frame, called "link", which dissipates the earthquake input energy by means of the inelastic cyclic shearing and bending. The assumed value of the q-factor in this case is almost the same than the one corresponding to MRF, depending on the location of the diagonal members (Mazzolani and Piluso, 1994, and Bruneau et al. 1998). The common types of EBF are classified as in Figure 4:

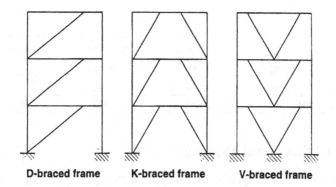

D-braced frame   K-braced frame   V-braced frame

Figure 4. Typical configurations of excentrically braced frames (EBF)

In order to provide lateral stiffness comparable to that of CBF and to offer high energy dissipation capacity during large inelastic deformations, short links, having the flanges connected to the columns by means of complete penetration welded joints, are to be preferred. This is because experimental results put in evidence that the links in shear possess a greater rotation capacity than the ones in bending (AISC, 1997).

A comparison between MRF and braced frames (CBF and EBF) can be qualitatively given on the base of the requirements which a seismic resistant structure has to satisfy: strength and stiffness against moderate ground motions with a small return period; strength, ductility and energy dissipation capacity against severe earthquakes with a great return period. It can be synthesised as in Table 1 (Mazzolani, 1995a).

Table 1

	Strength	Stiffness	Ductility
MRF	good	poor	good
CBF	good	good	poor
EBF	good	good	good

The ratio q_{BF}/q_{MRF} of the behaviour factor of braced frames (BF) versus the one of the MRF of ductile type, are given in Table 2, presenting a quantitative interpretation of the above remarks (Mazzolani, 1999a). According to the AISC provisions, concentrically braced frames (CBF) are subdivided into ordinary (OCBF) and special (SCBF).

Table 2. The q_{BF} / q_{MRF} ratio

		CBF		EBF	
		X - brace	V – brace	K-brace	D & V-brace
AISC (1997)	SCBF	0,75	0,75	1,00	0,875
	OCBF	0,75	0,40		
EUROCODE 8 (1994)		0,666	0,333	1,00	1,00

5   Improved Structural Typologies

A new family of structural typologies dissipating energy in bending could be conceived starting from two bays CBF, which in practice is the most economic system (Mazzolani et al., 1995b). Simply varying the distance "b" between the central columns B and C of Figure 5, structures with different values of stiffness and ductility can be obtained, ranging from CBF to MRF. As far as "b" increases, stiffness decreases, but ductility and energy dissipation increase, due to the presence of a central weak beams, where two plastic hinges form at the ends.

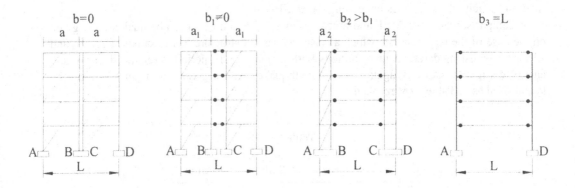

Figure 5. New typologies of frames: from a two bays CBF to a single bay MRF.

In this way a rational improvement of the poor ductility of a CBF can be achieved not so much by ductilising the system itself, but conceiving a system composed by two or more rigid CBF subsystems, connected by means of weak beams (Mazzolani et al., 1995b and 1995c).

The structural solutions ranging between the CBF and the MRF are called dual structures, because horizontal loading are resisted in part by moment resisting frames and in part by bracing systems acting in the same plane.

A similar evolution can be achieved starting from a classical one bay CBF system, with e=0, which is transformed in a MRF passing trough EBF systems, characterised by different dimensions of the links, from short to long (Figure 6).

In conclusion, the generation of such structural systems offers the possibility to obtain a wide range of structures, laying between a very rigid CBF up to a very ductile MRF, which can be characterised by a given combination of stiffness and ductility, for the same strength requirement.

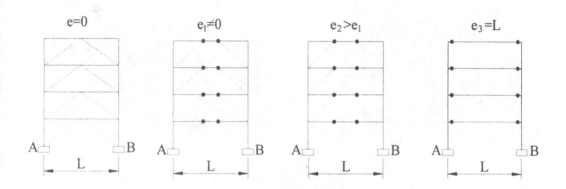

Figure 6. New typologies of frames: from a single bay CBF to a single bay MRF.

6 Analysis of Damage

It has been reported (Bertero et al., 1994) that, during the above-mentioned seismic events (see Section 1), many "pathological" failures occurred in steel buildings: structural designers expected that steel elements in building frames would be fully yield and finally collapsed in a ductile manner, after absorbing part of the energy input of strong ground motions; contrary, it was particularly astonishing to observe typical low-energy failures.

The damage can be characterised at different structural levels: material, cross-sections, members and connections (Mazzolani, 1999b).

As it is well known, steel is a very ductile material, but a big loss of ductility has been evidenced by fractures formed in the member far from the welded connections, crossing the entire sections, as it is showed in Figure 7a,b. This happened in the columns of the Ashiyahama apartment building in Kobe during the Hyogoken – Nanbu earthquake in 1995, but it must be pointed out that the material has been submitted to extremely severe loading conditions, due to very strong vertical quakes with very high velocity of propagation in presence of low temperature (it was 6 o'clock a.m. in winter).

At the cross-section level, a loss of ductility can occur due to local buckling phenomena. This happened in some bridge piles during the Kobe earthquake, under form of panel buckling (Figure 8a) and of "elephant foot buckled shape", typical for circular hollow sections (Figure 9a,b,c). In both cases the very high vertical components seems to be the main responsible.

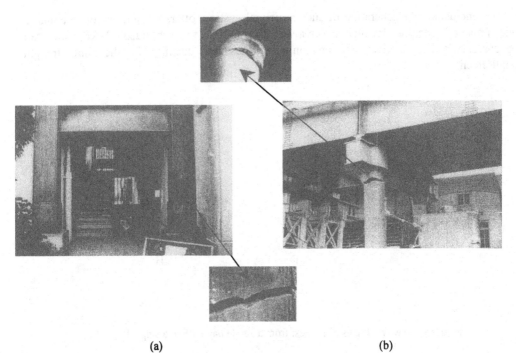

(a)   (b)

Figure 7. Material brittle fracture

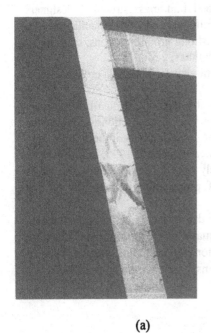

(a)   (b)

Figure 8. Cross-section failures

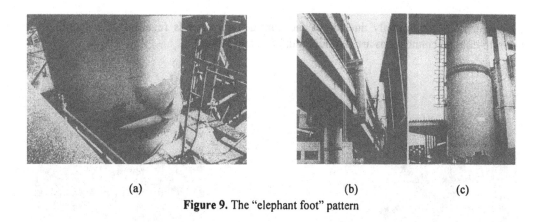

(a)   (b)   (c)

Figure 9. The "elephant foot" pattern

The local buckling in the column of the Pino Soarez building during the Mexico City earthquake (1985) is mainly due to the bad execution of weldings, which form the square hollow section (see Figure 8b).

Member ductility is conditioned by global buckling phenomena as it can be seen in Figure 10, where the overall buckling of bracing members produced large permanent deformations, leading to a change in the hysteresis loops and a consequent reduction in energy absorption capacity.

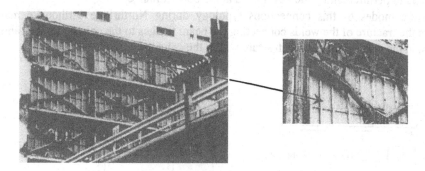

Figure 10. Member failures

But in general damage is mainly located in the connections. Typical are the failure of column base plates produced by the rupture of anchor bolts or the fracture of members at the bolted connections due to the presence of holes which generate a section with reduced resistance (Figure 11a). The fracture of welded beam-to-column connections in moment resisting frames is undoubtedly the most widespread type of failure occurred in steel structures during both Northridge and Hyogoken-Nanbu earthquakes (Figure 11b). However, it has been observed (Akiyama and Yamada, 1995) that the fracture mechanisms were different between the two earthquakes, mainly due to the fact that different trends regarding the detailing of beam-to-column connections are developed in USA and Japan.

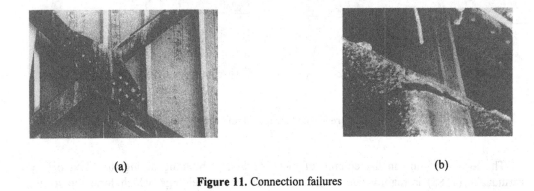

(a)   (b)

Figure 11. Connection failures

In fact according to the current U.S. practice for seismic resistant frames, both beams and columns are H-shaped (Figure 12a). The beam web is field bolted to a single plate shear tab which is shop welded to the column. The beam flanges are field welded to the column using complete penetration welds. Web copes are required to accommodate the backup plate at the top flange and to permit making the bevel weld at the bottom flange.

The failure modes of this connections typology during Northridge earthquake mainly consisted in the fracture of the welds connecting the beam flanges to the column flanges and, in some cases, in the propagation of this fracture within the column.

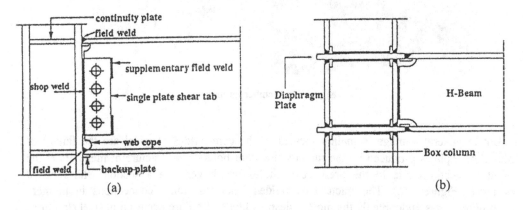

(a)   (b)

Figure 12. Beam- to column connections for MR steel frames, according to the current U.S. (a) and Japanese (b) practice.

Contrary, the current Japanese practice uses H-shaped beams which are field welded to box-columns (Figure 12b). The stresses of the beam flanges are safely transmitted by means of continuity plates welded through the box-columns. The transmission of the beam web stresses involves the out-of-plane bending resistance of the column wall plates and, therefore, is not perfect. This causes high stress concentration at the ends of beam flanges, which is further promoted by the web copes introduced for an easier welding (Mazzolani, 1998).

7   Lessons to Be Learned

The recent investigations have allowed to identify some causes of damage in steel buildings.

On one hand, it can be ascribed to the use of field welding so that poor workmanship is solely to blame and, therefore, it is necessary to tighten the site supervision and to improve the welding details and procedures; on the other hand, damage causes can be attributed to an excess of seismic loading and to defective design guidance leading to a rotation ductility supply in the connections lower than the earthquake imposed demand (Elnashai, 1994).

Some other factors influencing fracture modality are related to material properties, temperature, strain rate, joint geometry, plate thickness and so forth.

More extensive considerations can be done (Mazzolani, 1995c):

- at the material level, damage can be imputed to the concurrence of high velocity of load application and very low temperatures. These conditions are not reproducible by means of the usual laboratory tests and until now they have been ignored by the codes.
- local buckling of cross-sections in Kobe can be attributed to the important vertical component of the ground motions, which normally are underestimate or sometimes neglected by seismic codes.
- a distinction must be done between "far-field" earthquakes which are characterised by the cyclic alternation of action and "near-field" earthquake where the impulsive feature of the ground motion is predominant (see Chapter 2), as it occurred in Kobe.

It has also to be remembered that, regarding the strength requirements, the American code UBC 91 provides for special moments resisting frames a reduction factor R_w equal to 12, which is equivalent to a value of the European q-factor equal to 8. On the contrary, Eurocode 8 provides a q-factor value equal to 6 and the Japanese code a structural coefficient D_s equal to 0.25 which corresponds to $q = 4$. Therefore, concerning the damage experienced by steel buildings during the Northridge earthquake, it could be attributed to an excess of local ductility demand due to the high earthquake intensity compared with the adopted design level.

Furthermore, in U.S. practice the beam flange is welded to the column flange by provisionally using bolted connection in the beam web during the erection. In such a connection, the bending moment at the beam web is hardly transmitted to the column, resulting in the stress concentration in flanges at the end of beams. This normally occurs in a beam-to-column connection because, while the stress transmission between the beam flange and the column is completely made through the diaphragm plate, the bending moment in the web of the beam cannot be transmitted completely to the column, since the stress transmission is made through the out-of-plane bending of the column flange. The diaphragm plate is usually thicker and wider than the beam flange and, therefore, the fracture develops on the side of the beam (Mazzolani, 1998).

These effects were increased by the fact that the weak-beam type is preferred both in Japan and USA as a yield-mechanism. In addition the use of compact beam sections, with small width to thickness ratio, in order to avoid local buckling, gives rise to stress concentration in the beam flanges and brittle mode of failure in the connection. Besides, defects in material of the heavy column must be an incentive for the propagation of brittle cracks on the side of the column.

Both in USA and in Japan extensive programs of experimental tests has been carried out on beam-to-column subassemblages, in bi-dimensional and tri-dimensional frames.

It has been evidenced (Bertero et al., 1994) that the types of failure occurred in welded beam-to-column connections during the earthquakes have been already observed in laboratory experiments.

Numerical analyses of the seismic response of steel framed buildings damaged during Northridge earthquake have pointed out that there were several ground motions, recorded during this earthquake, able to significantly lead the structure into the inelastic range. In many cases, the plastic rotation demand at the beam ends exceeded 0.02 rad; therefore, on the base of the available experimental data and codes provisions (UBC 91 and AIJ 90), it is clear that the cracking occurring in the connections cannot be considered unusual.

In the experimental tests carried out in USA, a plastic rotation supply equal to 0.02 rad has been used as a benchmark to judge the seismic performance of beam-to-column connections, because it was believed to be sufficient to withstand severe earthquakes. As the recent experience demonstrated that this limit value can be exceeded, it is clear that the attention should be focuses on the design value of the q-factor which could be reduced in order to limit plastic rotation demands occurring during severe earthquakes or, as an alternative, on the improvement of the seismic performances of dissipative zones.

The SAC Steel Project has been developed to derive new design procedures accounting for the lessons learned from the Northridge earthquake. In particular, the structural design philosophy, as well as the characteristics of welding and of the structural details, the velocity of load application and the influence of the earthquake vertical component are the main issues under investigation within the SAC Project.

The final results of SAC are expected for the end of 1999. Some more refined conclusions will be drawn for middle 2000 and a special session will be devoted to the SAC official presentation during the STESSA 2000 Conference in Montreal (August, 2000).

Also in Europe, from the observation of damage in steel structure connections after the earthquakes of Northridge and Kobe it seemed that there is an urgent need to investigate new topics such as the influence of the strain rate on the cyclic behaviour of beam-to-column joints and more in general to review the whole background of modern seismic codes in order to grasp the design rules which failed (Mazzolani, 1998). In particular this revision should be aimed at the improvement of Eurocode 8, whose application will be widespread in the next years within the Mediterranean Countries.

The following question can raise: «Can the results of the American and Japanese "on field" experience be applied to European practice ? ».

It has to be considered that the steel grade, the chemical composition and the mechanical characteristics of the steel can be different. Also the welding technique can be different. In addition, different strength requirements and different levels of seismic input lead to different plastic rotation demands.

In this context, a group of 8 European Countries (Italy, Romania, Greece, Portugal, France, Belgium, Bulgaria, Slovenia) developed a joint research project (INCO-COPERNICUS), co-called RECOS ("Reliability of Connections of Steel Frame Buildings in Seismic Areas"), sponsored by the European Commission (Mazzolani, 1999c). The program of the joint project has been established in order to provide an answer to the above questions, by accomplishing the following objectives: a) Analysis and synthesis of research results, including Code Provisions, in relation with the evidence of Northridge and Kobe earthquakes. Particular attention has been devoted also to those research results which have not accounted for in preparing the provisions of the modern seismic codes. b) Assessment of new criteria for selecting the behaviour factor for different structural typologies and definition of the corresponding range of validity. In particular, in this field, the need to provide simplified methods for evaluating the q-factor is felt more and more urgent. The aim is to provide the designer with an operative tool, which allows him to be aware of the inelastic performances of the designed structure. c) Identification and evaluation of the structural characteristics of connections influencing the seismic response of steel buildings. Therefore, the research has been devoted to the strength and stiffness evaluation of moment resistant connections and to the prediction of their degradation under cyclic loads, by considering also the effect of strain rate and temperature. Low cycle fatigue has been also investigated. d) Definition of criteria for designing and detailing beam-to-column connections for seismic resistant structures, considering also the seismicity of the site (far-source and near-source effects).

After 30 months of activity, the RECOS project has been recently completed (Mazzolani, 1999d) and the main results are collected into a volume (Mazzolani, 2000), which is now in press.

References

AIJ (1990). Standard for Limit State Design of Steel Structures (Draft). Architectural Institute of Japan (English version, October 1992).

AISC (1997). Seismic Provisions for Structural Steel Buildings.

Akiyama H., Yamada S. (1995). Damage of steel buildings in the Hyogoken-Nanbu Earthquake.In *Proceedings of EASEC '95*. Gold Coast, Australia.

Bertero, V.V. (1996). The Need for Multi-Level Seismic Design Criteria. In *Proceedings of the 11th World Conference on Earthquake Engineering*. Acapulco.

Bertero V. V., Anderson J. C., Krawinkler H. (1994). Performance of steel building structures during the Northridge Earthquake. In *Earthquake Engineering Research Center, Report No. UBC/EERC-94/09*, University of California, Berkeley.

Bertero, R.D., Bertero, V.V. and Teran-Gilmore A. (1996). Performance-Based Earthquake-Resistant Design Based on Comprehensive Design Philosophy and Energy Concept. In *Proceedings of the 11th World Conference on Earthquake Engineering*. Acapulco.

Bruneau M., Uang C.M. and Whittaker A. (1998). *Ductile design of steel structures*. McGraw-Hill.

Commission of the European Communities (1994). Eurocode 8: Structures in Seismic Regions, ENV.

ECCS (1988). European Recommendations for Steel Structures in Seismic Zones, doc. n. 54.

Elnashai A. (1994). Comment on the Performance of Steel Structures in the Northridge (Southern California) Earthquake of January 1994. In *New Steel Construction*, Vol. 2, n. 5, October.

Mazzolani F. M. (1988). The ECCS activity in the field of recommendations for steel seismic resistant structures. In *Proceedings of the 9th World Conference on Earthquake Engineering, WCEE.* Tokyo, Kyoto.

Mazzolani, F.M. (1991a). Design of Seismic Resistant Steel Structures. In *Proceedings of International Conference on Steel and Aluminium Structures*, Singapore.

Mazzolani, F.M. (1991b). The European Recommendations for Steel Structures in Seismic Areas: Principles and Design. In *Proceedings of Annual Technical Session of SSRC (Structural Stability Research Council)*, Chicago.

Mazzolani F. M. (1992). Background document of EUROCODE 8, chapter 3: Steel Structures. In *Proceedings of the 1st State of Art Workshop COST C1*, Strasbourg.

Mazzolani F. M. (1995a). Design of seismic resistant steel structures. In *Proceedings of the 10th ECEE.* Vienna 1994, published by Balkema, Rotterdam.

Mazzolani F. M. (1995b). Eurocode 8 - chapter "Steel": background and remarks. In *Proceedings of 10th European Conference on Earthquake Engineering, ECEE.* Vienna, 1994, published by Balkema, Rotterdam.

Mazzolani, F.M. (1995c). Some simple considerations arising from Japanese presentation on the damage caused by the Hanshin earthquake. In *Stability of Steel Structures (ed. M. Ivanyi), SSRC Colloquium*, 21-23 September, Budapest, Akademiai Kiado, Vol. 2, 1007-1010.

Mazzolani F. M. (1998). Design of steel structures in seismic regions: the paramount influence of connections. In *Proceedings of the COST C1 International Conference on "Control of the semi-rigid behaviour of civil engineering structural connections".* Liege, September 17-18.

Mazzolani F.M. (1999a). Principles of design of seismic resistant steel structures. In *Proceedings of the National Conference on Metal Structures.* Ljiubljana, May 20.

Mazzolani F.M. (1999b). Design and construction of steelworks in seismic zones. In *Proceedings of the XVII C.T.A. Congress*, Napoli, October 3-6.

Mazzolani F. M. (1999c) Reliability of moment resistant connections of steel building frames in seismic areas: the first year of activity of the RECOS project. In *Proceedings of the 2nd European Conference on Steel Structures EUROSTEEL.* Prague, May 26-29.

Mazzolani F.M. (1999d). Reliability of moment resistant connections of steel building frames in seismic areas. In *Proceedings of the International Seminar on Seismic Engineering for Tomorrow*, in honor of professor Hiroshi Akiyama. Tokyo, Japan, November 26.

Mazzolani F.M. (edr., 2000). Moment resisting connections of steel frames in seismic areas: design and reliability. Published by E & FN SPON, London, in press.

Mazzolani, F.M., Georgescu, D., Astaneh-Asl, A. (1995a). Safety Levels in Seismic Design. In *Proceedings of the 1st International Workshop on "Behaviour of Steel Structures in Seismic Areas" STESSA ' 94*, Timisora, Romania, June 1994, Mazzolani F. M., Gioncu V. editors, published by E & FN SPON an Imprint of Chapman & Hall, London.

Mazzolani, F.M., Georgescu, D., Astaneh-Asl, A. (1995b). Remarks on behaviour of concentrically and eccentrically braced steel frames. In *Proceedings of the 1st International Workshop on "Behaviour of Steel Structures in Seismic Areas" STESSA ' 94*, Timisora, Romania, June 1994, Mazzolani F. M., Gioncu V. editors, published by E & FN SPON an Imprint of Chapman & Hall, London.

Mazzolani, F. M., Georgescu D., Cretu D. (1995c). On the ductility of concentrically braced frames. In *Proceedings of the C.T.A (Congresso dei Tecnici dell'Acciaio) Congress*, Riva del Garda.

Mazzolani, F.M. and Piluso, V. (1994). Manual on Design of Steel Structures in Seismic Zones. In *European Convention on Constructional Steelwork, ECCS-TC13.* doc. n. 76.

Mazzolani, F.M. and Piluso, V. (1995). Failure mode and ductility control of seismic resistant MR-frames. In *Costruzioni Metalliche n. 2.*

Mazzolani, F.M. and Piluso, V. (1996). *Theory and Design of Seismic Resistant Steel Frames.* E & FN SPON, an Imprint of Chapman & Hall, London.

Mazzolani, F.M., and-Piluso, V. (1997). A simple approach for evaluating performance levels of moment-resisting steel frames. In *Proceedings of the International Workshop on "Seismic Design Methodologies for the Next Generation of Codes".* Bled, June, published by Balkema, Rotterdam.

SEAOC, Structural Engineers Association of California (1995). VISION 2000: Performance Based Seismic Engineering of Buildings.

STESSA '94, (1995). *Proceedings of the 1st International Workshop on "Behaviour of Steel Structures in Seismic Areas".* Timisoara, Mazzolani F. M., Gioncu V. editors, published by E & FN SPON, an Imprint of Chapman & Hall, London.

STESSA '97 (1997). *Proceedings of the 2nd International Workshop on "Behaviour of Steel Structures in Seismic Areas",* Kyoto, Mazzolani F. M., Akiyama H. editors, 10/17, Salerno.

UBC (1991). Uniform Building Code. In *International Conference of Building Officials.* Whittier, CA.

UBC (1997). Uniform Building Code. In *International Conference of Building Officials.* Whittier, CA.

CHAPTER 2

DESIGN CRITERIA FOR SEISMIC RESISTANT STEEL STRUCTURES

V. Gioncu

Politechnica University Timisoara, Timisoara, Romania

Abstract. The paper presents the state-of –the art in the design of the steel structures, in the light of the lessons learned from the last great earthquakes: Michoacan (1985), Loma Prieta (1989), Northridge (1994) and Kobe (1995). In the introduction, the very recent progress in the conception and design are presented, showing the challenge for future research works. The design criteria as multi-level design approaches and rigidity, stiffness and ductility demands are detailed, emphasizing the unsolved problems. New aspects for ground motion modeling, as the differences between near-source and far-source earthquakes are examined. The response of the structure to these near-source earthquake is examined, taking into account the influence of superior vibration modes, velocity and strain-rate, vertical components, etc. The final conclusions consider that now is the right moment to introduce some new provisions in the design codes, in order to fill the gap that exists between the accumulated knowledge and design codes.

1. INTRODUCTION

1.1 The nature of the problem

The main purpose of structural design is to produce a suitable structure, in which one must consider not only the initial cost, but also the cost of maintenance, damage and failure, together with the benefits derived from the structure function. Thus, the optimum design of structure requires a clear understanding of the role of each of above aspects and therefore, it requires general view on the total process.

This objective can be achieved by the engineer, involved in designing of a specific building, without great problems for conventional actions as dead, live wind, and snow loads, but with difficulties for exceptional loads produced by natural disasters as hurricanes, floods, earthquakes, etc. Among these natural disasters, earthquakes were responsible for almost 60 percent of deaths and damage. Every strong earthquake is responsible of large losses of lives and of goods. In the past, during each event, the number of deaths was always very high, but it continues to be also very important at the time being. The analysis of these events shows that not only the ground motion severity is the main responsible of these losses in live. Indeed, the main feature of earthquake is that the most of the human and economic losses are not due to the earthquake mechanism, but to failure of human facilities, buildings, bridges, transport systems, dams, etc, which have been designed and constructed for the comfort of human being. So, in many cases of building collapses, the builders are thyself the producer of killing tools. The more important losses in life are always concentrated in the poor zones with ancient buildings or with very poorly built constructions. In exchange, the important economic losses are localized in rich zones with modern buildings, where, even if there are not many building collapses, the produced damage is very important and the cost of repairs very high.

It can be observed that buildings, which are poorly designed and constructed, suffer much more damage in moderate earthquakes, than well-designed and constructed buildings in strong earthquakes. Moreover, these last buildings should easily be able to withstand severe earthquakes with no loss of life and without severe damage. This observation is encouraging, because it seems that the earthquake problems are solvable and depend only on the results obtained in research works, their implementation in design practice and administration attitude regarding the measures to reduce the seismic effects. The duties of professionals are to detect the errors that have contributed to the collapse and to improve the conceptual design and quality of building constructions.

Generally, the engineering approach to design is quantitative and the structural members must be sized to have resistance greater than the actions caused by these events. But the design for the largest credibly imagined loads, resulting from the strongest expected earthquake on the structure site, is unreasonable and economically unacceptable. The design requirements are set at given level smaller than this associated with the largest possible loads. So, the structures occasionally fail to perform their intended function under these events that exceed the design values and consequently, they may suffer local damage, by the loss of resistance in a single member or in a small portion of structure. But a properly designed structure must preserve the general integrity, which is the quality of being able to sustain local damage, the structure as a whole remaining stable. This purpose can be achieved by an arrangement of structural elements that gives stability of the entire structural system. In the case of frequent minor earthquake it is required that structural and non-structural damage is prevented.

Unfortunately, in the last time this design philosophy fall behind the economic losses during recent seismic events, partially a product of the inadequate performance of modern buildings. Unfortunately very eloquent case is offered by the steel structures. For a long time it was generally accepted in design practice that steel is an excellent material for seismic resistant structures, thanks to its performances in strength and ductility at the level of material. But very serious alarm signals about this optimistic view arise after the last severe earthquakes of Michoacan (1985), Northridge (1994) and Kobe (1995). Beside a lot of steel constructions, which have shown a good performance, at the same time, a lot of others exhibited a very bad behaviour. For these last, the actual behaviour of joints, members and structures has been very different from the designed one, so the traditionally good performance of steel structures under severe earthquakes has been recognized as a dogmatic principle not always respected in the reality. Because in many cases the damage arose also when both design and detailing have been performed in perfect accordance with the code provisions, it means that something new happened, which was not foreseen in the design practice (Mazzolani, 1995). The engineers and scientists want to know exactly the reasons of this poor behaviour:

-inaccuracy of ground motion modelling ?
-modification of material qualities during severe earthquakes ?
-shortcoming in design concept, especially concerning the use of the simplified design spectrum method ?
-insufficient code provisions concerning ductility demands ?
-shortcoming in accuracy of constructional details ?

Today, concerning the measures necessary to eliminate the possibility of similar damage occurrence during the future strong earthquakes, the world of specialists is divided. Some of them consider that the actual design philosophy is proper and only some improvement of constructional details, especially concerning welded joints, is enough. But in the same time, there are many other specialists who consider that the above mentioned questions are real problems in the design and some pressing modifications in concept are required.

So, it is the moment to present the recent progress in design philosophy and the challenge for future research works, in order to improve the code provisions.

1.2 Main lessons after the last strong earthquakes

A dramatic increasing in the size of losses resulting from earthquake catastrophes has marked the period of 1985-1995. The first event of this decade was the Michoacan earthquake (1985) followed by Loma Prieta (1989), Northridge (1994) and Kobe (1995) events. If the economic losses in period 1965-1975 and 1975-1985 were only 10 and 67 billion US$, respectively, in the period 1985-1995 the losses have increased at 182 billion US$ (Fig. 1). These losses are certainly enormous, but its size pales against the possible losses of 100-150 billion US$ if "The Big One" occurs at San Francisco or 800-1,200 billion US$ at Tokyo if the 1923 earthquake is happened today (Berz and Smolka, 1995).

The reason for this remarkable increase in economic losses can be defined as follows:

(i) Due to the growth of world population, the accumulation of population and industrialization of high-risk regions is a today phenomenon. As a consequence of these developments, the occurrence of probability that a great event strikes an important city is drastically increasing. For instance the 1812 largest earthquake ever to hit the USA (M> 8) were concentrated in the New Madrid seismic zone, near to today Memphis City, Tennessee. But the damages were very low, because the area was relatively inhabited, the city Memphis being founded until several years later. How would the situation differ today? The Memphis area is now highly populated area and an earthquake of the similar magnitude is able to cause damage in the ten billions of dollars. Unfortunately, in future strong earthquakes will strike many of similar urbanized areas with very costly losses.

(ii) The vast majority of buildings in the seismic zones do not conform to current standards appropriate for the earthquakes expected in their areas, being designed and erected before or during the early development of seismic design provisions. A proper seismic rehabilitation program is required by this situation. Unfortunately, between earthquake professionals and politics an important gap exists and too little financial effort was paid on this program. These building will be the sitting targets of the coming strong earthquake.

(iii) For a long time the main purpose of seismic design of new buildings was the protection of public from loss of life or serious injuries and the prevention of buildings from collapse under the maximum intensity earthquake. Only the second goal was the reduction of property damages. Just on the last time the specialists have been awarded that to fulfil only the condition of live protection is economically unacceptable. They realized that it is necessary to pay more attention to the reduction of damage of all building elements for all the range of earthquake intensities. Consequently, a multi-level design approach has been developed (Bertero, 1996).

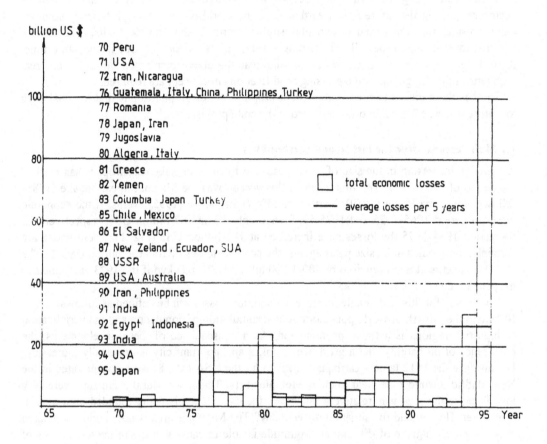

Fig. 1 Cost of losses due to earthquake from 1966...1995

(iv) Each event is basically unique, offering new surprises in the vulnerability of buildings affected by earthquakes and showing the great complexity of the phenomenon. Referring to steel structures, for a long time no serious damage during some major earthquakes were recorded, so the opinion that the steel structures are very safe solution for seismic areas have been consolidated among the structural designers. But the 1985 Michoacan earthquake produced in Mexico City the first collapse of a high steel building, due to the site conditions and lack of sufficient ductility. The 1994 Northridge and 1995 Kobe earthquakes produced many failures in steel moment-resisting frames, especially in beam-to-column joints. These damage in steel structures have shown that the present code provisions are insufficiently to prevent these failure, being the start of a large activity in research works, in which the control of ductility, both for members and joints, for severe conditions plays a leader role (Gioncu, 1999).

(v) In spite of the progress obtained in the research works between theory and practice a very important gap exists today. This is due to the fact that always the implementation of advanced new concepts is constrained by the needs of the profession to keep the design process as simple as possible. Contrary to the very great variability of the phenomenon, the current design methodology based on using the design spectra, contains only very few design parameters (acceleration level, earthquake and structural periods), insufficient for a proper description of the seismic loading. In the past, due to the reduced number of records during severe earthquakes (the famous El Centro records obtained in 1940 where for long time was the alone information about the time history ground motions of an earthquake) the developed design methods were mainly based on hypothesis with few possibilities to be verified. During the very last years, due to the development of a large network of instrumentation all over the world, there are a large amount of measurements of the ground motions for different distances from the sources and for different site conditions. The analysis of this new information has emphasized the great diversity of ground motion typologies. Between these ones, the most important are the differences between near-source and far-source recorded ground motions, which are not considered in the structure design.

(vi) Due to the development of computer science, today is not a problem to perform a nonlinear time history analyses. But the real problem is the option for an accelerogram, which adequately represents the expected earthquake at the structure site. As a result of the extensive monitoring of the areas with high seismic risk today it is possible to dispose of large number of records. All these records have random characteristics due to the particularities of source, magnitude, travel path, soil conditions, influence of neighboring buildings, etc. The choosing is complicated by the remark that at the same site and as the result of the same source, the ground motions may be very different in characteristics for different events.It is very difficult to select for analysis the recorded accelerogram having similar characteristics with those that will take place at the site of the analyzed structure. So, the blind confidence on computer results can be a source of potential damage.

(vii) Due to the large variability of the accelerogram characteristics (peaks, periods, patterns, duration, etc) able to influence the response of a structure, the response values show a very large scattering. Due to the randomness of these results it is very difficult to emphasize the main factors influencing the structure behaviour. In all the problems of the structure mechanics the studies begin with a deterministic concept to ascertain the theory. Only in the second step, when the determinist response is very well defined, the studies are extended to the random aspects. Contradictory with this general rule, the seismic analysis, using recorded accelerograms and time-history computer programs, has jumped the deterministic step, passing directly to the randomness approach. This fact complicates the discerning the main characteristics of the structure response. All these pointed out aspects show very clear that, aiming to explain the actual response of the structure, it is imperatively required to use in practice some artificial generated accelerograms. The control parameters of these accelerograms should be chosen with regard to the main characteristics of the expected earthquake on the site, neglecting secondary factors.

adequate member dimensions are derived, by considering the critical sections in the members selected for the eventual development of plastic hinges. The consideration of the overstrenght of these critical sections, where plastic hinges occur, is an important feature of the capacity design method. Finally, other members or part of members are designed to resist in elastic range, considering the overstrength of adjacent potential plastic hinges. This procedure ensures that the system may dissipate seismic energy with some local damage, but without global collapse.

Final detailing. The suitable achievement of design conceptions mainly depends on the simplicity of detailing of members, connections and supports. The respect of detailing requirement assures a good behaviour of structure during severe earthquakes, according to the design conception. The failures produced during the ground motions indicate more deficiencies in structure detailing than in structure analysis. However, the code provisions contain only some few details directly involved in the protection against local damage of members and connections. After the last strong earthquakes, a great amount of information has been obtained and a real progress in improving the detail conception is now possible.

(iii) Progress in construction. Design and construction are two phases intimately related of the birth of a building. A good design conception is effective only if also the building erection is qualitatively good. After each earthquake, field inspection has revealed that a large percentage of damage and failure has been due to the poor quality control of structural materials and/or to poor workmanship.

(i) Quality assurance during construction, referring to the rules for verifications of material qualities and proper workmanship. For instance, the analysis of structural steel specimen tests shows considerable variation in material characteristics. In view of this variability, many present seismic code provisions specify only the minimum value for yield stress, what can lead to an unsafe design because a random overstrength distribution can modify the global ductility. An upper bound for yield stress must be given in the code and more severe control of variation factor is required.

(ii) Monitoring and maintenance. In many cases the damage or failure of buildings may be attributed to a lack in monitoring and improper maintenance. Due to this fact the deterioration of mechanical properties of material and elements undermines the seismic response of the structure. The progress consists in the elaboration of severe rules for monitoring and maintenance of building during their life.

(iii) Refurbishment, repair and strengthening. The building may suffer some changing in function, which claims some modification of structure system. If this changing is performed without some aseismic rules, the modified structure may be victim of future earthquakes. On the other hand, there are a lot of buildings, which were built many years ago, before the introduction of seismic design. Today, the building industry is looking with particular interest at the restoration, repairing and consolidation of old buildings. The future of this industry will be extensively concerned with the refurbishment and strengthening of the existing buildings, many of them being very important from historical and architectural interest. In all cases steel is an ideal material for refurbishment and an important progress is marked in last time in use some specific technologies.

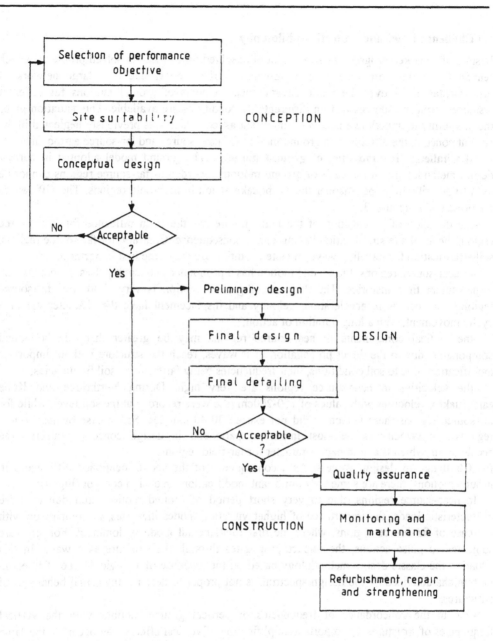

Fig. 2 Seismic protection philosophies

1.4 Challenge in seismic protection philosophy

In spite of a marked progress, there are a lot of unsolved or insufficient solved problems, which remain to be the main objectives of research works. Today, due to a large network of instrumentation all over the world several measurements of ground motions for different distances from the sources and on different site conditions are available. This situation gives the possibility to underline a new very important aspect which was previously neglected in the current concept: the difference in ground motion for near-source and far-source earthquakes.

(i) **Challenge in modelling of ground motions.** The ground model adopted in current design methodology on the basis of ground motions recorded in far-source regions cannot be used to describe in proper manner the earthquake action in near-field regions. The differences are presented in Figure. 3:

-the direction of propagation of the fault rupture has the main influence for near-source region, the local site stratification having minor consequence. Contrary, for far-source regions, soil stratification for traveling wave and site condition are the of the first importance;

-in near-source regions, the ground motion has a pronounced coherent pulses in velocity and displacement time histories. The duration of ground motion is very short. For far-source regions, the records in acceleration, velocity and displacement have the characteristic of a cyclic movement, with a long duration of action;

-the vertical components in near-source regions may be greater than the horizontal components, due to the direct propagation of P waves, reach the structure without important modifications due to soil conditions, their frequencies being far from the soil frequencies;

-the velocities in near-source regions are very high. During Northridge and Kobe earthquakes, velocities with values of 150-200cm /sec were recorded at the soil level, while for far-source regions these velocities did not exceed 30-40 cm/ sec. So, in case of near-source regions, the velocity is the most important parameter in design concept, replacing the acceleration, which is a dominant parameter for far-field regions.

(ii) **Challenge in design process.** As a consequence of the above mentioned differences in ground motions, there are some very important modifications in design concept (Fig. 4):

In near-source regions, due to very short period of ground motions and due to pulse characteristic loads, the importance of higher vibration modes increases, in comparison with the case of far-source regions, where the first fundamental mode is dominant. For structure subjected to pulse actions, the impact propagates through the structure as a wave. In this situation the classic design methodology based on the response of a single-degree-of freedom system characterized by the design spectrum is not proper to describe the actual behaviour of structures;

-due to the concordance of frequencies of vertical ground motions with the vertical frequencies of structures, an important amplification of vertical effects may occur. In the same time, taking into account the reduced possibilities of plastic deformations and damping under vertical displacements, the vertical behaviour can be of first importance for structures in near-field regions. The combination of vertical and horizontal components produces an increasing of axial forces in columns, an as a consequence, the increasing of second order effects;

-due to the pulse characteristic of actions, developed with great velocity and to lack of important restoring forces, the ductility response may be very different from the usual cases.

So, the use of inelastic properties of structures for seismic energy dissipation must be very carefully examined. In the same time, short duration of ground motions is a favourable factor;

-due to great velocity of seismic actions, an increasing of yield strength occurs, what means a significant decreasing of available local ductility. Due to this increasing in demand, as an effect of the pulse characteristic of loads, therefor with the decreasing of response, due to the effect of high velocity, the balance demand-response can be broken.

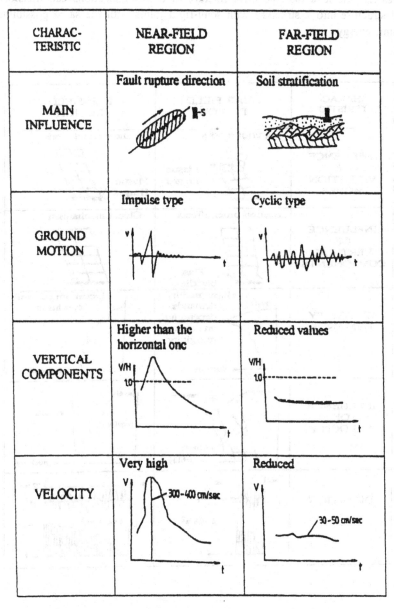

CHARAC-TERISTIC	NEAR-FIELD REGION	FAR-FIELD REGION
MAIN INFLUENCE	Fault rupture direction	Soil stratification
GROUND MOTION	Impulse type	Cyclic type
VERTICAL COMPONENTS	Higher than the horizontal one	Reduced values
VELOCITY	Very high	Reduced

Fig. 3 Near-source vs far-source ground motion features

The need to determine the ductilities in function of the velocity of actions is a pressing challenge for research works. If it is not possible to take advantage of the plastic behaviour of structures, due to this high velocity, it is necessary to consider the variation of energy dissipation through ductile fracture. The fact that many steel structures, which were damaged by fracture of connections during the the Northridge and Kobe earthquakes without global collapse, gives rise the idea that the local fracture of these connections can transform the original rigid structure into a structure with semi-rigid joints with the same possibilities to dissipate seismic energy.

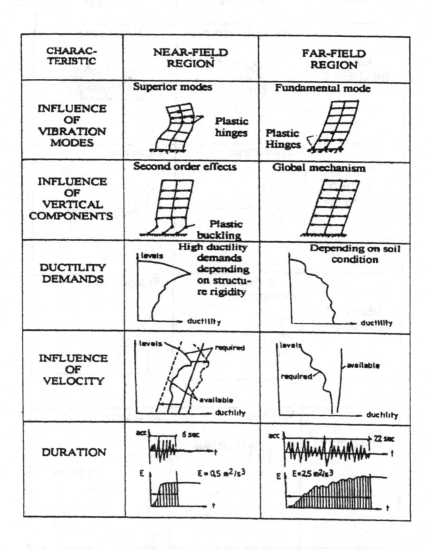

Fig. 4 Near-source vs far-source structure behaviour

The positive result of this weakening is the reduction of seismic actions at a level which can be supported by the damaged structure, taking into account that the duration of an earthquake is very short. This is not the case of far-source earthquakes, for which the effect of long duration can induce the collapse of the structure;

-pictures showing failures of structural members, in which details were grossly wrong, are very frequent in the post-earthquake reports. Very often, these mistakes were due to the fact that the detailed analytical calculations were not accompanied by a consistent set of structural details. This reality claims that special parts of codes must be elaborated in order to give constructional requirements for details. After the joint damage produced during the last earthquakes, when welded connections behaved very badly, the challenge in construction is to establish some provisions to improve the joint behaviour and to eliminate any source of brittle fracture.

2. SEISMIC DESIGN CRITERIA

2.1 Multi-level design approaches

The need to erect buildings in seismic areas has led to the development of a particular design philosophy. The basic principle of this philosophy consists in considering that it is not economically justified that, in a seismic active area, all structures should be designed to survive the strongest possible ground motion, without any damage. It is more reasonable to take the point of view that the structures must exceed a moderate earthquake without damage. In a rare event of very strong ground motion, damage would be tolerated as long as the structure collapse is prevented.

The code provisions for earthquake design cannot guarantee a safe structure or no damage during a very strong earthquake. Each building has some weakness not covered by the design and when a strong earthquake occurs, this weak point show up a serious damage.

So, the first step in the basic design philosophy is to define acceptable damage level due to an earthquake and this is the purpose of the design code. There is no general agreement yet on this damage level, but there are some general accepted criteria for determining these performances (Krawinkler, 1995):

(i) Life safety, which is primary requirement. The loss of life and the injuries in a building due to an earthquake are usually caused by the collapse of the building components. The evaluation of the number of deaths and injuries as an economic damage in an optimization process, poses very difficult ethical problems.

(ii) Collapse prevention, which is directly related to the prevention of loss of lives, injuries and damage of the contents of buildings. The structure can undergo important damage during the major earthquakes, but it must stand after the ground motion.

(iii) Reparable damage. A distinction is made between structural damage which cannot be repaired and damage which can be repaired. Irreparable damage is very much subject to individual engineering judgement of experts. The damage refers both to structure and to non-structural elements.

(iv) Acceptable business interruption, which can be appropriated by the owner. If the owner of a building wishes to avoid the cost of the interruptions, it is necessary to do more than the minimum requirement of codes. By using stronger and stiffer design, it is possible to reduce, or even eliminate, the building function interruption after a strong earthquake, but this means that more expensive structure result.

The main goal of seismic design and requirement is to protect life and structure collapse. However, the last earthquakes with extensive non-structural element collapses, interruption of functionality for many buildings, evacuation of people, losses in work places for varying periods of time, monetary losses, have shown that the above mentioned goal is not sufficient for a proper design methodology. So, the concept of more than one level of earthquake intensity must be adopted as a basic design philosophy.

A seismic territory can be subjected to low, moderate, or severe earthquake. The building may cross these events undamaged, can undergo slight, moderate or heavy damage, may be partial destroyed or can collapse. These levels of damage depend on the earthquake intensities and structure conformation. The low intensity earthquakes occur very frequently, the moderate earthquakes more rarely, while the strong earthquakes may occur once or maximum two times during the structure life. It is also possible that a devastating earthquake do not affect structure never during its life.

In these conditions, the checks required to guarantee a good behaviour of a structure during a seismic attack, must be examined at the light of a multi-level design approach. Thus, it is very rational to establish some limit states, as a function on the probability of a damage occurrence, both for structure and non-structural elements (Bertero, 1996). These limit states are presented in Fig. 5, in function of both structure and non-structural elements. In the seismic load-top sway displacement curve, there are three very important points: limit of elastic behaviour without any damage, limit of damage with major damage and limit of collapse, for which the structure is at the threshold of breakdown. In function of taking different limit states for structure and non-structural elements, some multi-level approaches are possible.

In spite of this very rational consideration, current code methodology is based on just one level criterion. A review of 41 codes elaborated all over the world, 38 are based on just one level, the principal design being concentrated on strength requirements (Bertero, 1997a).

Due to this code shortcomings, in the very last period some very important new conception are proposed, introducing four, three, or two design levels.

(i)Four levels. The performance-based seismic engineering elaborated by Vision 2000 Committee of SEAOC (Bertero, 1996) consists in a selection of appropriate parameters, so that at specified levels of ground motion and with defined levels of reliability, the structure will not be damaged, beyond certain limit states. Four specific limit states, or performance levels, have been defined, as a combination of damage of structure and non-structural elements and building contents (Fig. 5):

-fully operational, FO, or serviceability limit state, for which facility continues in operation without damage and only some minor damage in contents can occurs. The structure and non-structural elements work in elastic range;

-operational, O, or functional limit state, for which facility continues in operation without structural damage and with minor damage in non-structural elements. The structure works in elastic range, but for non-structural elements some partial plastic deformations occur;

-life safety, LS, or damageability limit state, for which the life safety is substantially protected, damage of structure is extensive, but reparable, and some non-structural elements are collapsed. The structure works in plastic field, numerous plastic hinges occur, the non-structural elements are out of work or partially damaged;

-near collapse, NC, or ultimate limit state, when life safety is risky, damage is severe, but the structure collapse is prevented. The structure is transformed in a plastic mechanism, which works near collapse. All the non-structural elements are collapsed.

These performance levels, or limit states, are connected with the specified probability of occurrence, given by the return period of each source. The correspondence return period-performance levels is a very disputable problem, depending on the socio-economic conditions of each country. But some medium values are general accepted for the following earthquake types:

-frequent, with a return period of 8-10 years;
-occasional, with a return period of 20-30 years;
-rare, with a return period of 450 years;
-very rare, being the maximum value, which can occurs, generally for over 970 years.

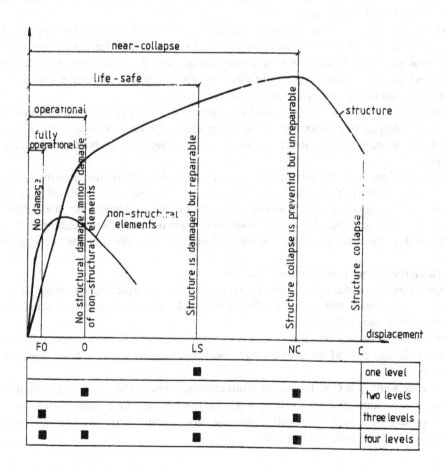

Fig. 5 Limit states

If for rare and very rare earthquakes there are no problems in definitions, for frequent and occasional earthquakes there are many proposals, depending on density of buildings in some zones. So, for California, Bertero, (1997b), proposed for return periods 47 years in the case of for frequent earthquakes and 72 years for occasional earthquakes.

The relation between the four limit states and the four probabilities of occurrence is presented in Fig. 6, as a performance objective matrix. The minimum objectives of an earthquake design illustrated in figure along the diagonal line. The unacceptable performance corresponding to situation under these minimum objectives. At the same time, there are enhanced objectives, if the owner consents a supplementary payment for providing better performance or lower risk than the one corresponding to the minimum objectives (Bertero, 1997a).

(ii) Three levels. Because it is very questionable to ask design engineers to perform so many verifications, it is more rational to introduce no more than three levels of verification (Mazzolani and Piluso, 1993, 1996):

-serviceability limit state, for frequent earthquakes with 10 years of return period. The corresponding design earthquake is called as service earthquake. This limit state imposes that non-structural elements should suffer minimum damage and the discomfort for inhabitants should be reduced to the minimum. So, for this level, the structure must remain within elastic range or may suffer unimportant plastic deformations;

-damageability limit state, for occasional earthquakes with 50 years of return period. This limit state considers an earthquake intensity which produces damage in non-structural elements and moderate damage in the structure, which can be repaired without great technical difficulties;

-survivability limit state, for earthquakes which may rarely occur, representing the strongest possible ground shaking, for which a return period of 450 years is considered. For these earthquakes both structural and non-structural damage are expected, but the safety of inhabitants has to be guaranteed, because no structure crash occurs. In many cases the damage is so ample that the structure cannot be repaired and the demolition is the recommended solution.

(iii) Two levels. If two levels are used those are :

-serviceability limit state, with a return period of 10 years, for which structures are designed to remain elastic or with minor plastic deformations and the non-structural elements remain undamaged or with minor damage;

-ultimate limit state, with a return period of 450 years, for which the structures exploit their capability to deform beyond the elastic range, the non-structural elements being partially or totally damaged.

In Eurocode 8 and U.B.C 97 the accelerations corresponding to the serviceability limit state are given as a fraction of corresponding ones for the ultimate limit state. Generally, this methodology cannot assure a controlled damage, also because the determination of this relationship is not clearly answered in code.

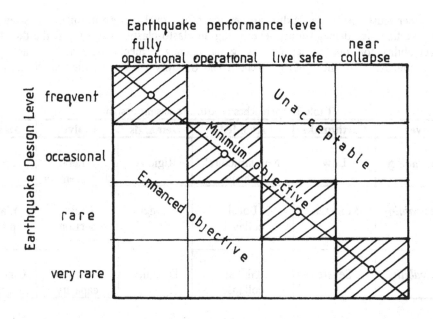

Fig. 6 Earthquake performance level

To be effective for design, these performance levels must be translated in seismic action values in term of design magnitude or accelerations. For this, it is necessary to dispose of a recurrence relation. For Vrancea and Banat earthquakes (Romania), this relation and the corresponding curves are plotted in Fig. 7, in case of maximum credible magnitude by 8.0 and 6.3, respectively. The following magnitudes for different limit states are presented in Table 1. Between these proposals, the verification for three level seems to be the more reasonable for practice. The approaches of this verification is presented in Table 2 (Filiatraut, 1996).

Table 1: Seismic magnitude at different levels

No. levels	Level	Vrancea	Banat
4	Fully operational	6.24	4.25
	Operational	6.64	4.70
	Life safety	7.90	5.90
	Near collapse	8.0	6.30
3	Serviceability	6.24	4.25
	Damageability	7.10	4.90
	Survivability	7.90	5.90
2	Serviceability	6.24	4.25
	Ultimate	7.90	5.90

The designer must verify the rigidity for serviceability level in elastic range, the strength of structure sections for damageability level using an elasto-plastic analysis and the ductility at the survivability for rotation capacity using the cinematic mechanisms at local and global structure collapses. So, the rigidity, strength and ductility triad is verified at different load levels.

Table 2: Approaches of structure seismic design

Level	Earthquake	Avoided	Demands	Analysis	Method
Serviceability	Low	Non-structural damage	Rigidity	Lateral displacement	Elastic
Damageability	Moderate	Local collapse	Strength	Cross-section capacity	Elasto-plastic
Survivability	Severe	Global collapse	Ductility	Rotation capacity	Cinematic mechanism

One of the most important problems of the multi-level design approach is the optimization of solutions. If the design is performed without any conception, it is possible that one of the limit state dominates the sizing. A structural solution able to satisfy two or more requirements represents the case of optimum solution.

2.2 Rigidity design criterion

The verification of structure rigidity is performed to avoid important damage of non-structural elements or building contents under minor earthquakes and to assure that the structure behaviour remain elastically. Damage in non-structural elements can be related directly to the level of distortion occurring during the earthquake. In the language of the structural engineer, this distortion is measured in the terms of **inter-storey drift**, that is, the lateral displacement of one floor relative to an adjacent floor. Inter-storey drift is usually expressed as a fraction or percentage of the inter-storey height.

In order to protect the non-structural elements, the rigidity design criterion is laid out through a required available formulation:

$$Required\ rigidity \leq Available\ rigidity \qquad (1a)$$

The relationship may be used also in the form of:

$$Determined\ inter\text{-}storey\ drift \leq Limit\ inter\text{-}storey\ drift \qquad (1b)$$

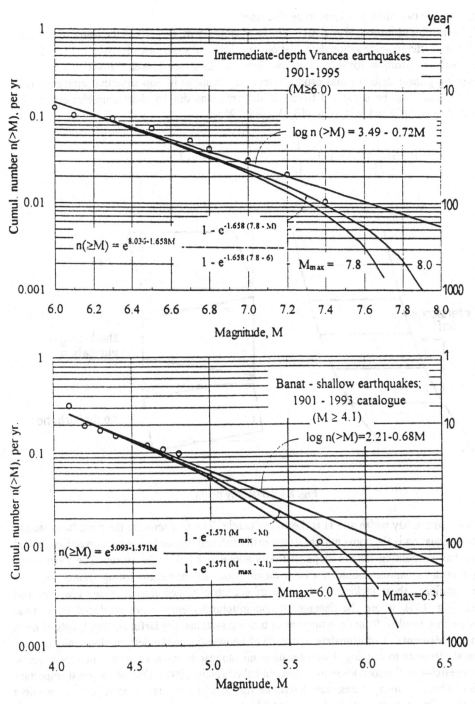

Fig. 7 Return periods

Thus, there are two main problems to be discussed:

(i) **Definition of inter-storey drift**. The inter-storey drift, (Fig. 8), is the result of a shear racking or shortening-elongation deformation, but there are some cases for which the drift occurs with no distortion, for significant foundation rocking (see Tower of Pisa).

Because only shear deformations produce important damage in non-structural elements, (Fig. 9a), these one must be separated from the deformations due to shortening-elongation. The tangential storey drift index is, (Fig. 9b), (Mayes, 1995):

$$R_T = \frac{1}{H}(u_3 - u_1) + \frac{1}{L}(v_6 - v_8 - v_2 - v_1) \qquad (2)$$

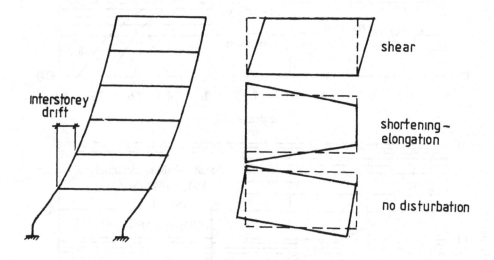

Fig. 8 Inter-storey drifts

in which L is the bay width and H is the storey height. The first term on the right-hand side of the above equation is the conventional storey drift ratio and the second is the correction applied to each bay accounting for the slope of floors above and below the storey. The shortening-elongation deformations should not be a major issue for moment-resisting frames, but can cause significant shear racking in the case of eccentric braced frames, (Fig. 10a), or dual frames, (Fig. 10b), where the damage of non-structural elements are produced by vertical displacements, resulted from deformations of bracing systems. **(ii) Deformation limits of non-structural elements.** Unfortunately, the stage of knowledge on the deformation limits of non-structural elements to racking distortion does not appear to have advanced in recent years. Little research-based information has been published (Nair, 1995). One of the most important criterion which, in many cases, can determine the structure sizing, is based on the obsolete information On most modern buildings nowadays an attempt is made to isolate the exterior cladding from the deformation of the structure due to seismic loading. The interior walls are detailed in such a way as to isolate them from structure deformations. If this were done, the deformation tolerance of the non-structural elements would not be a fact in establishing the lateral rigidity criterion for structure and only the building contents protection remain a criterion for inter-storey drift limitation.

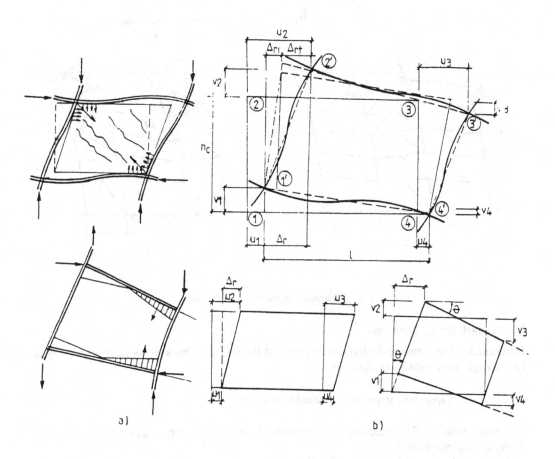

Fig. 9 Tangential inter-storey drift

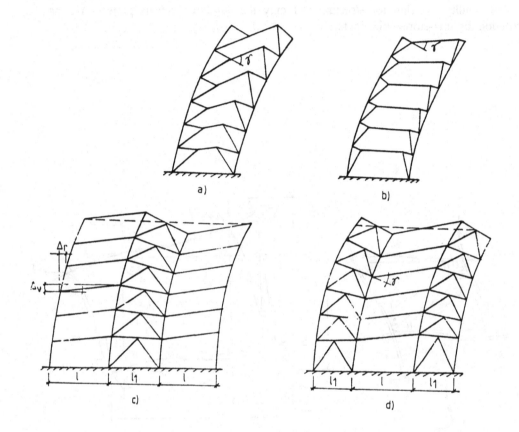

Fig. 10 Displacements in braced frames

2.3 Strength design criterion

Traditionally, the structure design has emphasized the strength checking as their primary goal. The strength must satisfy the equation:

$$Required\ strength \leq Available\ strength \tag{3}$$

The two terms of this equation are determined from the seismic actions and structure configuration, respectively.

   (i) Analysis of required strength. The stress due to seismic forces can be evaluated by means of different methods, which provide different degree of accuracy:

- *Equivalent static analysis* based on the assumption that the structural behaviour is governed by the first vibration mode. The characteristics of ground motions are described by means of linear elastic spectrum, Fig. 11a. For the inelastic deformations the design spectra are obtained by scaling the elastic spectra by means of a reduction factor, namely q-factor, which takes into account the energy dissipation capacity of the structure. The method has many shortcomings, being based on the grossly simplified physical models, so its use is limited only for simple structures. In order to improve this method efforts are paid for defining and refining the reduction factor q, which is regarded as a magic factor able to solve all difficult problems (Aribert and Grecea, 1997).

- *Push over analysis*. Following this method the structure is subjected to static incremental lateral loads which are distributed along the height of the structure, Fig. 11b. A nonlinear inelastic static analysis is performed and a load-deflection response curve is obtained. The method is relatively simple to be performed, but contains many limiting assumptions: the variation of loading pattern, the influence of higher modes and the effect of resonance. In spite of these deficiencies, the push-over method can provide a reasonable estimation of the global structure capacity, especially for structures which behave according to first mode.

- *Time-history analysis*. This method is based on the direct numerical simulation of the motion differential equations, using real or simulated ground motions. In this analysis the elaso-plastic deformations of the structure members are considered. The variations of displacement at diverse levels of a frame are presented in Fig. 11c. This is the only method able to describe the actual behaviour of a structure during an earthquake. But the great problem of this method is the choice of a proper accelerogram record, as it is presented in Section 1.3.

(ii) **Analysis of available strength**. The steps in the determining the available strength are the followings:

- *Establishing the suitable plastic mechanism*, between the beam hinge model (BH), column hinge model (CH), or weak storey model (WS), (Fig. 12a). The most suitable plastic mechanism is the beam hinge model and the obtaining this mechanism type must be the main objective of the design strategy;

- *Identifying the critical sections*, named as dissipative zones where the plastic hinges have the task to dissipate the seismic energy input. These zones may be the ends of members or the joints, (Fig. 12b). For all these sections, the determination of internal axial forces, bending moment and shear forces, taking into account the plastic behaviour of sections, plays a very important role in a structure design.

- *Sizing and detailing the plastic regions*, in order to assessing the critical sections to have the capacity to dissipate the seismic energy. At this step it is very important to consider the over-strength of these critical sections due to some factors like the variation of material characteristics, hardening effect, over-sizing the cross-section, redistribution of internal forces, etc.

- *Sizing and detailing the elastic regions*, which must exhibit a nominal strength greater than ones in plastic hinges, taking into account the possible over-strength. The aim of this step is to assure that no accidental plastic hinges can occur in some regions, which are not designed for this possibility. For instance, if it is considered in analysis that plastic hinges occur only in beams, it is very important to provide the joints with sufficient strength, so that no plastic hinges accidentally form there, (Fig. 12c).

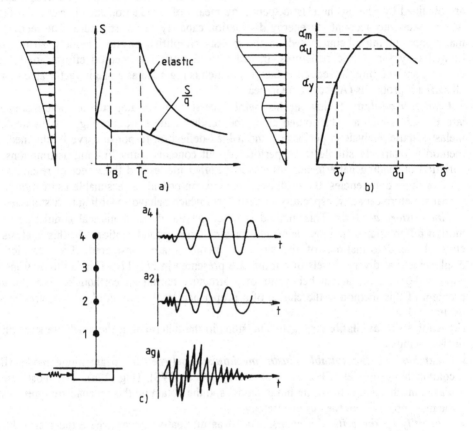

Fig. 11 Method of analysis

2.4 Ductility design criteria

The ductility is the capacity of structure to undergo high plastic deformations in some predetermined locations, ensuring a protection against structure collapse during severe earthquakes.

Recent development of advanced design concepts, as the ones included in the capacity design method (Bachmann et al, 1995), is based on the objective to provide a structure with sufficient ductility, in the same way as for strength and stiffness. For this, a comprehensive and transparent methodology for direct ductility control must be developed, with the same rigorous as for resistance and displacement checking. This purpose may be achieved if it is satisfied the criterion, Fig. 13:

$$\text{\textit{Required ductility} ≤ \textit{Available ductility}}$$ (4a)

where the required ductility is determined from the global behaviour of structure and the available ductility from the local plastic deformation. So, the ductility design criteria may be transformed in the relation:

$$\text{\textit{Required global ductility} ≤ \textit{Available local ductility}}$$ (4b)

These two ductilities are defined as:
  Global ductility or *displacement ductility*, determined at the level of full structure:

$$\mu_\delta = \frac{\delta_u}{\delta_y}$$ (5)

where δ_u, δ_y, are the elasto-plastic and the yielding top sway displacements, respectively.

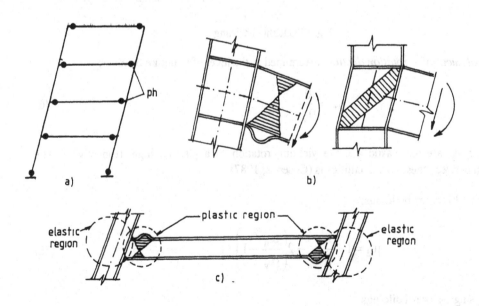

Fig. 12 Analysis of available strength

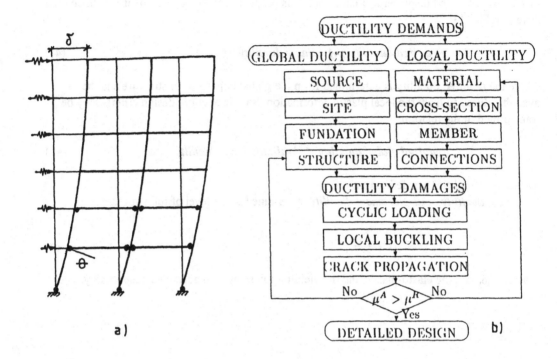

Fig. 13 Ductility checking

Local ductility or *rotation ductility* determined at the level of structure member:

$$\mu_r = \frac{\theta}{\theta_y} \tag{6}$$

where θ, θ_y, are the plastic and the yielding rotation of a plastic hinge, respectively. The relation between these two ductilities is (Cosenza, 1987):

-for multi-storey buildings:

$$\mu_d = 1 + \frac{2}{3}\mu_r - 2\left(\frac{\alpha_u}{\alpha_y} - 1\right) \tag{7a}$$

-for single-storey buildings:

$$\mu_d \approx 1 + \mu_r \tag{7b}$$

In relation (7a), α_u , and α_y are the multiplier α of horizontal forces for ultimate and for first yielding limit state, respectively.

For global ductility the hierarchy is at the level of source, site, foundation and structure, while for local ductility, cross-section, member and connections, are the main factors. The correlation between these two ductilities is performed at the level of ductility damages. These last are influenced by the cyclic characteristics of earthquake motions and the member collapse by local buckling or crack propagation.

For assessment of global ductility it is necessary to gather information on characteristics of possible earthquakes. A detailed description of these earthquakes is out of our scope, so only the aspects directly connected to the problems of structure ductility will be presented. With this purpose it is necessary to underline the importance of engineering judgment from a practical point of view and the collaboration of seismologists, geotechnists and structural engineers, for establishing the most characteristic design earthquakes. Source, site, foundation and structure directly influence the global ductility, (Fig. 14) (Gioncu, 1997).

(i) Global ductility. The estimation of required global ductility is based on the ultimate top sway displacement of structure. Unfortunately, there are no standard definition, accepted by all the specialists, with respect to this ultimate displacement. There are some different approaches, among them the following being the most reasonable:

- **Approximate estimation** is given by the relationship between the global ductility, μ_d and the behaviour factor q. It is generally accepted that:
-for short vibration period:

$$q = (2\mu_d - 1)^{1/2} \qquad (8a)$$

-for moderate and long periods

$$q = \mu_d \qquad (8b)$$

GLOBAL DUCTILITY	
GROUND MOTIONS	**STRUCTURE RESPONSE**
Source	Foundations
• **Earthquake type**	• **foundation type**
• **focal depth**	• **base isolation**
Distance from source	Structure system
• **near and far-field**	• **structure type**
• **attenuation**	• **collapse mechanism**
Site	Non-structural elements
• **soil profile**	• **interaction**
• **amplification**	• **damage limits**
• **duration**	• **collapse limits**
REQUIRED DUCTILITY	

Fig. 14 Factors influencing global ductility

- **Push over analysis,** Fig. 15a, in which the structure is subjected to incremental lateral loads, using a predetermined load patterns of horizontal forces. The ultimate global ductility can be defined as for α_m, at the maximum value, or taking into account of the post-mechanism branch, for $0.9\alpha_m$. By considering that for seismic action, where the direction of forces is variable, the plastic mechanism does not represent the collapse of structures, the second definition seems to be more reasonable.
- **Time-history analysis,** Fig. 15b, using a recorded or artificial accelerogram. The plastic rotation of these plastic hinges represents the required ductilities of beams and columns.

(ii) Local ductility. The local ductility can be defined at the level of material, cross-section, member, or connection. The factors influencing this ductility and the results which must be used in design practice are presented in Fig. 16 (Gioncu, 1997). The estimation of local ductility is based on the ultimate rotation capacity of plastic hinge, where the moment rotation curve plays a leader role. Unfortunately, as for global ductility, there are no standard definition accepted by all specialists, with respect to ultimate rotation.

There are three different approaches, (Fig. 17):

- The rotation capacity is defined related to the maximum moment M_{max}.
- The rotation capacity is determined in the lowering post-buckling curve at the intersection with the theoretical full plastic moment. This definition is given in the Background Document 5.02 Eurocode 3. In the same document of Eurocode 8 there is also a proposal to use a reduced plastic moment $Mp / 1.1 \approx 0.90 M_p$.
- The rotation capacity is considered in a more general way; in addition to rotation capacity corresponding to maximum moment, the slope of the descending part of the moment-rotation curve is calculated.

In order to improve these definitions, Gioncu and Mazzolani (1995a) have shown that the values reffering to M_{max} give unacceptable very low values for rotation capacity. The use of reduced values of M_{max} is a very disputable proposal, because the evaluation of M_{max} contains much incertitude. The proposal to consider also the slope of descending path is a very promising definition, because it is very important to know the member behaviour after ultimate rotation. A slow slope allows a supplementary ductility, while an abrupt slope is the indicator of a bad post-buckling behaviour. But this procedure remains as a guideline for the future research works, due to the difficulties to be used now in design practice. In these conditions the second definition proposed by Eurocode 8, seems to be the more reliable. The formula to calculate the **available rotation capacity**, as a measure of the local ductility, is given by, (Fig. 17):

$$\mu_r = \frac{\theta_{pu}}{\theta_p} = \frac{\theta_u}{\theta_p} - 1 \tag{9}$$

where θ_{pu} is the ultimate plastic rotation, θ_p is the rotation corresponding to the first plastic hinge and θ_u is the total ultimate rotation. For high moment variation in plastic range, the ultimate rotation can be obtained without any problem at the intersection with the full plastic moment M_p. But in the case of quasi-uniform moments, where, in many cases, the maximum moment does not reach the full plastic moment, it is possible to use a reduced reference plastic moment, for instance $0.9M_p$.

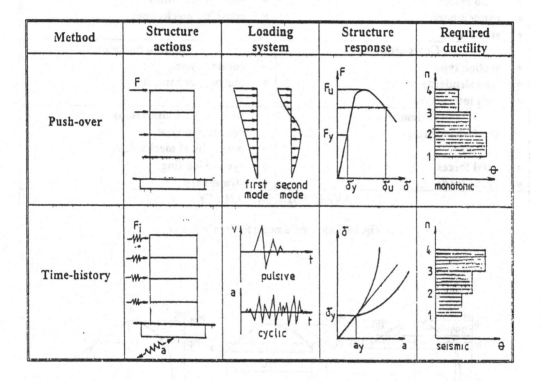

Method	Structure actions	Loading system	Structure response	Required ductility
Push-over		first second mode mode		monotonic
Time-history		pulsive cyclic		seismic

Fig. 15 Global ductility analysis

The problem of rotation capacity recently demonstrated a primary interest, as it is testified by several papers, presenting different methods:

-using of FEM, the ABAQUS and PROFIL computer programs being the most known;

-integrating the moment-curvature relationship; in this method only the rotation capacity corresponding to maximum moment is obtained;

-interpretation of the collapse plastic mechanism, coming from experimental evidence;

-using the effective width method;

-statistical analysis of experimental data of full-scale member tests;

-statistical analysis of numerical tests.

Among these procedures, certainly the best one is the finite element method, but, due to the necessity of a very dense network in the buckled zone the computer time and cost are very high. In this conditions, the method of plastic mechanism seems to be the most adequate for determining the rotation capacity for design purposes.

LOCAL DUCTILITY	
ELEMENT	**JOINTS**
Material	**Joint panel**
• steel grade • yield ratio • randomness • strain-rate	• joint panel type • shear mechanism • crushing mechanism
Cross-section	**Column flanges**
• section type • wall slenderness • wall interactions	• column type • plastic mechanism
Members	**Connections**
• strain-hardening • buckling • axial forces • cyclic loads	• connection type • plastic local mechanism • cycling loading • strain-rate
AVAILABLE DUCTILITY	

Fig. 16 Factors influencing the local ductility

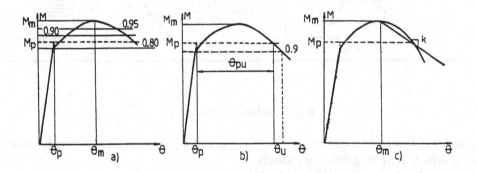

Fig. 17 Definition of available local ductility

This method has been proved to be a very efficient procedure by the studies of Timisoara research team. A computer program, DUCTROT (with the variants 93, 95, 96, 97 and now 99) has been elaborated (Petcu and Gioncu, 1997). Based on local plastic mechanism, (Fig. 18a), a post-buckling path is determined, allowing the calculation of ultimate plastic rotation, (Fig. 18b). The computed results have been compared with over 60 experimental tests, collected from technical literature, showing a good correspondence.

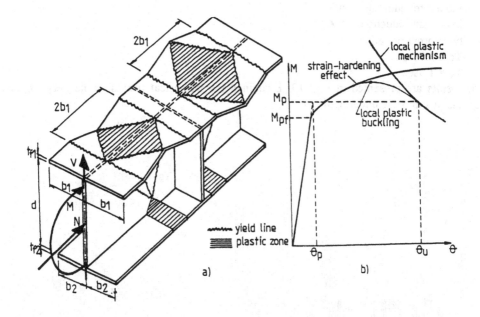

Fig. 18 Local plastic mechanism

2.5 Correlation between design criteria

The 3 and 6 levels MRFs was designed to satisfy the rigidity, strength and ductility criteria. For loading parameters the following values are used (Tirca et al, 1997):

- α, depending one site zone:
  -zones with low seismicity: $\alpha = 0.12g$; $0.16g$;
  -zones with medium seismicity: $\alpha = 0.20g$, $0.24g$;
  -zones with high seismicity: $\alpha = 0.28g$, $0.32g$.
- q, considering the frame ductility:
  -low ductility, $q = 4$;
  -medium ductility, $q = 6$;
  -high ductility, $q = 8$.
- two plastic mechanism types are analysed:
  -local plastic mechanism (L);
  -global plastic mechanism (G).
- for intr-storey drift criterion, a design spectrum for serviceability limit state are used with the values for reduction coeficient v :
  -limited reduction: $v = 2$;

-moderate reduction: $v = 3$;

-important reduction: $v = 4$.

• two corner periods:

-Tc = 0.7sec;

-Tc = 1.5sec.

The results are presented in Fig. 19. Examined the numerical results, some very important conclusions must be underlined:

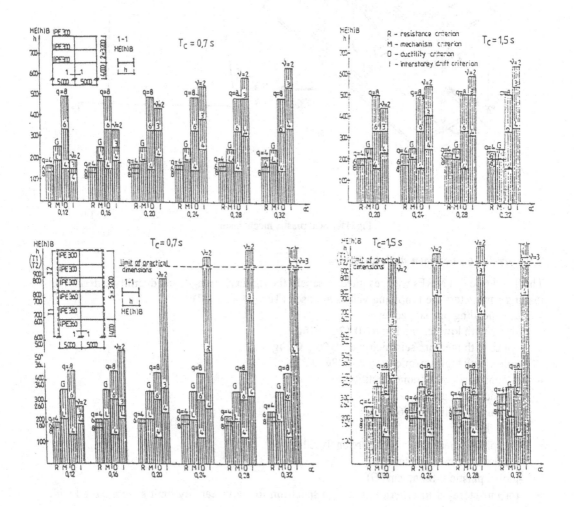

Fig. 19 Correlation between design criteria

-the resistance criterion is not an important requirement for frames with moderate height, but becomes decisive for tall frames;

-the mechanism criterion depends only on storey number and the beam bearing capacity, and generally, it is exceed by the other criteria. For zones with low seismicity the global mechanism criterion is too drastically , and a local mechanism, associated with a reduced behaviour factor q is recommended;

-the ductility criterion is essentially influenced by the value of q-factor. For high values, the demand for ductility is very high, so a balance between these two factors for ductility and action reduction must be considered. For frames with reduced number of stories it is recommendable to use a reduced value q = 4. Only for tall buildings it is suitable to be considered the high values, q = 6 or 8;

-the inter-storey drift criterion depends on the limit of the corner period (Tc) but, essentially on the value of the reduction factor v, which defines the service seism. Up to the present no semnificant studies exist for determining this very important coefficient. In EC-8 it is proposed v = 2, but, if the collapse earthquake is defined for 475 years, and the service one for 10 years it seems that this value is too small and the influence of this criterion on the frame dimensions is very important. Therefore, value v =3...4 are more resonable;

Reaserch works considering more reasonable and diversified limits for the inter-storey drift are also required. In the future more attention mast be paid in this aspect, because the inter-storey drift criterion plays a leading role in the frame sizing;

-for structures in zones with low seismicity the inter-storey drift criterion is not decisive, and the choosing of value for v coefficient is no very important. Using a reduced value for behaviour factor q = 4, associated with a local plastic mechanism is the best solution;

-for structures in zones with medium seismicity, one can see values q = 6...8, for which the global mechanism must be considered. For inter-storey drift criterion it is recommended to use v = 3...4;

-for structures in zones with high seismicity, the resistance and inter-storey drift criteria become very important. It is recommandable to use high value for q-factor, q = 8, associated with global mechanism, and values v = 3...4.

3.   PROGRESS IN GROUND-MOTION VALUATION

3.1 Source characteristics

A great amount of information concerning the feature of earthquakes is collected and important databases are operative. In the same time, important activity in macro and microzonation has been carried out all over the world to identify and characterize all potential sources of ground motions. For the structural engineers the interest of these results is focused in the source characteristics with direct influence on seismic action. There are the source mechanism and the spatial description (Lam, 1996).

The **source mechanism** may be, (Fig. 20):

-intrreplate mechanisms, produced by sudden relative movement of two adjacent tectonic plates at their boundaries. Very large magnitude and large natural vibration periods and duration characterize these interplate earthquakes;

-intraplate mechanisms, associated with relative slip across geological fault within a tectonic plate. Such earthquake types are generally smaller magnitude, natural vibration periods and duration.

The Californian, Japanese and New Zealand earthquakes are interplate ground motions, while European (with the exception of some Romanian and Greek sources), Canadian and Australian earthquakes are intraplate ground motions.

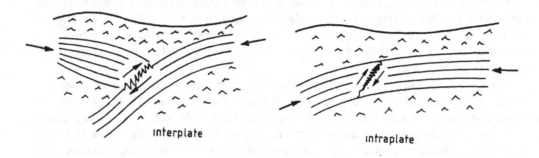

interplate   intraplate

Fig. 20 Mechanism types

The spatial description refers to the focal depth and directionality, (Fig. 21). Focal depth has a considerabie influence on the earthquake behaviour, the area affected being directly related to this depth. There are the following earthquake types:
- shallow crustal earthquakes, having the hypocenter situated at the deep till 25Km;
- crustal earthquakes, with depth from 25 to 70 Km;
- intermediate earthquakes, with depth from 70 to 300 Km;
- profound earthquakes, having depth over 300 Km.

The importance of source depth is underlined by the attenuation law, (Fig. 22). The attenuation is very important for crustal earthquakes and reduced for deep earthquakes. So, the surface earthquakes have influence on the reduced area around the epicentre, while for deep earthquakes the area are very large. For Europe and Mediterranean areas the shallow crustal events are the most frequent event, over 85 percent being ranged till 15 Km (Ambraseys and Free, 1997), (Fig. 23).

3.2 Near-source and far-source ground motions

During the past 20 years an ever-increasing database of recorded earthquakes has indicated that characteristics of ground motions can vary significantly between recording stations that are located in the same area. This is particularly true for recording stations located near the epicentral region. There are also very important differences between characteristics of ground motions recorded in epicentral areas and at same distance from the epicentre. So, in the last time, two main region with different ground motions are considered, (Fig. 24):

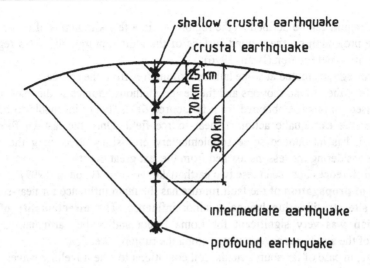

Fig. 21 Source depth

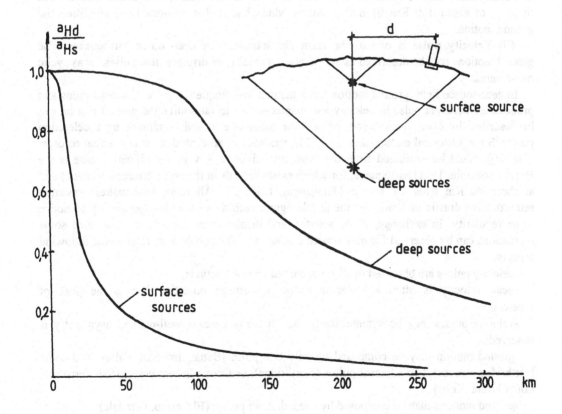

Fig. 22 Attenuation law

-near-source region can be defined as the region within a few kilometres of either the surface rupture or the projection on the ground surface of the fault rupture zone. This region is also referred as the near-field region (Iwan, 1996);

-far-source region situated at some hundred kilometres from the source;

Unfortunately, the ground motions and the design methods adopted in the majority of codes are mainly based on records obtained from far-source fields, being incapable to describe in a proper manner the earthquake action in near-source field. Only the last Uniform Building Code, UBC'97, has introduced some supplementary provisions concerning the near-source earthquakes, considering the lessons learned from the last great events.

The main differences between these two earthquake types are (Gioncu, 1999):

(i) **Direction of propagation** of the fault rupture has the main influence for near-source fields, the **local site stratification** having a minor influence. **The directionality of the wave propagation** was very significant for Loma Prieta and Kobe earthquakes, where **the damages** of the structures are connected along the rupture direction.

In exchange, in case of far-source fields, soil conditions for the travelling waves and the site stratification are of the first importance. The peak acceleration amplification at 400 kilometres from the source in the case of the Michoacan earthquake is a very good example for this aspect. The same effect may be observed in the case of Vrancea earthquake, recorded at Bucharest at distance of about 160 Km from the source, where bad soil conditions have amplified the ground motion.

(ii) **Velocity pulse** is one of the main characteristics of near-source earthquakes. The ground motions may be described as acceleration, velocity, or displacement pulses, or as cyclic movements.

In near-source field, ground motion has a distinct low-frequency pulse in accelerations and pronounced coherent pulse in velocity and displacement. In far-source the ground motion can be described by cyclic movements or in some cases due to soil conditions by acceleration pulses. Some historical earthquakes, (Fig. 25), recorded in epicentral areas and actual records, (Fig. 26), could be examined from the mentioned figures. A very semnificative case is the Banat (Romania, 1991) earthquake, for which exists records in the epicentre zone (Banloc) and at about 40 Km from the source (Timisoara), (Fig. 27). All these earthquakes present a semnificative drastic variation for the accelerogram records without any possibility to notice some regularity. In exchange, if the velocity and displacement records are examined some regulations can be observed for near-source earthquake. The conclusions refer to the following aspects:

-velocity pulses are observed in all the recorded ground motions;

-peak velocity is often a better indicator of damage potential than is the peak of acceleration;

-velocity pulses may be symmetrically, but in many cases a well-marked asymmetry is observed;

-ground motion may be composed by only one pulse (Banat, Imperial Valley, and some Northridge records), two adjacent pulses (Northridge), or three adjacent pulses (San Fernando, Loma Prieta, Kobe);

-ground motions may be composed by some distinct pulses (El Centro, Ferndale).

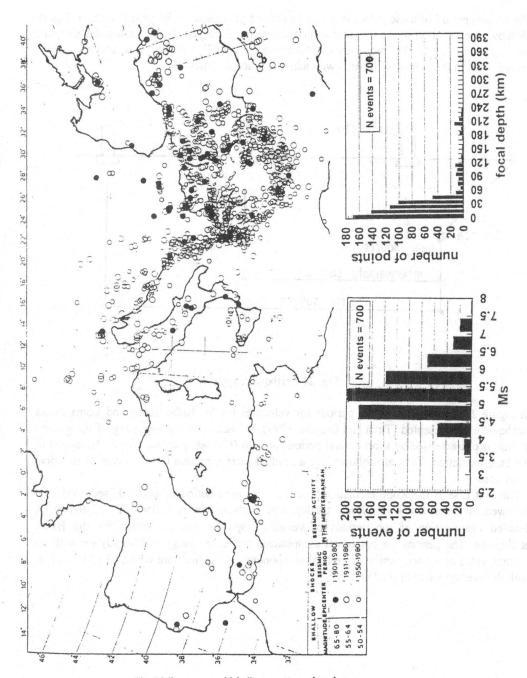

Fig. 23 European and Mediterranean earthquakes

The natural period of these pulses is a very important parameter of the ground motion. Figure 28 shows the variation of the frequencies in acceleration during the Kobe earthquake. One can see large variation of the frequency (Mohammadioun, 1997), but the pulse periods do not exceed 1 sec in the portion of records with maximum accelerations.

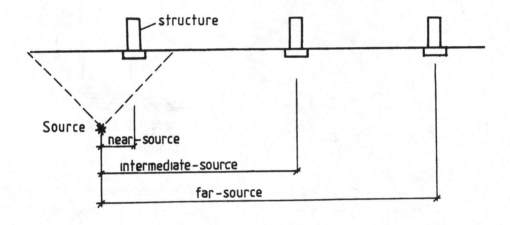

Fig. 24 Earthquake types

In Figure 29 the variation of the periods for velocities for the Kobe, Banat and Loma Prieta earthquake are presented (Tirca and Gioncu, 1999). In each case, the beginning of the ground motion is characterised by short natural periods of 0.20-0.30 sec. periods, which correspond to the base excitation. They are followed by a gradual increasing due to the effects of the local conditions.

The first case is the Kobe earthquake, with a maximum period of about 0.7sec., which is followed by Banat and Loma Prieta earthquakes, with maximum periods of 0.3-0.4sec. A detailed examination of the periods of velocity impulses was performed for the Banat earthquake. The periods for horizontal components are of the range (0.25-0.80) sec with an average value of 0.4 sec, while for vertical components, the periods are within (0.15-0.50) sec, with the average value of 0.22 sec.

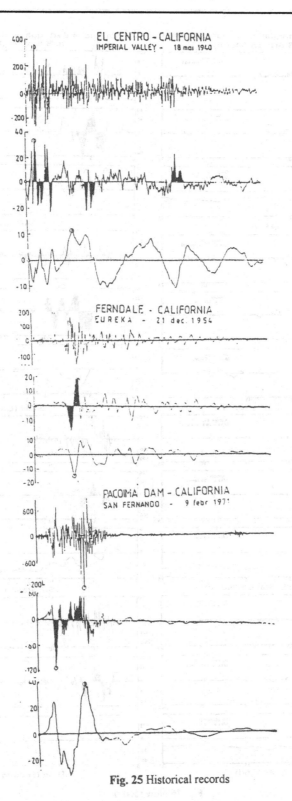

Fig. 25 Historical records

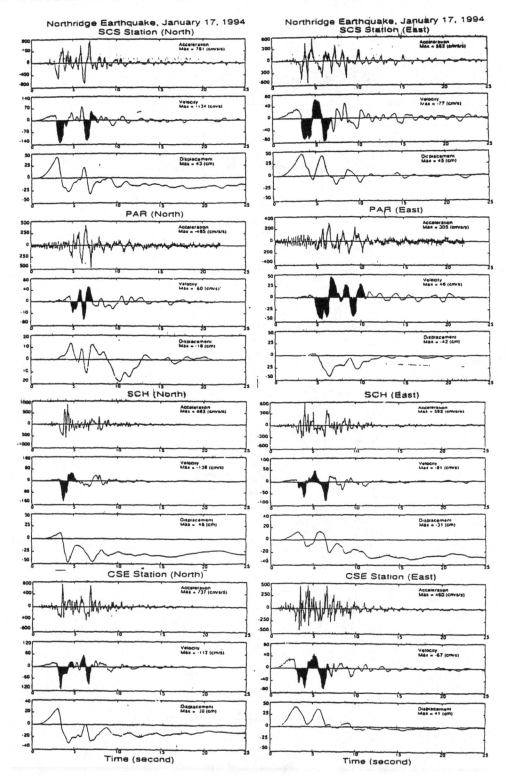

Fig. 26 New records

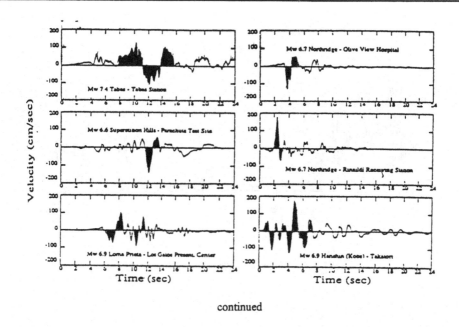

continued

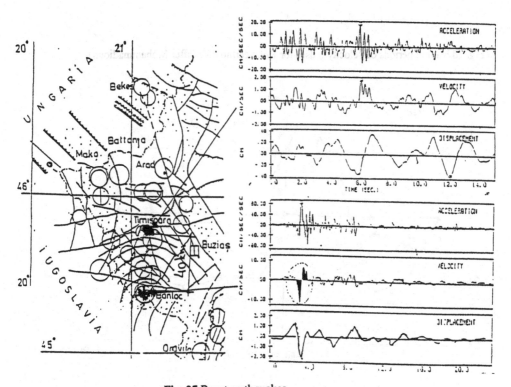

Fig. 27 Banat earthquakes

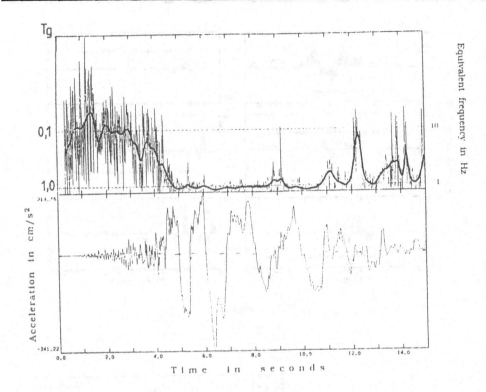

Fig. 28 Natural periods of velocity pulses for Kobe earthquake (after Mohammadioun)

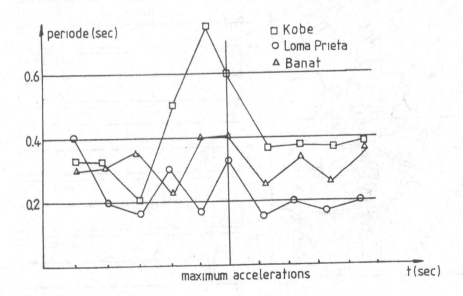

Fig. 29 Variation of natural period

During some recorded ground motions (Landers, 1993), the ground displacements were very important, being very directionally associated with the fault-rupture progress (Hall, 1999). These displacements pulses can be one of the most damaging earthquakes, if the structure is situated on the rupture line.

In exchange, for far-source records, cyclic ground motions may be noticed. But in some cases, when the soil conditions are very bad, acceleration pulses being observed in recorded accelerograms. These were the cases of Bucharest, (Fig. 30) and Mexico City. The distances of epicentre were about 160 Km for Bucharest earthquake and 400 km for Mexico City earthquake.

(iii) The **vertical components** in the near-source field could be greater than the horizontal ones, due to the direct propagation of the P-waves. Figure 31 shows the ratio of vertical to horizontal spectra for the Northridge earthquake (Hudson et al, 1996). Similar aspects were related to other earthquakes (Table 3).

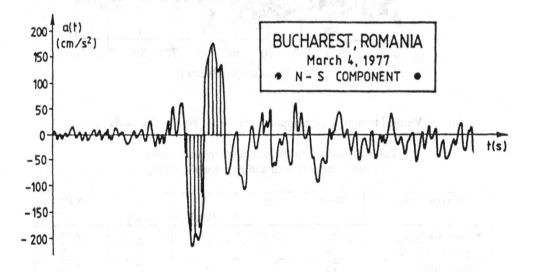

Fig. 30 Bucharest 1997 earthquake

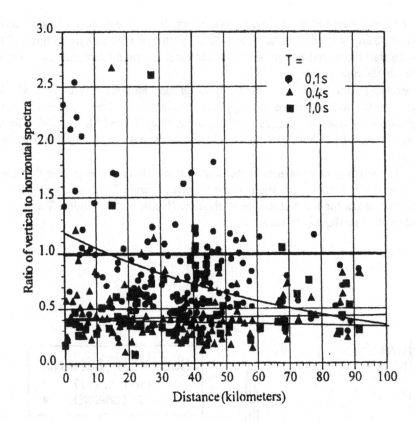

Fig. 31 Vertical to horizontal component ratios for Northridge earthquake

Table 3: Vertical to horizontal component ratios
(data from El Nashai and Papazouglu, 1997)

Earthquake	Date	M	d [km]	h [km]	V/H
Nahani, Canada	85	6.79	8	6	2.15
Coyote Lake, USA	79	5.72	4	6	1.78
Gazli, USSR	76	7.10	19	10	2.28
Imperial Valley, USA	79	6.86	1	8	1.85
Imperial Valley, USA	79	6.86	1	8	3.70
Imperial Valley, USA	79	6.86	6	8	0.62
Imperial Valley, USA	79	6.86	4	8	1.17
Imperial Valley, USA	79	6.86	4	8	1.53
Imperial Valley, USA	79	6.86	5	8	1.75
Kobe, Japan	95	7.20	18	14	0.54
Kobe, Japan	95	7.20	20	14	1.96

Earthquake	Date	M	d [km]	h [km]	V/H
Kobe, Japan	95	7.20	25	14	1.56
Loma Prieta, USA	89	7.17	0.1	17	0.92
Loma Prieta, USA	89	7.17	14	17	1.35
Loma Prieta, USA	89	7.17	16	17	0.91
Loma Prieta, USA	89	7.17	9	17	1.09
Loma Prieta, USA	89	7.17	7	17	1.32
Morgan Hill, USA	84	6.17	15	8	1.87
Morgan Hill, USA	84	6.17	14	8	3.94
Morgan Hill, USA	84	6.17	14	8	1.86
Morgan Hill, USA	84	6.17	17	8	1.76
Montenegru, Yug.	79	7.04	21	10	1.74
Northridge, USA	94	6.70	6	18	1.79
Northridge, USA	94	6.70	12	18	0.94
Northridge, USA	94	6.70	9	18	0.89
Whittier, USA	87	6.00	3	14	0.87
Whittier, USA	87	6.00	4	14	1.65
Whittier, USA	87	6.00	11	14	1.38

The attenuation of the vertical to horizontal components in function of the epicentral distance is presented in Figure 32 (El Nashai and Papazouglu, 1997). One can notice a very important reduction of this ratio with the distance.

(iv) The **velocities** in near-source field are very high. During the Northridge and Kobe earthquakes, values of 177 cm/sec were recorded at the soil level (Table 4). The spectral velocities are drown in Figure 33 showing for Northridge and Kobe earthquakes an increasing till the values of 400 - 500 cm/sec for short periods. The velocities increase dramatically near the epicenter (Fig. 34).

So, in the case of near-source field, the velocity is the most important parameter in the design concept, replacing the acceleration, which is the dominant parameter for the far-source field.

For far-source earthquakes, the velocities of horizontal and vertical components are more reduced as values than in the near-source area, not exceeding 30 – 40 cm/sec.

4. PROGRESS IN STEEL STRUCTURE RESPONSE ANALYSIS

4.1 Artificial generated accelerograms for pulse actions

As a consequence of the above mentioned differences in the ground motions, there are some important modifications in the behaviour of the structures subjected by near-source earthquakes versus structures acted by far-source ground motions.

The seismic response of a structure with a complex behaviour can be correctly calculated only through direct non-linear dynamic time-history analysis, where the structural model is subjected to a base acceleration input. The used accelerograms have to represent adequately the characteristics of the expected earthquake at the site of the structure.

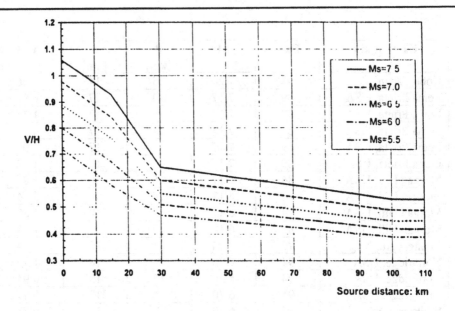

Fig. 32 Attenuation of V/ H ratio with the epicentral distance

Table 4: Peak near-source ground motions
(Data from Hall, 1995)

Earthquake	Date	Mg	Distance [km]	Acceleration [g]	Velocity [cm/s]	Displace-ment [cm]
Imperial Valley						
El Centro Array 7			1	0.65	110	41
El Centro Array 6	1979	6.5	1	1.74	110	55
Bonds Corner			4	0.81	44	15
El Centro Array 5			4	0.56	87	52
El Centro Array 8			4	0.64	53	29
Loma Prieta						
Los Gatos Presentation Center	1989	6.9	0	0.62	102	40
Lexington Dam			5	0.44	120	32
Landers						
Lucerne	1993	7.2	1	0.90	142	255
Northridge						
Rinaldi Receiving Station			0	0.85	177	50
Sylmar Converter Station			0	0.90	129	50
Los Angeles Dam			0	0.32	79	22
Sepulveda Vetrans Hospital			0	0.94	75	15
Jensen Filtration Plant	1994	6.7	0	0.85	103	38
Sylmar County Hospital			2	0.91	134	44
Van Nuys (hotel)			2	0.47	48	13
Arleta fire station			4	0.59	44	15
Newhall fire station			5	0.63	101	36
Tarzana nursery			5	1.82		
Hanshin						
Kobe (JMA)	1995	6.9	0	0.85	105	26
Kobe University			0	0.31	55	18
Takatori			0		176	

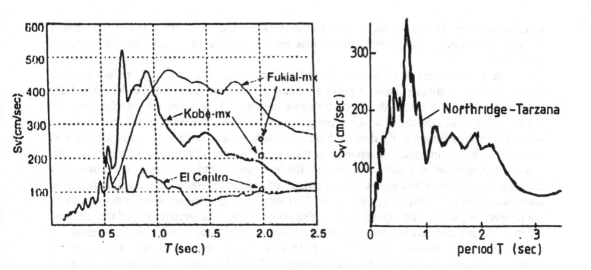

Fig. 33 Velocity response spectra for Northridge and Kobe earthquakes

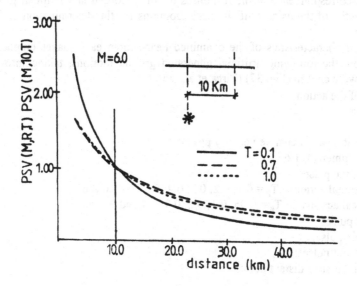

Fig. 34 Velocity of vertical components near to source

As a result of the extensive monitoring of the areas with high seismic risk (see USA-California and Japan) it was possible to collect a large number of records, in recent years. All these records have random characteristics. The design practice consists in the chose of the most appropriate ground motion type and to perform a time-history analysis, obtaining the randomness values, involved by the used acceleration records. This procedure has some major disadvantages, presented in the introduction.

So, as an alternative to earthquake recordings, the use of artificial generated accelerograms, with respect to certain given restraints, has been allowed by some recent codes. These control parameters should be chosen with regard to the characteristics of the expected earthquakes at the site, depending on the generic mechanisms and the position of the site, in relation to the focal area.

In order to control the possibility to use artificial accelerograms for the determination of the behaviour of the structure during a pulse earthquake, the response should be determined for time fragments in the recorded accelerograms. Figure 35 presents the accelerogram of Bucharest 77, where the duration of the main important 17 seconds was divided into four time intervals (denoted from a to d): three of 5 seconds each one and another of 2 seconds (Ifrim et al, 1986). The 'a' interval corresponds to the beginning, 'b', represents the major ground motion characterized by a pulse acceleration, the intervals 'c' and 'd' are the consequences of the main shock. Figure 36a shows the importance of each accelerogram fragment, considering each interval as an independent earthquake. It is very clear that the response is dominated by the interval 'b', which correspond to the pulse acceleration. The superposing of the response for the different intervals is presented in figure 36b. Curve A corresponds to the interval 'a', curve B, to 'a'+'b', curve C, to 'a'+'b'+'c' and the curve D, to the sum of the four intervals 'a' to 'd'. One can see that for periods ranged between 0.2 and 1.6 seconds, the curves B, C and D practically coincide. The simulation of the structural response, using a base harmonic excitation with one and two pulses corresponding to the interval 'b', is also presented in Figure 36b. The good correspondence of the responses for the recorded and artificial ground motion shows the potential of the using artificial accelerograms for the determination of the response of the structure.

Considering the characteristics of the examined near-source earthquakes emphasized in the previous section, the following artificial generated ground motions, taking into account the pulse velocity were chosen (Fig. 37) (Tirca et al, 2000).

- direction of the action:
  - horizontal;
  - vertical;
- ratio of positive and negative velocity peaks:
  0.6; 1.0 (symmetry); 1.6;
- duration of the pulse:
  - for horizontal actions: $T_g = 0.1; 0.2; 0.3; 0.4; 0.5$ and 1.0 sec
  - for vertical actions: $T_g = 0.05; 0.1; 0.2; 0.3; 0.4$ and 0.5 sec
- number of pulses:
  - one single pulse;
  - two adjacent pulses;
  - two distinguished distant pulses.

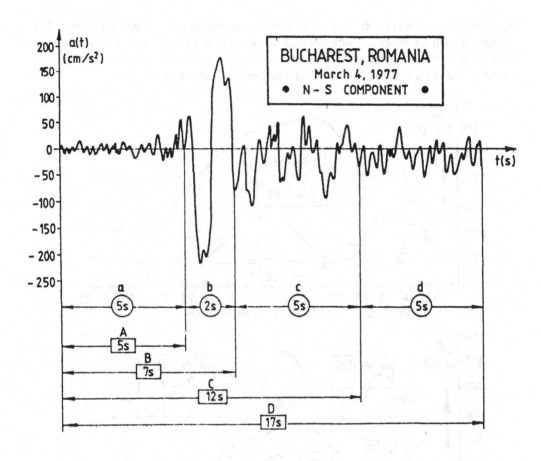

Fig. 35 Bucharest 77 accelerogram

4.2 Spectra for pulse actions

Using the artificial generated accelerations for pulse actions some spectra were obtained for different pulses periods and structure natural periods (Tirça et al, 2000).

(i) Spectra for horizontal components. Figure 37 shows the spectra for horizontal components. A comparison between one single pulse, two adjacent pulses and two distinguished pulses is presented in figure 38. One can see that the amplification is maximum for adjacent pulses. A comparison between the EC8 spectrum and the pulse action is shown in Figures 37 and 38. One can see that the EC8 value does not cover the high amplification in the range of low structure periods. In exchange, for the medium periods, the EC8 values are too large.

(ii) Spectra for vertical components. Spectra for vertical ground motions are presented in Figure 40, with a comparison with the EC8 proposals (Tirca and Gioncu, 2000). It is clear that the code provisions are not sufficient, because the maximum obtained amplification correspond exactly with the field of the vertical periods of the structure.

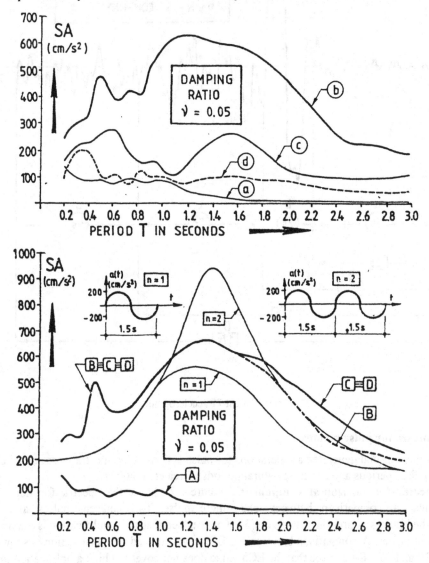

Fig. 36 Sequential seismic response spectra

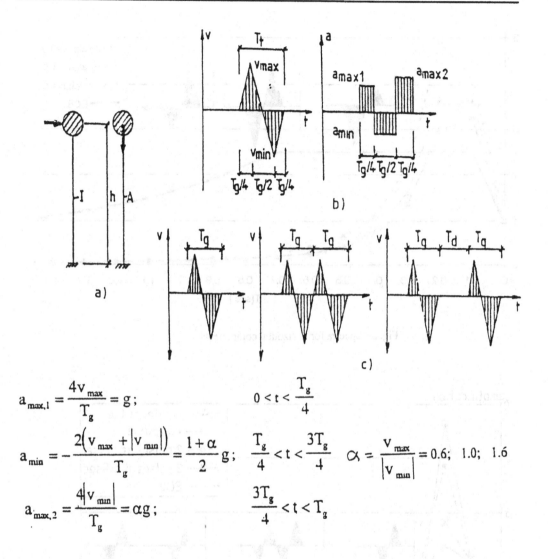

$$a_{max,1} = \frac{4v_{max}}{T_g} = g; \qquad 0 < t < \frac{T_g}{4}$$

$$a_{min} = -\frac{2\left(v_{max} + |v_{min}|\right)}{T_g} = \frac{1+\alpha}{2}g; \qquad \frac{T_g}{4} < t < \frac{3T_g}{4} \qquad \alpha = \frac{v_{max}}{|v_{min}|} = 0.6; \quad 1.0; \quad 1.6$$

$$a_{max,2} = \frac{4|v_{min}|}{T_g} = \alpha g; \qquad \frac{3T_g}{4} < t < T_g$$

Fig. 37 Artificial ground motions for near-source earthquakes

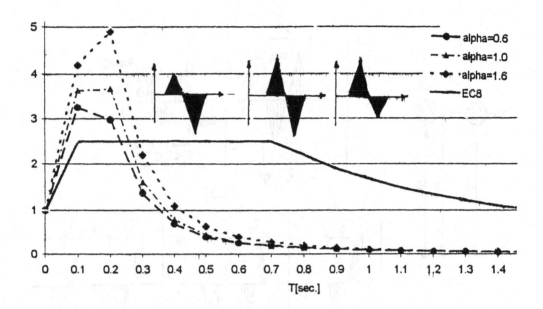

Fig. 38 Spectra for horizontal components

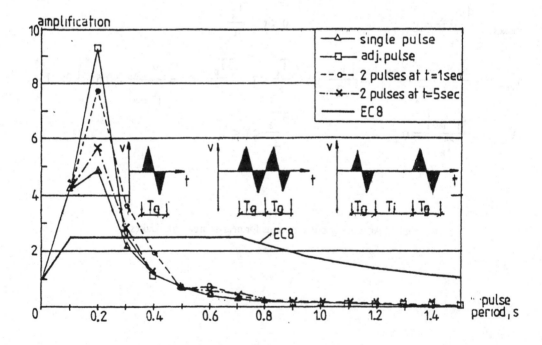

Fig. 39 Influence of multiple pulses

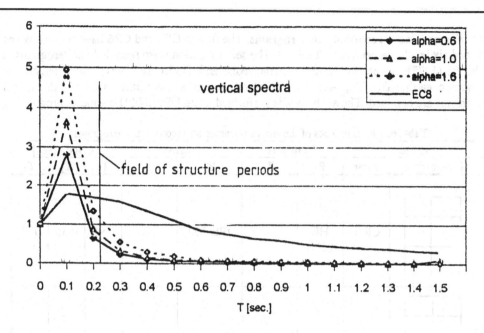

Fig. 40 Spectra for vertical components

4.3 Analysed structures

(i) **Frames under artificial ground motions.** The one span and six levels analysed frames are presented in Table 5 (Tirca and Gioncu, 2000). MRF1a is an ordinary moment resisting frame with variable cross-section, MRF1b, with constant cross-section for the columns, MRF2 is a special resisting frame, where the superior modes are not considered, while MRF3 considers the second vibration mode. The collapse mechanism of MRF1 is a story one, MRF2 and MRF3 develop global collapse mechanisms. The natural period of the three frames is also presented. The used accelerograms are of artificial pulse type, with peak values of 0.35g, 0.25g and 0.15g, corresponding to high, moderate and low seismic actions and with pulse periods varying from 0.3 to 2.0 sec.

Table 5: Characteristics of analysed frames for pulse actions

Frame		Columns		Beams		Periods [sec]		
Geometry	Type	P_1	P_2	P_1	P_2	T_1	T_2	T_3
	MRF1a	HE240B	HE200B	IPE360	IPE300	1.81	0.65	0.40
	MRF1b	HE240B	HE240B	IPE360	IPE300	1.80	0.64	0.39
	MRF2	HE260B	HE220B	IPE360	IPE300	1.72	0.61	0.35
	MRF3	HE450B	HE400B	IPE360	IPE300	1.35	0.40	0.18

(ii) **Frames under recorded accelerograms.** The frames CR3 and CR6 have two spans and three and six levels, respectively (Table 6). The seismic action were recorded accelerograms at Bucharest (1977), typical for far-field earthquakes, and Timisoara (1991), characteristic for near-source earthquake (Fig. 41). The acceleration peaks were increasing till the global mechanisms are obtained. The analysis was performed using DRAIN-2D computer program.

Table 6: Characteristics of the analysed frames for recorded accelerograms

Frame		Columns		Beams		Periods [sec]		
Geometry	Type	P_1	P_2	P_1	P_2	T_1	T_2	T_3
	CR 3	HE 260B	-	IPE 300	-	0.95	0.30	0.12
P_2 ── P_1	CR6	HE 360B	HE 320B	IPE 360	IPE 300	1.30	0.48	0.20

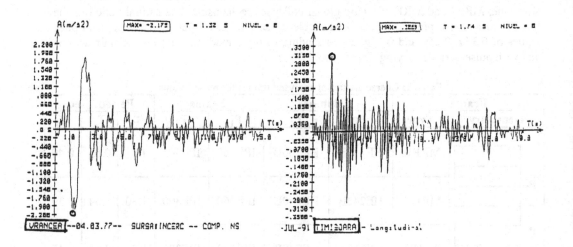

Fig. 41 Used accelerogram type

4.4 Seismic base shear forces

The seismic base shear forces for pulse actions are provided in Fig. 42, in comparison with the values obtained from EC-8 provisions (with $a_g = 0.35g$, and $T_c = 0.7sec$). Generally, the shear forces for pulse action are 2...3 times greater than the one given by EC-8. The differences increase with the increasing of column rigidity. For MRF 1a, which is a frame with reducing of column cross-sections in the top part of the structure, one can see a collapse for pulse periods corresponding to the second vibration mode.

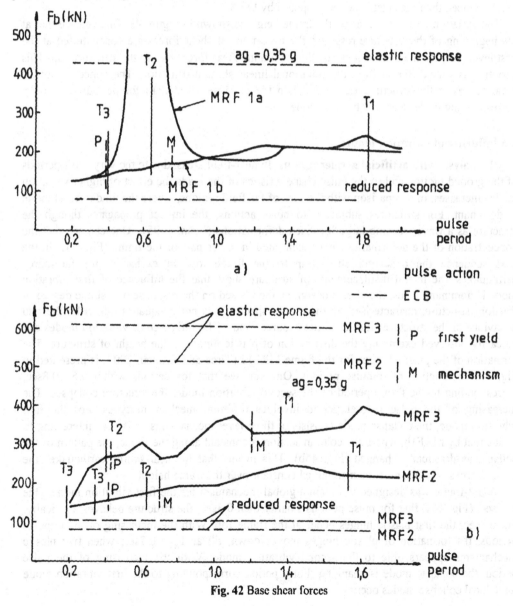

Fig. 42 Base shear forces

In exchange, for MRF1b, where the cross-sections of columns are constant along height, the collapse is eliminated. In all the cases the base shear forces increase with the increasing of pulse period.

On the figure there are plotted the periods corresponding to the occurrence of first plastic hinge and the formation of the collapse mechanism. For pulse periods about 0.40 sec the structures work elastically at values much lower than the elastic values given by EC 8. The differences between time-history and code values show: (i) the proposed behaviour factor is too large for the long pulse ground motions, (ii) the elastic values for short pulse ground motions are very much reduced than the elastic values proposed by EC-8.

The variation of shear forces on the frame height is shown in Figure 43. One can see that at the beginning of the structure response the maximum of shear forces are concentrated at the first level, but, for the next sequences, the maximum shear force rises to the structure top. This variation is very different from the traditional linear shear distribution. The concentration of shear forces at the structure top can explain the collapse of MRF1a for periods 0.5...0.8s, corresponding to the second vibration mode.

4.5 Influence of vibration modes

(i) Analysis with artificial accelerograms. In near-field areas, due to the very short periods of the ground motion and to the pulse characteristics of the loads the effect of higher vibration modes increases, in comparison with the case of far-field regions, where the fundamental mode is dominant. For structures subjected to pulse actions, the impact propagates though the structure as a wave, causing large localised deformations (Iwan, 1995). One can see that the concentration of the deformations is concentrated in lower part of the frame (Fig. 44). In the next sequence the concentration go up to the frame top. In exchange, for far-source earthquakes, the lateral displacements of structure show that the influence of first vibration mode is dominant. Thus, the classic design method based on the response of a single degree of freedom structure, characterised by the design spectrum, is not adequate to describe the real behaviour of the structure situated in near-source fields. The influence of vibration modes can be easily observed examining the distribution of plastic hinges in the height of structure. The formation of the plastic hinges for the frame MRF1a (Tirca and Gioncu, 2000) is presented in Fig. 45a in function of impulse periods. One can see that for periods within 0.5...0.8sec, corresponding to the frame period of the second vibration mode, the structure collapses. The increasing of the pulse natural period involves different mechanism types. For the first vibration mode, the collapse mode consists of three level mechanisms. It is very interesting to notice that by MRF1b, where the column section is constant along the frame, the pattern of the collapse is drastically changed (Fig. 45b). This means that for near-source earthquakes the column cross-sections must be maintained constant over the frame height.

MRF2-frame was designed to obtain a global mechanism for the first vibration mode. One can see, (Fig. 46a) that for pulse periods smaller than 0.3sec, the structure behaves elastically. For 0.4sec, the first plastic hinges occur at the frame top. With the increasing of the impulse periods, the formation of plastic hinges moves down, till at $T_g = 0.7$sec, when two plastic mechanisms appears, due to the second vibration mode. With the increasing of the pulse period, the collapse mode is changing. For a period corresponding to the first vibration mode three local collapse modes occur.

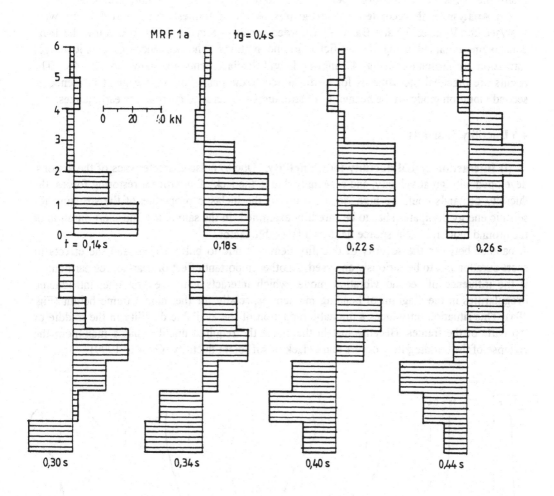

Fig. 43 Shear forces for MRF1a

MRF3 is sized to eliminate the local plastic mechanisms (Fig. 46b) One can see that the first plastic hinge occurs at the pulse periods corresponding to the second vibration mode and the global mechanism is reached for a period closed to the value of the first vibration mode.

(ii) Analysis with recorded accelerograms. A lot of frames with 3 and 6 levels were analysed for Vrancea 77 and Banat 91 recorded accelerograms in order to confirm the main conclusions obtained using the artificial ground motions. The sequences of plastic hinges formation are presented in Fig. 47 and the lateral displacements are shown in Fig. 48. The results are practical the same as for artificial accelerograms, showing the great influence of second vibration mode on the behaviour of structures subjected to near-source earthquakes.

4.6 Ductility demands

(i) Interaction global ductility-local ductility. Due to pulse characteristics of the actions, developed with great velocity, and especially due to the lack of significant restoring forces, the ductility demands could be high, so the using of the inelastic properties of the structure for seismic energy dissipation has to be carefully examined. In the same time, the short duration of the ground motion in near source fields is a favorable factor.

A balance between the severity of ductility demands due to pulse actions and the effects of short duration has to be seriously analysed. Another important effect of near-source earthquake is the influence of second vibration mode, which interacts with the first one, introducing irregularities in the diagram of bending moment especially for the middle frame height (Fig. 49a). This situation introduces a dramatic reduction of the available ductility in the middle or top parts of the frames. Because just in this place the required ductility has a maximum the collapse of the building may occurs due to lack of sufficient ductility (Gioncu, 1999).

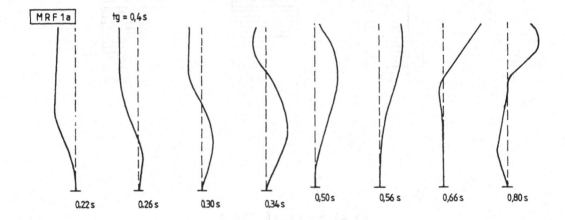

Fig. 44 Variation of lateral displacements

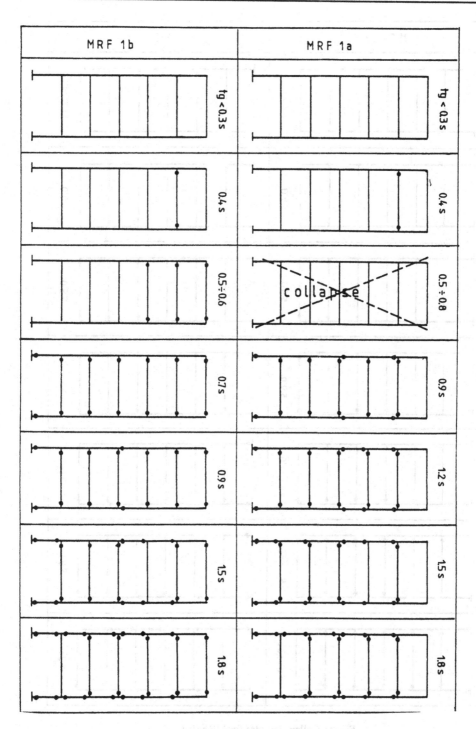

Fig. 45 Collapse modes for MRF 1a, b

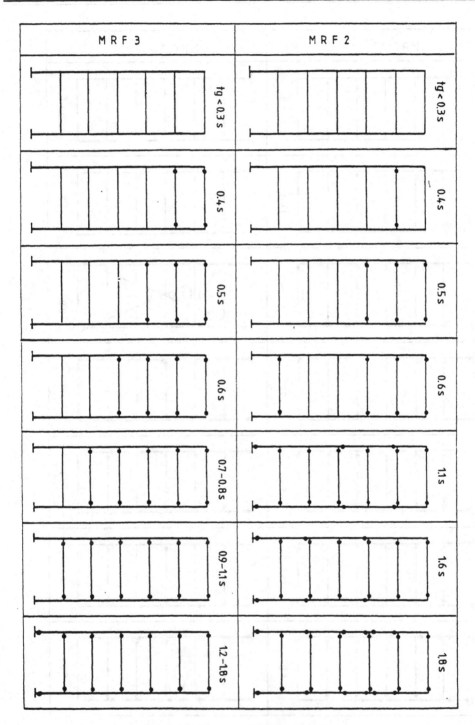

Fig. 46 Collapse modes for MRF 2, 3

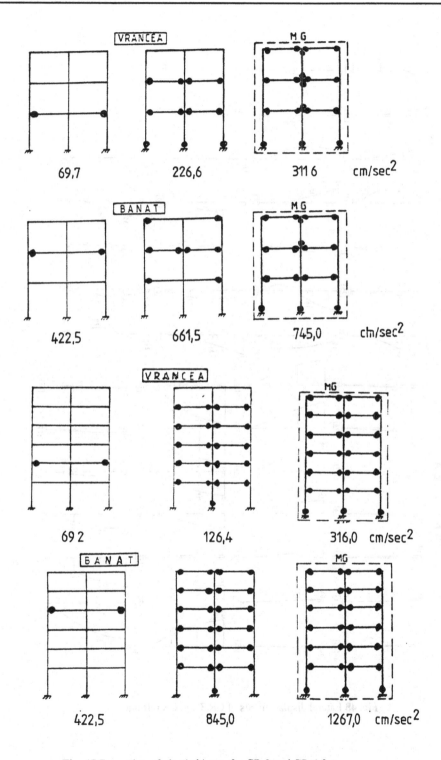

Fig. 47 Formation of plastic hinges for CR 3 and CR 6 frames

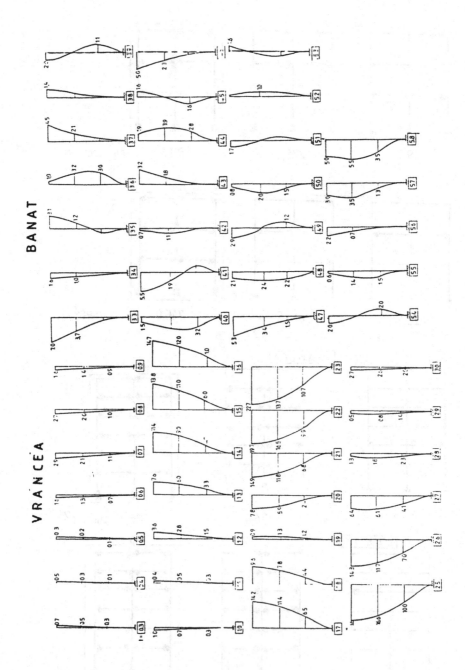

Fig. 48 Lateral displacements of CR 3 and CR 6 frames

This was a common phenomenon during the Kobe earthquake, where many buildings were damaged at the middle stories of structures. Fig. 49b shows the available and required ductilities in the case of far-source earthquakes where the required and available ductilities have the maximum values at the first storey.

(ii) **Global ductility for pulse loads**. Fig. 50 shows the influence of pulse period on the ductility demands (Tirca and Gioncu, 2000). For the periods near to second mode the maximum required ductility is concentrated at the top of structure, while for pulse periods near to first vibration mode, the maximum ductility demands occurs at the first levels. The required ductilities for beam and columns, resulting from the second vibration mode, are more reduced than the ones obtained from the first vibration mode (Fig. 50b). MRF3-frame is sized to eliminate the level plastic mechanism. One can see from Figure 51 that the first plastic hinges occur at the pulse period corresponding to the second vibration mode and the global mechanism is reached for a period close to the value of the first vibration mode. The maximum ductility demands is obtained at the top of the structure for all the pulse periods. The influence of number of adjacent pulses is presented in Fig. 52. One can observe that the required ductility at the top of frame increases drastically with the pulse number.

(iii) **Local ductility**. Beams, columns and joints compose a structure. In the capacity design method, some critical sections are chosen to form a suitable plastic mechanism, able to dissipate an important amount of the input energy. Generally it is considered that these sections are located at the beam-ends, where plastic hinges occur during a strong earthquake. But this end of beam is tied into a node, which connects also the column. So, the local plastic mechanisms in the structure can be localised not only at the beam or column end, but also at joints, or at both, member ends and joints (Fig. 53), depending on the earthquake characteristics and the node conformation. The modern codes impose that the plastic deformations occur only at the beam ends and the column bases without considering the joints, even it is well known that these could show, in some conditions, a stable behaviour. But in reality, the required conditions (the joint capacity must be 20% stronger than the adjacent members) does not assure the elastic behaviour of the joints, and as a consequence, the joint could be the weakest component of the node. So, the local ductility has to be defined at the level of the node, which is composed by the members and joints (Gioncu, 1999). Some definitions are necessary for the terms of node, joint and panel, because of indiscriminate everyday use in literature (Fig. 53c):

-panel zone is the portion of the web corresponding to the connection high;

-the joint is composed by the panel zone and connections, the last one representing the physical components which mechanically fasten the beams and columns;

-nodal zone covers the joint and the adjacent beam and column ends, where plastic deformations may occur.

The main hypothesis of the **component method** states that the overall behaviour of the node is dictated by the behaviour of the weakest component (Tschemmernegg et al, 1998). From the ductility point of view, the components may be classified as (Fig. 54):

-*high ductile component*, with an almost unlimited increasing of the deformation capacity. The collapse is due to a ductile fracture, when the ultimate strains are reached;

-*limited ductile component*, when the load-deformation curve presents a moderate decreasing in capacity, after the attainment of the maximum values. The collapse is produced by local plastic buckling, in this case;

-reduced ductile component, with a brittle fracture. Especially high strength bolts and some welding procedures present this behaviour type.

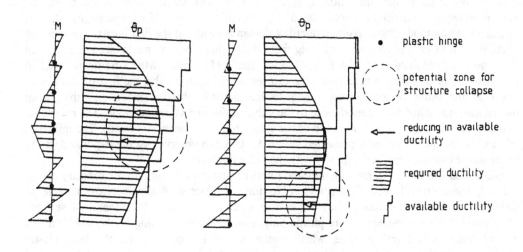

Fig. 49 Ductility demands

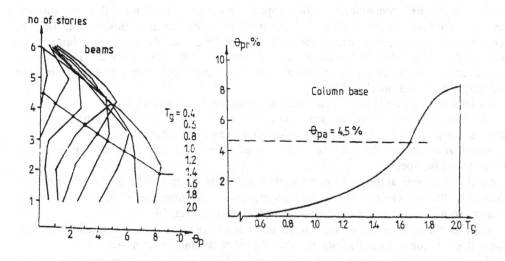

Fig. 50 Ductility demands for MRF2

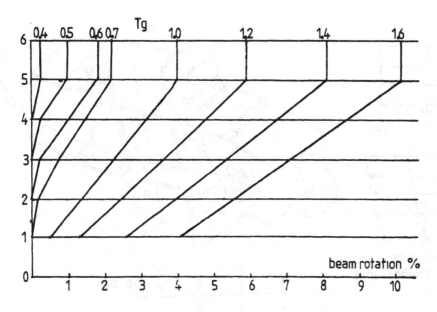

Fig. 51 Ductility demands for MRF 3

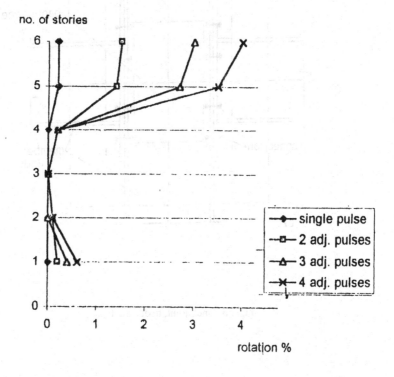

rotation %

Fig. 52 Influence of adjacent pulses on ductility demands

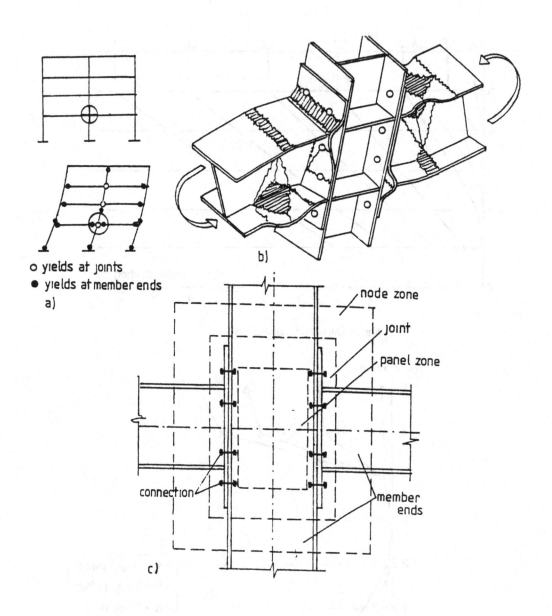

o yields at joints
● yields at member ends
  a)

b)

node zone

joint

panel zone

connection

member
ends

c)

Fig. 53 Panel, joint, node structure

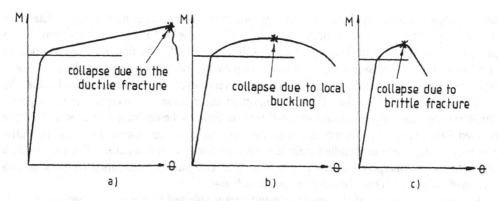

Fig. 54 Ductile and brittle collapse

The analysed elements are presented in Fig. 55, referring to:
-members with different cross-section types;
-welded, bolted and mixt joints

For all these elements the monotonic ductilities are determined, using the great facilities offered by the local plastic mechanism methodology (Gioncu and Petcu, 1997). Some of these local mechanisms for beams and joints are presented in Fig. 56. Using these plastic mechanisms, the rotation capacities are determined both for beams, columns and joints. The **DUCTROT MJN (DUCT**ility for **ROT**ation of Members, Joint and Nodes) computer program was elaborated at INCERC Timisoara. Some results for different geometrical parameters are presented in Fig. 57, in comparison with the EC-3 cross-section ductility classes and the member ductility classes proposed by Mazzolani and Piluso (1993, 1996). It is very clear that the member ductility classes is a most suitable proposal to describe the member ductility. These ductilities determined for monotonic loads must be affected by the influence of seismic actions. So, the required and available ductilities are modified in time due to ground motions and superposition of structure vibration modes (Fig. 58) In the same time erosion of available ductility is marked. This erosion depends on the characteristics of cyclic loading and local collapse mode by buckling or by fracture (Gioncu, 1997),(Fig. 59).

Influence of the cyclic loading. Some research works have classified the failure of the structural members during seismic actions as a low-cycle fatigue. During an earthquake, the structure resists hundreds of loading cycles, but only few cycles cause high plastic deformations. During an earthquake with short duration (fewer than 5 cycles) large plasticity occurs, while by a long duration (about 20 cycles), plastic excursions are induced. Therefore, classifying failures under repeated large deformations as belonging to the category of fatigue-failure is questionable. The low cycles with high plastic deformations cause accumulation of these deformations along the yield lines, inducing cracks or rupture in the deformed plates, rather than a reduction of the material strength, as by fatigue under high cycle loading. But the effects of seismic loads are more intricate, because the movement history is a chaotic one, depending on a great number of factors. The deformations induced by the seismic motion vary from one member to another and also from one earthquake to other. Therefore, it is extremely difficult to select a particular deformation time-history to generalise the earthquake-induced deformations. Considering these uncertainties, one must adopt a caution approach to determine the member rotation capacity under seismic actions. The behaviour of **I-sections** has a particular feature by cyclic loading (Anastasiadis et al, 2000). The first semi-cycle produces buckling of the compression flange and the section rotates around a point located in, or near to opposite flange. As a result, the tension forces are very small. By the reversal semi-cycle, the compressed flange buckles also. The most important observation is concerned that the opposite flange remains unchanged because of small tension force, it being incapable to straighten the buckled flange (Fig. 60). Therefore, during the next cycle, the section works as having initial geometrical imperfections resulted from the previous cycle. In this manner, after each cycle a new rotation is superposed over the previous one. The plastic mechanism consists of two deformed shapes, each one for the corresponding flange.

The inelastic rotation of the **joints** is significantly affected by the cyclic loading. As it is presented in Figure 60b, the first semi-cycle produces a plastic displacement of the end plate, when the beam rotates around a point located in opposite position. For the unloading field, at zero force, some residual displacement occurs. The reversal semi-cycle produces a compression force, which for the maximum value reduces the residual displacement at zero. At the end of the first cycle, a residual displacement remains. The start of the second cycle is characterised by the presence of a residual displacement, so the needed force to reach a constant displacement is reduced in comparison with the first cycle. Results a continues decreasing of the joint moment capacity and an accumulation of residual rotations, which can reach the fracture rotation after several cycles.

As in the case of members, the fracture may occur at the first cycle (near-source earthquake with one important impulse), or after 2-3 cycles, if the ground motion is characterised by some adjacent impulses. In case of far-source earthquakes, characterised by cyclic loading, the fracture appears after 5-10 important cycles, producing plastic excursions in the joint elements.

Influence of the strain rate. In the case of near-source earthquakes, the velocity is very high (Gioncu, 2000) inducing very important strain-rates. In addition, the first important ground motion is the strongest, for these earthquakes. So, the member behaviour could be studied using pulse forces. It is well known that the strain-rate has the main influence on the increasing of the yield stress (Fig. 61), especially for values greater than 10^{-1}/s. In the same time, the increasing of the ultimate strength is moderate, consequently, the yield ratio has the tendency to reach the value of 1, with increasing of the strain-rate. So, a reduction of the ductility occurs, especially for strain rates greater than 10^{-1}/s, due to the increasing of the yield ratio. The influence of the yield ratio on the ductility was studied considering the fracture ductility for the rotation of a bended stub:

$$\mu_{\theta.fr} = \frac{\theta_{rf}}{\theta_p} = \frac{1}{2} \frac{E}{f_y} \left(\frac{1}{\rho_y} - 1 \right)^2 \left(\frac{c}{t} \right)^2 \varepsilon_u^2 \tag{10}$$

with θ_{rf}, calculated from a simplified approach.

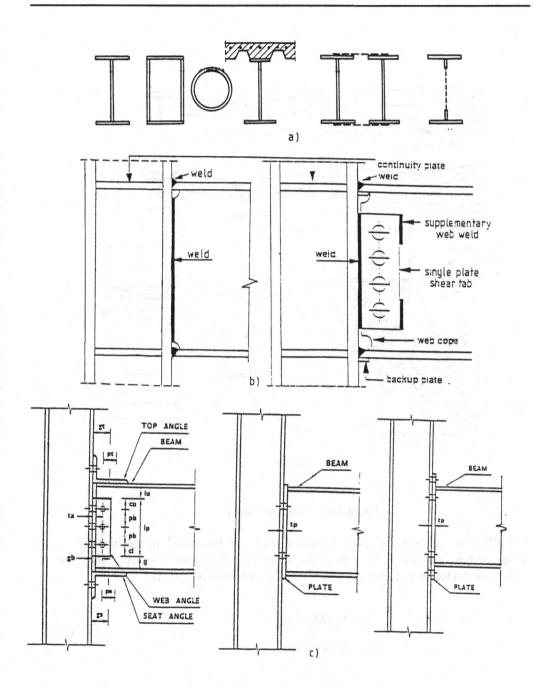

Fig. 55 Analysed elements

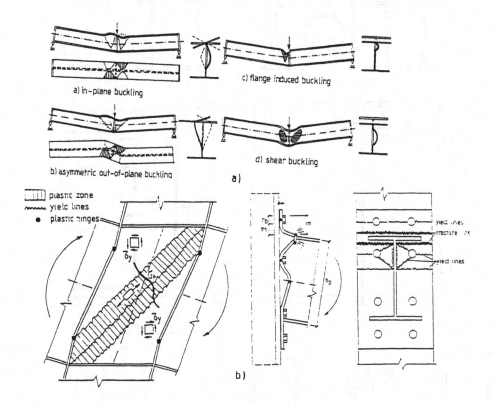

Fig. 56 Local plastic mechanism

The effect of high velocity cycling loading could be introduced in the analysis by increasing the yield ratio in function of the strain-rate ε. Using the Shorousian and Choi (1987) relation for the yield and ultimate stresses, for Fe 360 steel, results:

$$\rho_{ysr} = \frac{1.46 + 0.0925 \cdot \log \varepsilon}{1.15 + 0.0496 \cdot \log \varepsilon} \tag{11}$$

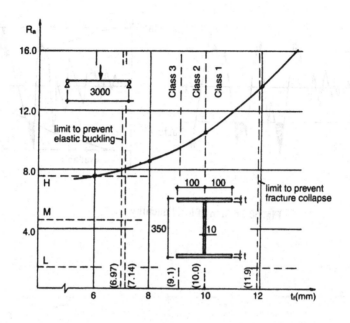

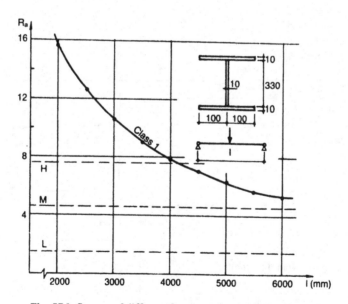

Fig. 57 Influence of different factors on local ductility

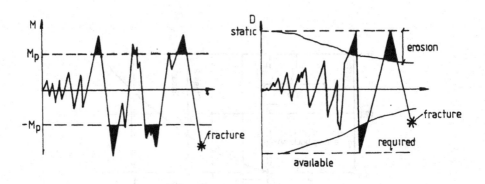

Fig. 58 Erosion of local ductility

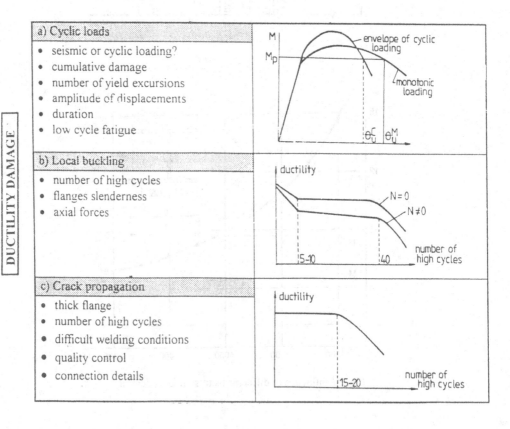

Fig. 59 Factors influencing the ductility damage

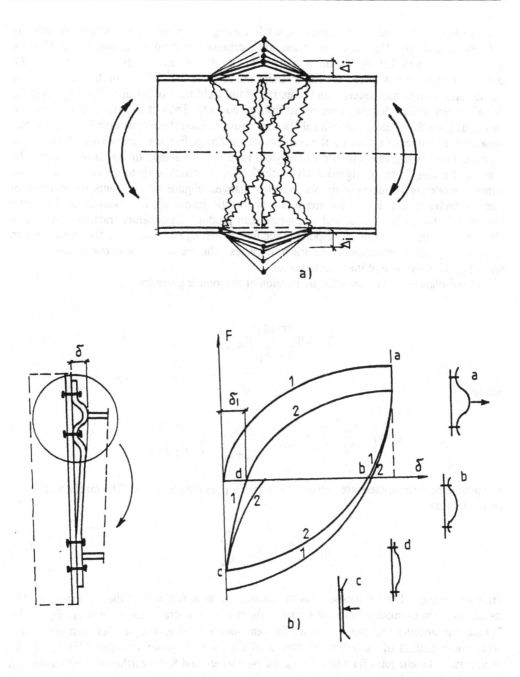

Fig. 60 Influence of cyclic loading

If the values obtained from monotonic loading correspond to the plastic ductility and the values resulted from (10) represent the fracture ductility, the interaction of the two curves, determined for different yield ratios, show two distinct domains. The ductility of the domain with low yield ratio is given by the plastic deformations, while for high yield ratio, the fracture dramatically reduces the ductility (Fig. 62). The strain-rate increases the yield stress and the yield ratio. Taking into account that the randomness of the yield ratio of usual steels is about 0.60...0.73, and that the yield stress increases with a strain-rate of about 10^{-1} - 10^{1}/ s in the field of strong earthquakes, results consequently an increasing of the yield ratio of about 0.75-0.95; this is the domain where plastic ductility changes to fracture ductility. Thus, it is possible that in case of near-source earthquakes, a brittle local fracture could replace the plastic ductility. Figure 62b presents the number of cycles till the reaching of the flange fracture for the case of increasing rotation. There is a great influence of the yield ratio and the flange thickness to remark. The number of cycles decreases significantly by thick flanges, which leads to the conclusion to use rather a moderate slenderness for the member flanges. Figure 62c presents the number of fracture cycles in function of the strain-rate and cyclic loads. So, the coupling of these two erosion effects, cyclic loading and strain-rate, can produce a premature fracture. Because of 10^{-1}...10^{1}/s of the strain-rate correspond to the field of strong near-source earthquakes, one can mark an important reduction in the fracture cycles. The fracture may occur at the first, or second cycle in the case of these earthquakes.

For end-plate joint type the ultimate rotation of the joint is given by:

$$\theta_u = \varphi \frac{m \cdot m_1}{h_b \cdot t_p} \overline{\varepsilon}_u \tag{12}$$

where:

$$\varphi = 2\left[2\frac{r_b}{m_1}\left(\frac{r_b}{m_1} + \frac{\rho_y}{1+\rho_y} \right)\frac{\overline{\varepsilon}_h}{\overline{\varepsilon}_h + \overline{\varepsilon}_u} + \frac{1-\rho_y}{\left(1+\rho_y\right)^2} \right] \tag{13}$$

The geometrical parameters are presented in Fig. 63, ρ_y is the yield ratio. The mechanical parameters are:

$$\overline{\varepsilon}_h = \frac{\varepsilon_h}{\varepsilon_y} \quad ; \quad \overline{\varepsilon}_u = \frac{\varepsilon_u}{\varepsilon_y} \tag{14a,b}$$

From the relation (12) one can see that the ultimate rotation depends on the yield ratio ρ_y. The reduction of the ultimate rotation of a joint in function of the strain-rate is shown in Figure 63. Taking into account that the strain-rate for near-source earthquakes lies between 10^{-1}....0.5· 10^{1}/s, the reduction of the ultimate rotation of the joints is about 40-50%. This can be an explanation of some joint fractures during the Northridge and Kobe earthquakes, characterized by very high-recorded velocities.

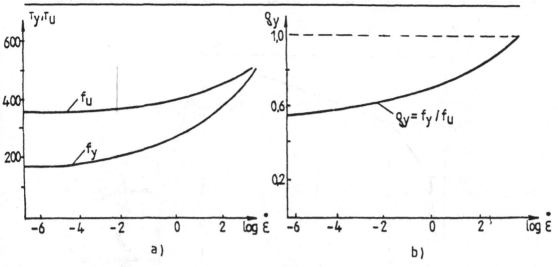

Fig. 61 Influence of strain–rate on steel characteristics

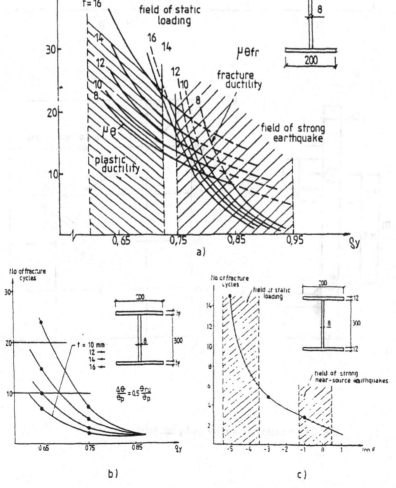

Fig. 62 Influence of strain-rate on beam behaviour

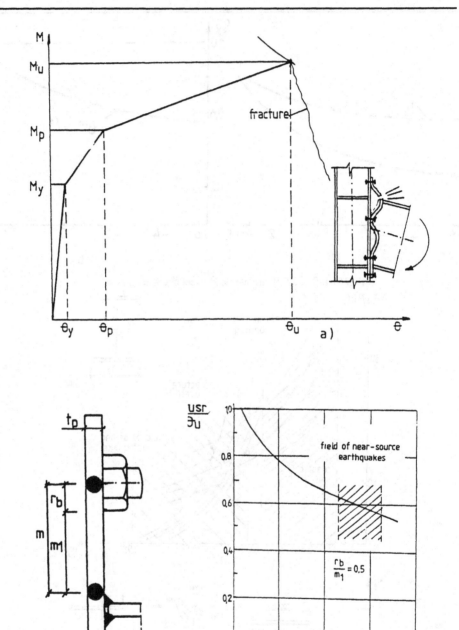

Fig. 63 Ultimate rotation for end-plate joint

4.7 Influence of vertical components

Of fundamental importance are the natural periods for vertical vibrations for those buildings, which are more stiffer in the axial than the transverse direction, and hence possess shorter periods in vertical direction (Papazoglou and El Nashai, 1996, Papaleontiou and Roesset, 1993), (Fig. 64a). Due to the fact that the amplification range for vertical strong-motion records is found to lie for period between 0.05 and 0.25, an important amplification of the vertical effects may occur. Naeim (1998) shows that during Northridge earthquake, the instrumented buildings indicated that peak vertical ground accelerations were amplified by the structures by a factor ranging from 1.1 to 6.4. In the same time, taking into account the reduced possibilities of plastic deformations and damping under vertical displacements, the vertical behaviour can be of first importance for the structures located in near-source fields. The combination of vertical and horizontal components produces an increasing of the axial forces in columns, (Fig. 64b), and consequently, the increasing of second order effects. Due to this effect, additional plastic hinges may occur in the columns of the first level.

The damping effects have reduced values for the vertical components, compared with the horizontal ones. A value of 2% was proposed for the viscous damping ratio. The reduced factor taking into account the effect of energy dissipation is smaller for vertical than for horizontal components; values of $q=1.0...1.5$ was proposed.

The vertical components can produce an important increasing of the axial forces, due to the high amplification and the reduction factors of damping and energy dissipation. This can give rise to a reduction of the rotation capacity of the columns and could extend the stability problems. All these aspects have to be introduced in the calculation of the available ductility of the columns.

(i) **Analysis with artificial accelerograms.** For the frame from Figure 65, a vertical, asymmetric ground motion pulse of $T_g=0.15$ sec was applied. The first natural period of the structure is 0.9 sec for the horizontal, and 0.081 sec for the vertical movement, which correspond to the range of high amplification. The first plastic hinge and the collapse mode occur for acceleration values of a_p = 1353 cm/s^2 and a_c = 4709 cm/s^2, and velocities v_b=54.1 cm/sec and vc = 147.2 cm/sec, respectivelly. These values are very high, exceeding the recorded vertical accelerations and velocities. So, the vertical ground motion has to be considered as a factor influencing the effects of the horizontal components only. It is interesting to notice that, even if the vertical components applied to the structure are the same for each column, the formation of the plastic hinges has a marked asymmetry. This is due to the analysis imperfections, but also to the structural imperfections, characteristically for the in site frame behaviour. A very important problem to be solved in the future is the asynchronism of the vertical movement (Fig. 66), which can introduce high internal forces, especially in beams and nodes.

(ii) **Analysis with recorded accelerations.** Papaleontiou and Roesset (1993), respectively, studied two structures with 4 and 16 levels, for the Loma Prieta-Capitola earthquake, considering the horizontal and vertical components and their simultaneous action (Fig. 67). The results show an important increasing of the axial forces in the bottom columns, which is greater for the structure with a less number of levels.

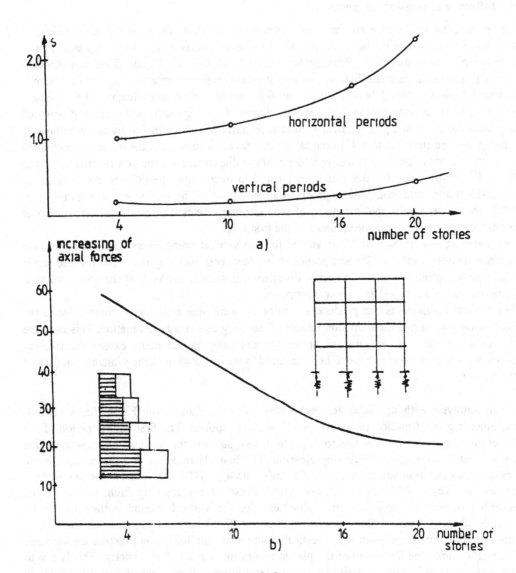

Fig. 64 Influence of vertical components

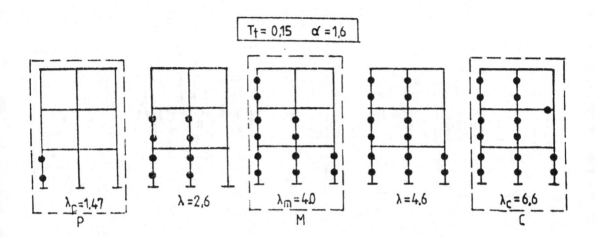

Fig. 65 Analysed frame for vertical pulse

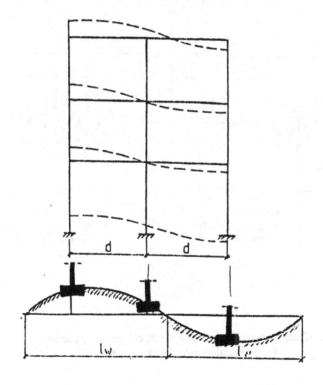

Fig. 66 Asynchronism of vertical movements

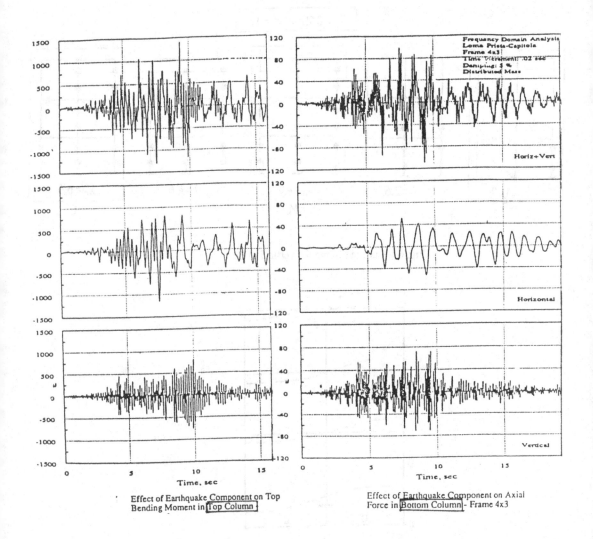

Fig. 67 Simultaneous horizontal and vertical component action
(after Papaleontiou and Roesset)

5 CONCLUSIONS

Recent earthquakes have demonstrated that the behaviour of many steel structures built in concordance with the existing codes was unacceptable. The earthquakes which have been hit the urban areas have demonstrated that the economic impact of physical damage, loss of functions and business interruptions were huge and some new conceptions in the structural design have to be considered.

The paper presents the last results of an intense activity in the research work all over the world concerning the improving of the design criteria for seismic resistant steel structures, showing that we have probably reached today the stage where the actual structural performance during strong motions can be satisfactorily explained. It is an ethical duty of specialists to find the way from the theory to practise, in order to fill the gap that exists between the accumulated knowledge and the design codes.

REFERENCES

Ambraseys N.N, Free M.W. (1997): Surface-wave magnitude calibration for European region earthquakes. Journal of Earthquake Engineering, Vol. 1, 1-22.

Anastasiadis A., Gioncu V., Mazzolani F.M. (2000): New trends in evaluation of available ductility of steel members. In Behaviour of Steel Structures in Seismic Areas, STESSA 2000, 21-24 August 2000, Montreal (manuscript).

Aribert J.M., Grecea D. (1997): A new method to evaluate the q-factor from elastic-plastic dynamic analysis and its application to steel structures. In Behaviour of Steel Structures in Seismic Areas, STESSA 97 (eds. F.M. Mazzolani and H. Akiyama), 3-8 August 1997, Kyoto, 10/17, Salerno, 382-393.

Bachmann H., Linde P., Werik Th. (1995): Capacity design and non-linear dynamic analysis of earthquake resistant structures. In 10th European Conference on Earthquake Engineering (ed. G. Duma), 28August-2 September 1994, Vienna, Balkema, Rotterdam, Vol. 1, 11-20.

Bertero V.V (1996): State-of-the-art report on design criteria. In 11th World Conference on Earthquake Engineering, 23-28 June 1996, Acapulco, CD-ROM 2005.

Bertero V.V (1997a): Codification, design and application. General report. In Behaviour of Steel Structures in Seismic Areas, STESSA 97 (eds. F.M. Mazzolani and H. Akiyama), 3-8 August 1997, Kyoto, 10/17, Salerno, 189-206.

Bertero V.V (1997b): Performance-based seismic engineering: A critical review of proposed guidelines. In Seismic Design Methodologies for the Next Generation of Codes (eds. P. Fajfar and H. Krawinkler), 24-27 June 1997, Bled, Balkema, Rotterdam, 1-31.

Bertz G., Smolka A. (1995): Urban earthquake potential: Economic and insurance aspects. In 10th European Conference on Earthquake Engineering (ed. G. Duma), 28August-2 September 1994, Vienna, Balkema, Rotterdam, Vol.2, 1127-1134.

Cosenza E. (1987): Duttilità globale delle strutture sismo-resistenti in accaio. PhD thesis, University of Napoli, Italy.

El Nashai A.S., Papazoglou A.J. (1997): Procedure and spectra for analysis of RC structures subjected to strong vertical earthquakes loads. Journal of Earthquake Engineering, Vol. 1, No 1, 121-155.

Filiatrault A. (1996): Elements de Genie Parasismique et de Calcul Dynamique des Structures. L'Ecole Polytechnique de Montreal.

Gioncu V. (1997): Ductility demands. General report. In Behaviour of Steel Structures in Seismic Areas, STESSA 97 (eds. F.M. Mazzolani and H. Akiyama), 3-8 August 1997, Kyoto, 10/17, Salerno, 279-302.

Gioncu V. (1999): Framed structures. Ductility and seismic response. General report. In Stability and Ductility of Steel Structures, 9-11 September 1999, Timisoara, Romania.

Gioncu V. (2000): Effect of strain-rate on the ductility of steel members. In Behaviour of Steel Structures in Seismic Areas, STESSA 2000, 21-24 August 2000, Montreal (manuscript).

Gioncu V., Petcu D. (1997): Available rotation capacity of wide-flange beams and beam-columns. Part I. Theoretical approaches. Part II. Experimental and numerical tests. Journal of Constructional Steel Research, Vol. 43, No. 1-3, 161-217, 219-244.

Hall J.F (1995a): Near-source ground motion and its effects in flexible buildings. Earthquake Spectra, Vol. 11, No 4, 569-605.

Hall J. F. (1995b): Parameters study of the response of moment-resisting steel frame buildings to near-source ground motions. Technical Report SAC 95-05, 1.1-1.83.

Hudson M.B., Skyers B.N., Lew M. (1996). Vertical strong motion characteristics of the Northridge earthquake. In 11th World Conference on Earthquake Engineering, 23-28 June 1996, Acapulco, CD-ROM 728.

Ifrim M. Macavei F., Demetriu S., Vlad I. (1986): Analysis of degradation process in structures during the earthquake. In 8th European Conference on Earthquake Engineering, Lisbon, 65/8-72/8.

Iwan W.D (1995): Drift demand spectra for selected Northridge sites. Technical Report SAC 95-05, 2.1-2.40.

Iwan W.D. (1997): The drift demand spectrum and its application to structural design and analysis. In 11th World Conference on Earthquake Engineering, 23-28 June 1996, Acapulco, CD-ROM 1116.

Law N., Wilson J., Hutchinson G. (1996): Building ductility demands. Interplate versus intraplate earthquakes. Earthquake and Structural Dynamics, Vol. 25, 965-985.

Maeys R.L. (1995): Interstorey drift design and damage control issues. The Structural Design of Tall Buildings, Vol. 4, 15-25.

Mazzolani F.M. (1995): Some simple considerations arising from Japanese presentation on the damage caused by the Hanshin earthquake. In Stability of Steel Structures (ed. M. Ivanyi), SSRC Colloquium, 21-23 September 1995, Budapest, Akademiai Kiado, Vol.2, 1007-1010.

Mazzolani F.M., Piluso V. (1993): Design of Steel Structures in Seismic Zones. Manual ECCS Document.

Mazzolani F.M., Piluso V. (1996): Theory and Design of Seismic Resistant Steel Frames. E & FN Spon, London.

Mohammadioun B. (1997): Nonlinear response of soils to horizontal and vertical bedrock: earthquake motion. Journal of Earthquake Engineering, Vol. 1, No.1, 93-119.

Naeim F. (1998): Research overview: Seismic response of structures. The structural Design of Tall Buildings, Vol. 7, 195-215.

Nair R. Shankar (1995): Stiffness and serviceability issues in tall steel buildings. In Habitat and the High-rise . tradition and Innovation (eds L.S. Beedle and D. Rice), 14-19 May 1995, Amsterdam, 1171-1183.

Papaleontiou C., Roesset J.M. (1993): Effect of vertical accelerations on the seismic response of frames. In Structural Dynamics. EYRODYN'93, Balkema, Rotterdam, 19-26.

Shorousian P., Choi K.B. (1987): Steel mechanical properties at different strain-rate. Journal of Structural Engineering, Vol. 113, No.4, 863-872.

Tirca L., Gioncu V., Mazzolani F.M. (1997): Influence of design criteria for multistorey steel MR frames. In Behaviour of Steel Structures in Seismic Areas, STESSA 97 (eds. F.M. Mazzolani and H. Akiyama), 3-8 August 1997, Kyoto, 10/17, Salerno, 266-275.

Tirca L., Gioncu V. (1999): Ductility demands for MRFs and LL-EBFs for different earthquake types. In Stability and Ductility of Steel Structures, 9-11 September 1999, Timisoara, Romania.

Tirca L., Mateescu G., Gioncu V. (2000): Artificial ground motions for design of MRFs subjected to pulse type earthquakes. In Behaviour of Steel Structures in Seismic Areas, STESSA 2000, 21-24 August 2000, Montreal (manuscript).

Tirca L., Gioncu V. (2000): Behaviour of MRFs subjected to near-field earhquakes. In Behaviour of Steel Structures in Seismic Areas, STESSA 2000, 21-24 August 2000, Montreal (manuscript).

UBC (1997): Uniform Building Code, Division V, Soil Profile Types.

CHAPTER 3

METHOD BASED ON ENERGY CRITERIA

H. Akiyama
Nihon University, Tokyo, Japan

1. INTRODUCTION

1.1 Short History of Earthquake Resistant Design Method

Earthquake resistant design method in a sense of modern technology started immediately after the Great Kanto Earthquake (1923)(See Table 1.1). Since then, during three fourths century earthquake resistant design has made a great progress. As the scope of the applied structures was extended, the design ideas became versatile. As the electronic digital computers and apparatuses for field observation became available, analytical techniques for response analysis developed drastically and became versatile. On the other hand, successive attacks of big hazardous earthquakes made us continuously recognize the arrived level earthquake resistant design method to be still incomplete. The earthquake resistant design started as a measure by which a structure basically designed for gravity loading is to be supplied additionally with a resistance against earthquakes. Increased experiences of earthquake hazard and the progress of response analysis techniques have raized the importance of earthquake resistant design to such a extent that the design of structures is governed mostly by earthquake resistant design.

Under such a circumstance, a methodology to be perfectly free from dreadful hazard made by earthquakes has been sought after in realizing so-called "base-isolated structures." Such a methodology, however, had to wait the chance of germination for about a half century in the shade of a steady development of earthquake resistant design method. In order to establish a base-isolated structure, matured design theory and reliable structural element such as the laminated rubber bearing which appeared in 1970's had to be prepared as preconditions.

The development of earthquake resistant design and realization of base-isolated structured, however, did not developed favorably without any obstacles and each had to go through hardships. One of the reason is that the earthquake resistant design in Japan assumed from the start a character of law to be observed for the public welfare. On one hand, the earthquake resistant design must be rational one supported by scientific grounds and, on the other hand, it must be equipped with a lucid logic. The scientific fact and a logic, however, are not same. Incorrect recognition of a fact can be always stated logically. This basic contradiction lying under the earthquake resistant design has been the cause of a tenaciously made long dispute. This dispute is called the dispute about "flexible or stiff", which started immediately after the introduction of the lateral force method and raised a question on the recommended principle, at that time, that structure should be stiff. Although the lateral force method is superior in its practicability and logical consistency, it is not perfect scientifically.

The lateral force method has been established during the period of introducing nuclear power

Table 1.1 Historical Background

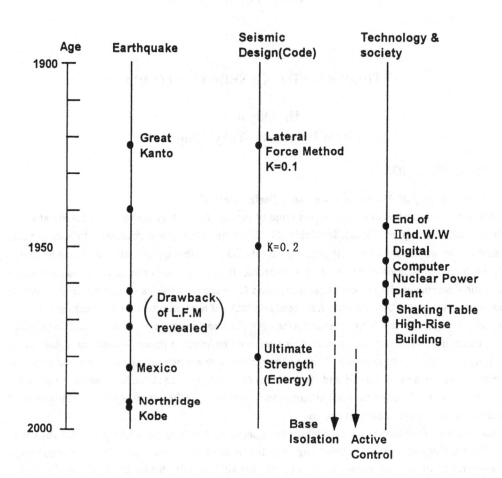

plants and drastic economic growth in Japan as a design methodology that the important structures should be designed stiff and strong. A high rate of economic growth in late 1960's promoted raising the height of buildings in urban areas, and the possibility of super high-rise buildings in highly seismic countries were discussed in the second dispute about "flexible or stiff". As a result, it was made clear that as the natural period of a structure increases, seismic force applied to the structure reduces reversely proportional to the natural period. Thus, the side of "flexible" triumphed.

As the development of computers and techniques of field observation brought about more detailed knowledge of structural behavior under earthquakes, also, it has been made clear that the structures with intermediate height can receive far stronger seismic forces than that prescribed by the lateral force method. Comings of the Niigata Earthquake (1963), the Tokachi-oki Earthquake (1968) and the Miyagiken-oki Earthquake (1973) made us realize the insufficiency of the lateral force method. Then, the third theme of the dispute was how to endow a structure with a sufficient energy absorption capacity.

In 1981, the Japanese building code was revised to cope with the energy absorption capacity.

In addition to the previous lateral force method, the energy absorption capacity was introduced as a basic performare which should be equipped with a structure to resist to earthquakes.

By this, the basic framework of seismic design in which the basis requirement of force by the lateral force method was supplemented with the requirement of deformation capacity, was established.

How to secure the deformation capacity, however, is a matter concerned with the nonlinear field beyond the elastic limit and can not be easily implemented even by applying advanced modern technologies. Even more as the analytical tools are well prepared, we should be confronted with the situation in which we are bewildered by getting too many options and losing objective measures to judge the validity of the results of analysis.

The main theme of the third dispute about "flexible or stiff" can be said to be a pragmatic problem about how to reflect the energy absorption   capacity in the inelastic range to the seismic design in general. At the same time, also, in this period, as an opponent to the earthquake resistant design which dashed forward to the way of complication, the base-isolated structure, which aims to escape fundamentally from the devastative damage, began to be seriously discussed of its real applicability in Japan.

The earthquake resistant structures are expected to develop both the strength and deformation capacity as required to resist earthquake. Therefore, it is inevitable that structural skeletons suffer some damage under an attack of an earthquake. Contrary to this, the base-isolated structure challenged the earthquake resistant structure, asserting that the structural skeletons should be made free of any damage. For this assertion, the circle of the earthquake resistant design hardly understood the true meaning, still keeping the standpoint that the possibility of base-isolated structures couldn't be proven in such a highly seismic country as Japan. On the other hand, in spite of tenacious resistance of the circle of the earthquake resistant design, the circle of the base-isolated structures at last succeeded to prove the applicability, based on the same ground as that on which the earthquake resistant design stands.

To be able to do so, the appearance of the excellent structural element such as "laminated rubber isolater" was inevitably necessary.

The principle of the base-isolated structure can be recognized as follows.

> The base-isolated structure is a dual structure, in which the building part forms a superstructure and the base-isolating layer and foundations form a substructure. The base-isolating layer is formed with a group of laminated rubber isolaters which acts as a flexible elastic part and with a group of dampers which acts as a stiff energy absorbing part. Since the rigidity of the superstructive is sufficiently stiff compared to that of the substructure, the base-isolated structure is identified to be a one-mass system.

The high vertical-load bearing capacity and the large elastic deformation capacity of the laminated rubber bearings enable the base-isolated structure to be equipped with a natural period longer than 4.0sec, and the base-isolating layer can behave as a highly efficient energy absorbing mechanism due to a favorable combination of elasticity of isolators and plasticity of dampers.

The base-isolated structure is equipped with simplicity and clarity which should be intrinsic in the true earthquake resistant structure.

Moreover, it is equipped with advantageous characteristics of the long-period structure, and the uncertainty in damage distributions is basically removed in the base-isolated structure in which the seismic energy input is totally absorbed in the base-isolating layer. Although the base-isolated structure just emerged as an opposing concept to the earthquake resistant structure, through the fourth dispute on "flexible or stiff", it was proven to be an excellent earthquake resistant structure. On the other hand, ordinary earthquake resistant structures, structural skeletons of which are designed primarily to support gravity loading and are utilized also to develop the strength and energy absorption capacity required to resist to earthquakes, have a fundamental deficiency to be simple and clear structures.

Just then, at dawn of January, 17th in 1995, The Hyogoken-Nambu Earthquake happened and taught us the real level of earthquake resistant structures at present.

The intensities of ground motions in the epicentral zone were partly far greater than that prescribed by Japanese building code. Observed various sorts of damage, however, made us feel that the goal of seismic design is still far away. Two base-isolated buildings constructed not so far from the epicentral zone were proven to have developed an anticipated performance.

The seismic design of 20th century is closing in rise and flourish of base-isolated structures. It is, however inconceivable that all buildings would turn to be base-isolated structures. The earthquake resistant structures will make a strenuous effort toward a breakthrough, awakened by superiority of base-isolated structures.

And also, the base-isolated structures would change qualitatively with an enormous increase of their application. In any way, the fifth dispute about "flexible or stiff" would develop on the axis of base-isolated structure.

This dispute must be a hard and desperate battle for the earthquake resistant structure.

1.2 Importance of Energy Concept [1)]

In order to know the process of collapse of structures subjected to earthquakes, a high nonlinearity of structures must be analysed by using the equation of motion. Although the obtained results are specific and isolated solutions, characterized strongly by specific conditions, these results can be synthesized through the equation of energy balance which is obtained by multiplying a displacement increment on both sides of the equation of motion and by integrating over the duration of the ground motion.

The reason is;

- Information is integrated through integration.

- The energy is a scaler composed by the product of force and deformation and is a suitable quantity for synthesis.

- The seismic energy input in total is a stable amount which depends mainly on the total mass and the fundamental natural period of a structure as predicted by Housner [2)].

As far as the total energy input is a stable constant quantity, the major concern in the seismic design should be focussed on the manner in which the input energy is distributed over a structure. The equation of motion for one-mass system is written as

$$M\ddot{y} + C\dot{y} + F(y) = -M\ddot{z}_0 \tag{1.1}$$

where M : mass

   y : relative displacement of the mass

   C : damping coefficient

   $F(y)$: restoring force

   $\ddot{z}_0$: ground acceleration

Eq(1.1) expresses a fundamental relationship which governs the vibrational response and any responses can be obtained by integrating the equation.

By multiplying dy to Eq(1.1) and integrating over the duration of ground motion, the equation of
energy balance is obtained as follows.

$$\int_0^{t_0} M\ddot{y}\dot{y}dt + \int_0^{t_0} C\dot{y}^2 dt + \int_{y(0)}^{y(t_0)} F(y)dy = -\int_0^{t_0} M\ddot{z}_0\dot{y}dt \tag{1.2}$$

Eq(1.2) can be written also as

$$W_e + W_p + W_h = E \tag{1.3}$$

    where W_e : elastic vibrational energy

             W_p : cumulative inelastic strain energy

             W_h : energy absorbed by damping

             E : total energy input exerted by an earthquake

Eqs.(1.1) and (1.2) are easily extended to apply to multi-mass systems by expressing related quantities with matrices and vectors.

Eq.(1.1) expresses a balance of force at an instant and the numerical integration of Eq.(1.1) yields
structural responses at any level of structural damage irrespective of structural states to be elastic
or elastic.

The obtained information is scattered and discrete one governed by specific conditions.

Only one guarantee for exactness of results obtained by Eq.(1.1) is given by a proper execution of
numerical calculation, and Eq.(1.1) does not speak much of the structural behavior under earthquakes.

On the other hand, Eq.(1.2) or Eq.(1.3), which is also an exact expression of structural responses,
can speak generally about various phases of structural behavior in terms of W_e, W_p and W_h.

Moreover, the stable nature of the total energy input stated below enhances the applicability of
Eqs.(1.2) and (1.3).

    The total energy input exerted by an earthquake is mainly governed by the total mass and the
    fundamental natural period of a structure and is hardly influenced by design parameters such
    as the strength distribution, mass distribution, and stiffness distribution.

As a basis of design consideration, deep understanding for the real behavior of structures under
earthquakes is indispensable. In order to store knowledge, it is effective to accumulate individual
results obtained by Eq.(1.1) by means of Eq.(1.3). Through this procedure, the relationship between
the seismic input and the structural response can be synthetically grasped and the practical measure to realize the structural performance which is aimed by structural designers can be explicitly
shown.

The design method based on Eq.(1.3) grounded by enormous amount of results of numerical analyses made by Eq.(1.1) can cope with the general design judgements. In this sense, this design
method can be called a general design method or a synthetic design method.

The required structural performance of a structure is originally claimed by the owner of the structure. Therefore, the design method can be stated by a language which can be understood not only
engineers but also laymen. Since the synthetic design method can be spoken by a plain language,
the significance of synthesis in constructing a system of technical language can not be disregarded.

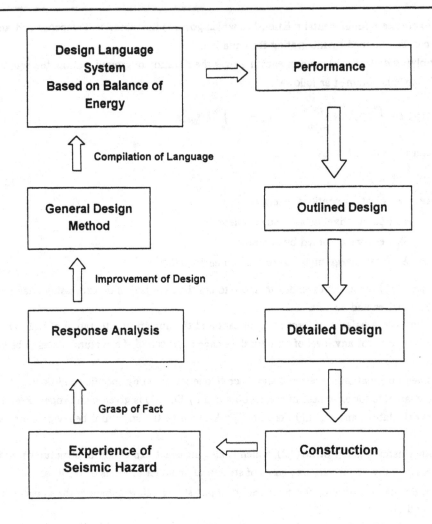

Fig.1.1 Development of Seismic Design

An outlined design can be made using the design language and the general design method as shown by Fig.1.1.

In the process of the detailed design, sometimes, the numerical response analysis is required to prove the fulfillness of structural performances.

The numerical response analysis is useful to improve the applicability of the general design method and, then, to improve the design language.

In such a manner, repetition of the analysis and synthesis and the accumulation of experiences of seismic hazard will lead to the progress of the earthquake resistant design method.

2 Energy Input in Single-Degree of Freedom System.

2.1 Equilibrium of Force and Equilibrium of Energy

The equation of equilibrium of the one-mass system shown in Fig.2.1(a) is expressed as

$$M\ddot{y} + C\dot{y} + F(y) = F_e \tag{2.1}$$

where M : mass

$C\dot{y}$: damping force

$F(y)$: restoring force

F_e : seismic force

z_0 : horizontal motion of the ground

y : displacement of the mass relative to the ground

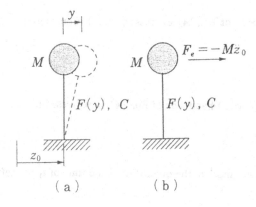

Fig.2.1 One-mass Vibrational System

Eq.(2.1) can be also applied to the system which stands on the fixed ground and is subjected to the force, F_e applied on the mass. The response of the system can be obtained by integrating Eq.(2.1). Whereas the analytical closed form of solution is obtained for elastic systems, Eq.(2.1) is generally solved by means of numerical analysis for the plastified systems. When the solution is obtained, the equilibrium of force stated by Fig2.1(b) can be transformed to the equilibrium of energy. The important response is the relative displacement, y. y causes strains in the system. Also, with respect to energy, the energy which is associated with strain is important in structural design point of view. Therefore, herewith, the equilibrium of energy is evaluated for the model shown by Fig.2.1(b), by multiplying dy ($= \dot{y}dt$) to Eq.(2.1).

The equilibrium equation is written as

$$M\int_0^t \ddot{y}\dot{y}dt + C\int_0^t \dot{y}^2 dt + \int_0^t F(y)\dot{y}dt = \int_0^t F_e\dot{y}dt \tag{2.2}$$

The constituent of Eq.(2.1) is discriminated as follows.

$$E(t) = \int_0^t F_e\dot{y}dt \tag{2.3}$$

$$W_e(t) + W_p(t) = M \int_0^t \ddot{y} \dot{y} dt + \int_0^t F(y) \dot{y} dt \tag{2.4}$$

$$W_h = C \int_0^t \dot{y}^2 dt \tag{2.5}$$

where $E(t)$: energy input at the time of t

$W_e(t)$: elastic vibration energy at the time of t

$W_p(t)$: cumulative inelastic strain energy at the time of t

$W_h(t)$: energy absorption due to damping at the time of t

The elastic vibration energy W_e can be written as

$$W_e(t) = W_{es}(t) + W_{ek}(t) \tag{2.6}$$

where $W_{es}(t)$: elastic strain energy

$W_{ek}(t)$: kinetic energy

The first term of the left hand side of Eq.(2.4) expresses, $W_{ek}(t)$. Considering $\dot{y}(0) = 0$, $W_{ek}(t)$ is written as

$$W_{ek}(t) = \frac{M\dot{y}^2(t)}{2} \tag{2.7}$$

Therefore, the second term of the right-hand side of Eq.(2.4) is expressed as

$$W_{es}(t) + W_p(t) = \int_0^t F(y) \dot{y} dt \tag{2.8}$$

t_0 being the duration of the ground motion, the quantities at the time of t_0 are defined as follows.

$$\left.\begin{array}{l} E = E(t_0) \\ W_e = W_e(t_0) \\ W_p = W_p(t_0) \\ W_h = W_h(t_0) \end{array}\right\} \tag{2.9}$$

where E : total energy input

W_e : elastic vibrational energy

W_p : cumulative inelastic strain energy

W_h : energy absorption due to damping

Eq.(2.2) is rewritten as

$$W_e + W_p + W_h = E \tag{2.10}$$

Eq.(2.10) is the fundamental equation on which the earthquake design method is constructed. Similar to Eq.(2.1), Eqs.(2.2) and (2.10) are equations of exact equilibrium. Eq.(2.1) provides structural responses. On the other hand, Eqs.(2.2) and (2.10) do not provide structural responses directly, but are very helpful to express and interpret structural responses.

The reasons are; whereas Eq.(2.1) expresses a state of equilibrium of force at an instant, Eqs.(2.2) and (2.10) provide an integrated information of vibrational state. Moreover, the effectiveness of Eq.(2.10) is guaranteed by the stability of the total energy input.

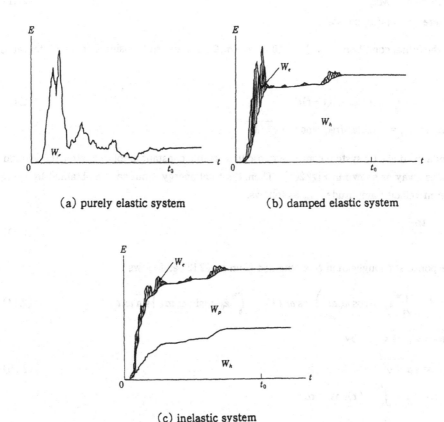

(a) purely elastic system   (b) damped elastic system

(c) inelastic system

Fig.2.2 Time History of Energy

Fig2.2 indicates the structural responses in time histories of energy. Fig2.2(a) indicates the case of an undamped system, in which the response is all elastic vibrational energy.

The energy input, $E(t)$ being always positive, is not necessarily increasing as time passes. Under t greater than t_0, $E(t)$ is kept to be E, since the ground motion which adds energy input does not exist any more. The purely elastic system continues to vibrate under a constant energy input, E.

Fig 2.2(b) is a case of damped elastic system. The energy absorption due to damping, $W_h(t)$ is monotonously increasing. Therefore, as the damping factor increases, $E(t)$ tends to be monotonously increasing. The displacement response, y, reaching the maximum at a time within t_0, rapidly reduces to zero.

Fig.2.2(c) is a case of elastic-plastic system. The cumulative inelastic strain energy, $W_p(t)$ is also monotonously increasing. The absorbed energy due to damping becomes smaller than that absorbed by the elastic system with same damping coefficient, C.

2.2 Fundamental Characteristics of Energy Input

2.2.1 Energy Input in Undamped Elastic System

The vibrational equation of the undamped elastic system is given by the following equation.

$$M\ddot{y} + ky = -M\ddot{z}_0 \tag{2.11}$$

where k : spring constant

Under the initial condition of $y(0) = \dot{y}(0) = 0$, Eq.(2.11) is solved by using Duhamel integral as follows.

$$\dot{y}(t) = -\int_0^t \ddot{z}_0(\tau)\cos \omega_0(t-\tau)d\tau \tag{2.12}$$

where $\omega_0 = $ circular frequency $= \sqrt{k/M}$

The undamped elastic system continue to oscillate with a constant amplitude after the ground motion fades away as shown in Fig2.2(a). Then, the total energy input can be obtained by using the maximum velocity amplitude, $\dot{y}_{max}$ as follows.

$$E = \frac{M\dot{y}_{max}^2}{2} \tag{2.13}$$

The response at t longer than t_0 is obtained from Eq.(2.12) as follows.

$$\dot{y} = \left(-\int_0^{t_0} \ddot{z}_0(\tau)\cos \omega_0 \tau d\tau\right)\cos \omega_0 t + \left(-\int_0^{t_0} \ddot{z}_0(\tau)\sin \omega_0 \tau d\tau\right)\sin \omega_0 t \tag{2.14}$$

Therefore $\dot{y}_{max}$ is given by

$$\dot{y}_{max} = \sqrt{a^2 + b^2} \tag{2.15}$$

where $a = \int_0^{t_0} \ddot{z}_0(\tau)\cos \omega_0 \tau d\tau$

$b = \int_0^{t_0} \ddot{z}_0(\tau)\sin \omega_0 \tau d\tau$

a and b in Eq.(2.15) are called Fourier integral. The meaning of this integral is to extract the wave component characterized by the circular frequency, ω_0 among wave components which composes the ground motion, $\ddot{z}_0$.

Therefore, Eq.(2.15) tells that the wave component which exerts energy input to an undamped elastic system is limited to a single wave component characterized by ω_0.

Vibration of the undamped elastic system becomes resonant under a stationary sinusoidal input with ω_0. This fact is common to the fact shown by Eq.(2.15) Thus, the undamped elastic system receives the energy very selectively.

The total energy input is transformed into an equivalent velocity by applying the following equation.

$$E = \frac{MV_E^2}{2}, \quad \left(V_E = \sqrt{\frac{2E}{M}}\right) \tag{2.16}$$

where V_E : equivalent velocity of the total energy input

V_E is the square root of the total energy input per unit mass, having an understandable dimension of velocity. The relationship between V_E and ω_0 (otherwise $f = \omega_0/2\pi$ or $T_0 = 2\pi/\omega_0$, f : natural frequency, $T_0 = $ natural period) is called energy spectrum. The energy spectrum of the undamped elastic system coincides with the so-called Fourier Spectrum.

2.2.2 Energy Input in Elastic Damped System

The second and third terms of the left-hand side of Eq.(2.1) are combined to express a quantity, R. The relationship between R and y is depicted schematically for the elastic undamped system, the damped elastic system and the inelastic system as shown in Fig.2.3. Fig.2.3(a) is for the case of purely elastic system. The $R-y$ relationship for the damped elastic system increases nonlinearity as C increases.

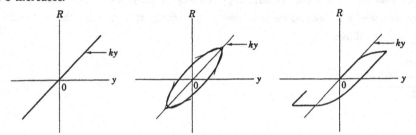

  (a) purely elastic system   (b) damped elastic system   (c) inelastic system

Fig.2.3 $R-y$ Relationship

The free vibration of a damped elastic system is expressed by

$$\ddot{y}+2h\omega_0\dot{y}+\omega_0^2 y = 0 \tag{2.18}$$

where $h = C/2M\omega_0$: damping constant

Under the initial condition of $y = y_0$ and $\dot{y} = 0$, the free vibration is obtained as

$$y = \frac{y_0}{\sqrt{1-h^2}}e^{-h_0\omega_0 t}\cos\left(\sqrt{1-h^2}\,\omega_0 t-\varepsilon\right), \quad \tan\varepsilon = \frac{h}{\sqrt{1-h^2}} \tag{2.19}$$

the period of free vibration is given by

$$T = \frac{2\pi}{\sqrt{1-h^2}}\sqrt{\frac{M}{k}} \tag{2.20}$$

The period given by Eq.(2.20) almost coincides with the natural period of the undamped elastic system, $T_0 = 2\pi\sqrt{M/k}$. The tangent slope in Fig.2.3(b), k_t, however, varies around k. The instantaneous period, T_i is defined to be

$$T_i = 2\pi\sqrt{\frac{M}{k_t}} \tag{2.21}$$

T_i lies in the range of

$$T_0-\Delta T \le T_i \le T_0+\Delta T \tag{2.22}$$

Δt which indicates the band of variation of T_i increases as h increases.

h and ΔT are roughly related by the following relationship [1].

$$\Delta T = 1.5 k T_0 \tag{2.23}$$

Therefore, the damped elastic system is characterized by a band of vibrational periods, while the purely elastic system is characterized by the single period of T_0. In this manner, as the vibrational system becomes complex, the wave components which supply energy to the system are plurallized. On the assumption that each wave component supplies energy evenly, the energy input in the damped elastic system can be calculated on the basis of the energy spectrum for the undamped elastic system, $_0V_E(T_0)$ as follows.

$$E = \frac{M \int_{T_0-\Delta T}^{T_0+\Delta T} {}_0V_E^2 dT}{2 \cdot 2\Delta T} \tag{2.24}$$

2.2.3 Energy Input in Inelastic System

The energy input for the system with $R = F(y)$ is discussed. The $R-y$ relationship is shown in Fig.2.3(c), where the initial slope is k and the instantaneous vibration period is elongated as the inelastic deformation develops. Then, the band of variation of vibrational period is expressed as follows, taking the width of band to be $2\Delta T$, similarly to the case of damped elastic system.

$$T_0 \leq T \leq T_0 + 2\Delta T \tag{2.25}$$

where T : substantial vibrational period

Therefore, the energy input in the inelastic system can be expressed as follows, similarly to the case of damped elastic system.

$$E = \frac{M \int_{T_0}^{T_0+2\Delta T} {}_0V_E^2(T)dT}{2 \cdot 2\Delta T} \tag{2.26}$$

Thus, the energy input of the damped elastic system and that of the inelastic system are basically identical and can be calculated through averaging the energy input of the purely elastic system within a band width of period which corresponds to the variation of substantial vibrational period of each system.

2.2.4 Shape of Energy Spectrum

In this chapter, the major characteristics stated in 2.2.1 to 2.2.3 are ascertained and observed minutely.

Used ground motion records are;

  • El centro record of the Imperial Vallay Earthquake (1940)
  • Hachinohe record of the Tokachioki Earthquake (1968)
  • Kobe Marine Observatory record of the Hyogoken-nanbu Earthquake (1995)

In Fig.2.4, the energy spectra for elastic systems are shown. It is clearly seen from the figure that as the damping is increased, the effect of averaging is deepened. The energy spectrum for the purely elastic system is highly dependent on the period. On the other hand, the energy spectrum

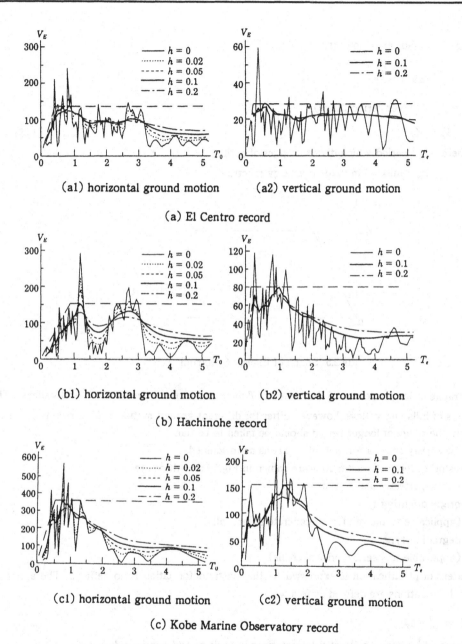

(a1) horizontal ground motion   (a2) vertical ground motion

(a) El Centro record

(b1) horizontal ground motion   (b2) vertical ground motion

(b) Hachinohe record

(c1) horizontal ground motion   (c2) vertical ground motion

(c) Kobe Marine Observatory record

Fig.2.4 Energy Spectra for Elastic System

for the highly damped systems is not so dependent on the period. In each figure, the bi-linear type of spectrum which envelopes the energy spectrum for the damping of $h = 0.1$ is depicted by a broken line. As is stated in 2.2.6, the energy spectrum for the damping of $h = 0.1$ has an importance to be a design spectrum. The envelope of the energy-spectrum with $h = 0.1$ has a shape as shown in Fig. 2.5. According to the range of period, the simplified energy spectrum can be described as follows.

$T \leq T_G$ (shorter period range)

$$V_E = aT = \frac{V_{Em}T}{T_G}$$ (2.27)

$T > T_G$ (longer period range)

$$V_E = V_{Em}$$ (2.28)

where T_G : period which divides ranges of period

V_{Em} : maximum value of energy spectrum

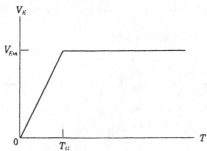

Fig.2.5 Shape of Energy Spectrum

In the range of longer period, V_E fluctuates as T increases, characterized by hills and valleys. The positions of hills and valleys, however, differ for different ground motions. As a result, a general shape in the range of longer period should be taken to be flat.

Next, the energy spectra for inelastic systems are discussed.

Used restoring force characteristics are shown in Fig.2.6 and these are;

• elastic-perfectly plastic type

• origin-orienting type

(applicable to the reinforced concrete shear walls)

• degrading type

(applicable to steel cylindrical shells)

The extent of plastification is expressed by the plastic deformation ratio defined. The apparent plastic deformations are defined as follows.

$$\delta_{pm}^{\pm} = \delta_m^{\pm} - \delta_Y$$ (2.29)

where $\delta_{pm}^{\pm}$: apparent plastic deformations in positive and negative directions

$\delta_m^{\pm}$: maximum deformations is positive and negative directions

The plastic deformation ratio is defined as follows.

$$\mu^{\pm} = \frac{\delta_{pm}^{\pm}}{\delta_Y}$$ (2.30)

$$\bar{\mu} = \frac{\mu^+ + \mu^-}{2}$$ (2.31)

where $\bar{\mu}$: average plastic deformation ratio

   $\mu^{\pm}$: plastic deformation ratios in positive and negative directions

In inelastic system, by adjusting the level of Q_Y, the aimed response of $\bar{\mu}$ can be easily obtained. First, $\bar{\mu}$ is fixed to be a certain value. Next, the response of $\bar{\mu}$ is calculated by Eq.(2.1). When the obtained response of $\bar{\mu}$ is smaller than the aimed value, the revised response of $\bar{\mu}$ closer to the aimed value is obtained by reducing Q_Y by a suitable amount. Thus, after several times of trials, the aimed value is obtained, since the total energy input is a very stable amount.

In Fig.2.7, the total energy energy input in the inelastic system obtained for a specified value of $\bar{\mu}$ is indicated. In Figs.2.7(a1), (b1) and (c1), V_E is depicted for T_0. A general tendency that the extent of averaging is deepened with the increase of $\bar{\mu}$ is clearly seen in reduction of undulation of energy inputs with the increase of $\bar{\mu}$.

In the range of shorter period, the averaging of energy spectrum, $_0V_E$ is made on the left hand side of T_0 as shown in Eq.(2.26). Thus, when the averaged value of V_E is depicted on the abscissa of T_0, it is anticipated that V_E increases with the increase of $\bar{\mu}$. Such a tendency is also clearly seen in Fig.2.7. In case of the elastic-perfectly plastic type of restoring force characteristics, this tendency is not remarkable. In other two cases, however, the increasing tendency of V_E is so remarkable that the energy spectrum for the elastic system with the damping of $h = 0.1$ can not be a spectrum enveloping the energy input for inelastic systems in the range of shorter period.

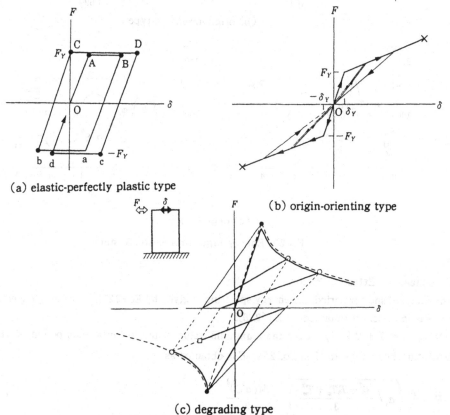

(a) elastic-perfectly plastic type

(b) origin-orienting type

(c) degrading type

Fig.2.6 Types of Restoring Force Characteristics

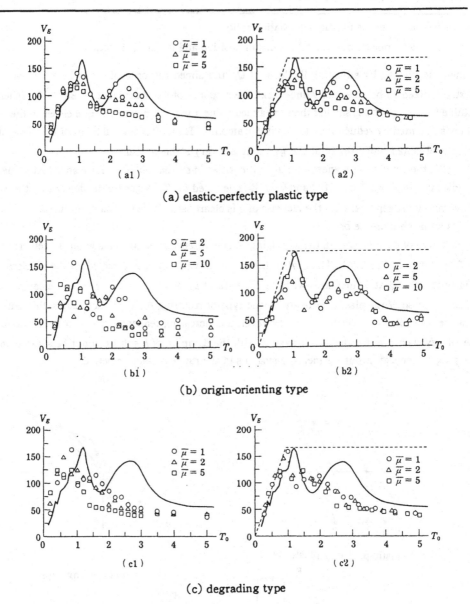

(a) elastic-perfectly plastic type

(b) origin-orienting type

(c) degrading type

Fig.2.7 Energy Input in Inelastic System

2.2.5 Concept of Effective Period

In the range of shorter period, the energy spectrum is given by Eq.(2.27). The energy spectrum for the inelastic system is obtained by Eq.(2.26).

Denoting $T_m = T_0 + 2\Delta T$, T_m means the maximum value of the instantaneous period of vibration.

Substituting Eq.(2.27) into V_E in Eq.(2.26), E is obtained as

$$E = \frac{M}{2}\left(a\sqrt{\frac{T_0^2 + T_0 T_m + T_m^2}{3}}\right)^2 = \frac{M(aT_e)^2}{2} \qquad (2.32)$$

$$\text{where} \quad T_e = \sqrt{\frac{T_0^2 + T_0 T_m + T_m^2}{3}} \tag{2.33}$$

Eq.(2.32) implies that the energy input in the inelastic system can be expressed by the same expression as Eq.(2.27) by applying the effective period, T_e in place of T_0.

As is shown in Fig.2.8, Eq.(2.33) can be approximated by

$$T_e = \frac{T_0 + T_m}{2} \tag{2.34}$$

Eq.(2.34) implies that the effective period can be given by the simple average of the natural period and the instantaneous maximum period. Considering that an energy spectrum can be represented by piecewise-linear relations, it is concluded that Eq.(2.34) can be applied to any shapes of energy spectrum.

2.2.6 Application of Effective Period

When the restoring force characteristics are to be described precisely, a standard load-deformation relationship is required. To be standard is identical to be well-definable. The best well-definable load-deformation relationship is the load-deformation relationship under monotonic loading. Thus, the monotonic load-deformation curve is indispensable to describe the restoring force characteristics. In Fig.2.9, a monotonic load deformation curve is schematically shown. Referring to a maximum response, μ and the monotonic load-deformation curve, the instantaneous rigidity of a system, k_s can be defined as follows.

$$k_s = \frac{qQ_Y}{(1+\mu)\delta_Y} \tag{2.35}$$

$$\text{where} \quad qQ_Y : \text{yield level, associated with } \mu$$

The period of vibration which corresponds to k_s, T_s is defined as follows.

$$T_s = 2\pi \sqrt{\frac{M}{k_s}} = T_0 \sqrt{\frac{1+\mu}{q}} \tag{2.36}$$

The maximum instantaneous period of vibration, T_m can be evaluated on the basis of T_s and is expressed in a following formula [4].

$$T_m = a_T T_s = a_T T_0 \sqrt{\frac{1+\mu}{q}} \tag{2.37}$$

$$\text{where} \quad a_T : \text{modifying constant}$$

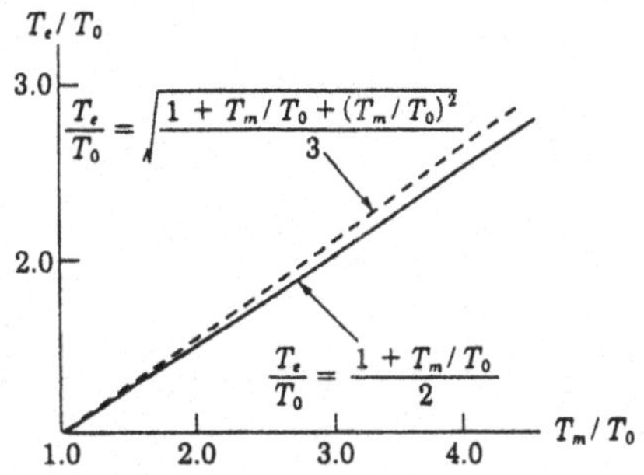

$$\frac{T_e}{T_0} = \sqrt{\frac{1 + T_m/T_0 + (T_m/T_0)^2}{3}}$$

$$\frac{T_e}{T_0} = \frac{1 + T_m/T_0}{2}$$

Fig.2.8 Effective Period

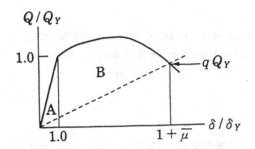

Fig.2.9 Load-Deformation Curve under Monotonic Loading

a_T is obtained as follows, according to the type restoring force characteristics.

For the elastic-perfectly plastic type:

$$a_T = \frac{1+\frac{\bar{\mu}}{8}}{\sqrt{1+\bar{\mu}}} \tag{2.38}$$

For the origin-orienting type and degrading type.

$$a_T = 1.0 \tag{2.39}$$

In Figs.2.8(a2), (b2) and (c2), the energy input in inelastic systems are indicated on the abscissa of T_e. It is clearly seen that the energy spectra for the elastic system with the damping of $h = 0.1$ can be representative of all cases of energy inputs of the inelastic systems, irrespective of the level of damage. Thus, the envelope spectrum shown by broken lines can be a design spectrum for general use.

3. Energy Input in Multi-Degree of Freedom System [3]

3.1 Energy Input in Elastic System

Based on the modal analysis, the total energy input in the continuous shear strut shown in Fig.3.1 which corresponds to a system with infinitely large number of masses is given by

$$E = \sum E_j$$

$$E_j = -\int_0^{t_0} \dot{z}_0 \dot{q}_j \left(\int_0^H m \phi_j dx \right) dt \tag{3.1}$$

where E_j : total energy input in jth mode

   q_j : time function

   ϕ_j : mode function

   m : distributed mass

   H : height of shear strut

   x : height from the ground

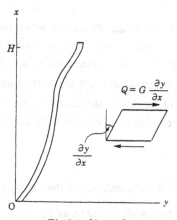

Fig.3.1 Shear Strut

Evaluating q_j and ϕ_j for various distribution of shear rigidity, the total energy input was found to be [1]

$$E > \frac{Mg^2 T^2}{4\pi^2} \cdot \frac{\alpha^2(0)}{2} \tag{3.2}$$

where $\alpha(0)$: base-shear coefficient = $Q_{max}(0)/Mg$

   $Q_{max}(0)$: maximum shear force response at the base

   M : total mass

   T : fundamental natural period

Therefore, the total energy input in the lastic system can be estimated conservatively by

$$E = \frac{Mg^2 T^2}{4\pi^2} \cdot \frac{\alpha^2(0)}{2} \tag{3.3}$$

The base-shear coefficient of structure, $\alpha(0)$ is governed by the fundamental natural period. Then, it can be said that the total energy input in elastic system is only governed by the total mass and the fundamental natural period of the system.

3.2 Energy Input in Inelastic System

It is shown that the total energy input in elastic systems is determined exclusively by the total mass and the fundamental natural period. The correspondence between many-mass inelastic systems and many-mass elastic systems is similar to that between one-mass inelastic systems and one-mass elastic systems. That is, plastification has the effect of expanding instantaneous periods. Eventually, the total energy input of an inelastic system governed by the natural period in the elastic range, T_0, and a period, T_1 greater than that in the elastic range is expressed by a mean value of energy inputs for an elastic system, the natural period of which drops in the range of $T_0 \leq T \leq T_1$. T_1 depends on the extent of plastification, becoming longer as plastification develops. In this chapter, the above-mentioned inference will be verified with some specific examples of multi-story inelastic systems. The multi-mass systems adopted are undamped, shear-type five-mass systems.

Because any modes higher than the fifth mode are of no significance in actual buildings, the five-mass system may represent multi-mass systems. A multi-mass vibrational system is specified by the mass ratio m_i/m_1, the yield-shear force ratio α_i/α_1, the stiffness ratio k_i/k_1, and a set of m_1, α_1, k_1. m_1 and k_1 are reduced to one parameter m_1/k_1, as m_1/k_1 is related to the fundamental natural period. The subscript, i, denoting the number of masses or stories, increases in ascending order with the height of a structure. The yield-shear coefficient, α_i is defined by the following relation.

$$\alpha_i = \frac{Q_{Y_i}}{\displaystyle\sum_{j=i}^{5} m_j g} \tag{3.4}$$

where Q_{Y_i} : yield-shear force coefficient of the ith story.

In order to distinguish distributions of parameters, such notations as M_n for the mass distribution, A_n for the yield-shear force distribution, and K_n for the stiffness distribution, are introduced. Thus, a vibrational system is denoted by M_n, A_n, K_n, α_1, and T.

Table 3.1 Parameters in Vibrational Systems

		i (Number of Stories or Masses)				
	Index	1	2	3	4	5
$\dfrac{m_i}{m_1}$	M_1	1.0	1.0	1.0	1.0	1.0
	M_2	1.0	0.333	0.333	0.333	0.333
	M_3	1.0	1.0	3.0	1.0	1.0
	M_4	1.0	1.0	1.0	1.0	3.0
$\dfrac{\alpha_i}{\alpha_1}$	A_1	1.0	1.10	1.25	1.565	2.0
	A_2	1.0	10.0	10.0	10.0	10.0
	A_3	1.0	10.0	10.0	10.0	1.0
	A_4	1.0	1.0	1.0	1.0	0.1
$\dfrac{\kappa_i}{\kappa_1}$	K_1	1.0	0.867	0.733	0.600	0.400
	K_2	1.0	0.820	0.640	0.500	0.200
	K_3	1.0	1.0	1.0	1.0	0.1

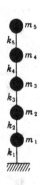

M_1 stands for the case of a uniform distribution of masses. M_2 expresses the case for which masses other than the first mass are one-third of the first mass. M_3 and M_4 correspond to the case in which one mass is greater than the others by three times.

A_1 expresses the case for which the yield-shear force distribution is controlled so that the cumulated ductility ratio, η_i, becomes nearly equal in all stories. η_i is defined by the accumulated inelastic horizontal deformation of the ith story divided by the elastic horizontal deformation of the ith story under a yield-shear force. The yield-shear force distribution for this case is obtained by a trial-and-error procedure of numerical analysis. A_2 stands for the case in which all the stories above the first story are ten times stronger than the first story. In this case, inelastic deformation can take place only in the first story. A_3 is the case for which only first and fifth stories are forced to succumb to inelastic deformation. In the case of A_4, only the fifth story behaves inelastically.

K_1 and K_2 are cases for which spring constants change linearly along the height, K_2 being equipped with a steeper change of stiffness distribution. In the case of K_3, the stiffness of the fifth story is one-tenth that of the other stories.

Actual buildings are conditioned almost by a set of cases (M_n, A_n, K_n) or (M_1, A_1, K_1). The applied restoring-force characteristics of each story are two typical types. One is the elastic-perfectly plastic type, in terms of story shear-force, Q_i, and story displacement, $\delta_i (= y_i - y_{i-1})$ (y_i being a horizontal displacement of the ith mass). Another type is the degrading type, as shown in Fig.3.2. Broken lines in this figure, denoted by OAE and OA'E', are the $Q_i - \delta_i$ relations under monotonic loading in the positive and negative directions, respectively. OAE and OA'E' are symmetric with respect to the origin, and under an arbitrary history of deformations it is assumed that the $Q_i - \delta_i$ relation is governed by the following law:

The slope in the elastic range is unchangeable, and the slope in the inelastic range, $dQ_i / d\delta_i$ is constant. A relation, which is obtained by connecting through parallel movement, piecewise load-deformation curve in the inelastic range under the positive sign of Q_i, coincides with the load-deformation curve under the same sign of monotonic loading. That is, moving parallel

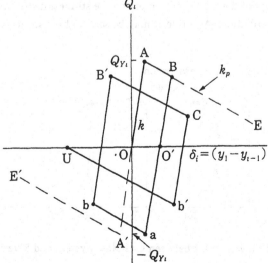

Fig.3.2 Restoring-Force Characteristics with Deterioration in Strength

to the segment B'C in Fig.3.2 and overlapping the point B' on B, the curve thus obtained, A−B(B')−C, agrees with the curve under monotonic loading. In the same manner, under the negative sign for Q_i, inelastic deformation increments connected sequentially provide the inelastic portion of the monotonic load deformation curve.

Under such a hysteretic law, inelastic deformation develops until point U in Fig.3.2 is reached, where with $Q_i = 0$, the restoring force is lost, and the system loses its resistance to the P−δ effect (due to vertical loading) and collapses. The aim of applying such degrading types of restoring-force characteristics is to discern whether the energy input in a system that collapses in a certain story is comparable to that in a system with elastic-perfectly plastic restoring-force characteristics. The accelerogram used is from El Centro.

Total energy input can be calculated by the following equation, using a velocity for each mass, $\dot{y}_i$, relative to the ground.

$$E = - \sum_{i=1}^{5} \left(m_i \int_0^{t_0} \ddot{z}_0 \dot{y}_i \, dt \right) \tag{3.5}$$

In Fig.3.3, the relation between the non-dimensionalized total energy input, A_E and α_1 is shown for a representative natural period. A_E is defined to be

$$A_E = E / \frac{Mg^2 T^2}{4\pi^2} \tag{3.6}$$

The solid line in the figure is the total energy input of undamped one-mass systems with restoring-force characteristics of the elastic-perfectly plastic type. It is clearly shown in the figure that the energy input into five-mass systems is fairly close to that in one-mass systems.

Similarly, the energy input in elastic-perfectly plastic systems is shown in Fig.3.4, in which the solid line indicates the responses of one-mass systems. $A_E - \alpha_1$ relations ($A_E - \alpha_5$ in the case of A_4 where only the fifth story can behave inelastically) for various sets of K_n, M_n, and A_n are compared. The aim is to discern the dependence of the total energy input on the stiffness distribution, the mass distribution, and the strength distribution. As a rule, it may be said that the stiffness distribution, the

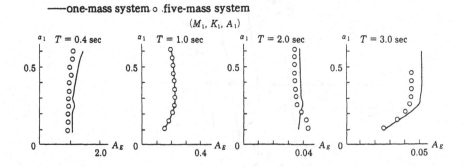

Fig.3.3 Comparison of Energy Input between One-Mass System and Five-Mass System

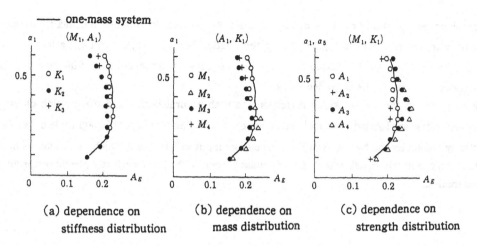

(a) dependence on   (b) dependence on   (c) dependence on
stiffness distribution   mass distribution   strength distribution

Fig.3.4 Energy Input in Five-Mass Systems

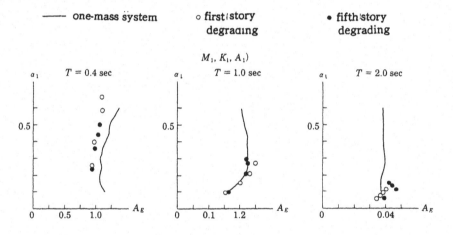

Fig.3.5 Energy Input in Degrading Systems

mass distribution, and the strength distribution are not influential with regard to the total energy input.

Fig.3.5 shows the response of a five-mass system that is equipped with the degrading type of restoring-force characteristics (as shown in Fig.3.2) in the first or fifth story, while the other stories are of the elastic-perfectly plastic type. This case is conditioned by (M_1, K_1, A_1). The ratios of the slope in the degrading range, k_p to that in elastic range, k were selected to be

$$\frac{k_p}{k} = -0.025, \ -0.05, \ -0.075, \ -0.1$$

Whether a story with a degrading type of restoring-force characteristics will collapse depends on the magnitude of the yield-shear force. The magnitude of the yield-shear force for each story is described by a_1. It is obvious that the larger a_1 becomes, the less the possibility of collapse. When the limit of a_1, which enables the system to remain uncollapsed is denoted by a_m, collapse does not occur under the condition $\alpha_m < \alpha_1$, α_1, at the collapse limit, is obtained by making α_1 increase gradually, and is shown in Fig.3.5. As for the relation between $|k_p|/k$ and α_m, a larger value of α_m is required to prevent collapse as $|k_p|/k$ increases, owing to the decrease of inelastic energy absorption capacity in the degraded story.

Four points in the figure, which correspond to each natural period and the position of each degraded story, are located upward from one another, in ascending order of magnitude of $|k_p|/k$. As clearly shown by the figure, even the total energy input in a building which just collapses in a certain story is nearly equal to that of a one-mass system with elastic-perfectly plastic restoring characteristics.

4. Estimate of Structural Damage

4.1 Expression of Damage

In the basic equation of earthquake resistant design based on the balance of energy shown by Eq(2.10), the term which corresponds to structural damage is W_p. Whereas the elastic deformation is restored to the non-stress state as the load is removed, the inelastic deformation remains unreleased and is monotonously accumulated until the collapse state is reached.

In this sense, the cumulated inelastic deformation or cumulated inelastic strain energy can be called damage and its quantity implies the degree of damage. W_p means the total sum of structural damage. When a structure is composed of many elements, W_p is generally written as

$$W_p = \sum W_{pi} \tag{4.1}$$

where W_{pi} : cumulative inelastic strain energy of ith element

Assuming a structure with a single element, W_p is expressed as follows.

$$W_p = Q_Y \delta_p = \eta Q_Y \delta_Y \tag{4.2}$$

where $\eta = \dfrac{W_p}{Q_Y \delta_p}$: cumulative inelastic deformation ratio

δ_p : cumulative inelastic deformation

η, being a nondimensionalised damage, is called cumulative inelastic deformation ratio. Assuming the elastic-perfectly plastic restoring force characteristics, η is written as

$$\eta = \frac{\delta_p^+ + \delta_p^-}{\delta_Y} = \eta^+ + \eta^- \tag{4.3}$$

where $\eta^+ = \delta_p^+ / \delta_Y$: cumulative inelastic deformation ratio in the positive direction (see Fig.4.1)

$\eta^- = \delta_p^- / \delta_Y$: cumulative inelastic deformation ratio in the negative direction

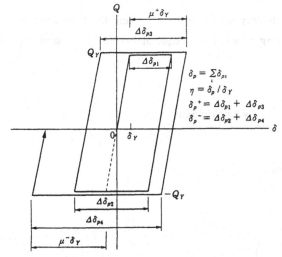

Fig.4.1 Restoring Force Characteristics of Elastic-Perfectly Plastic Type

In this case, η means the real cumulative inelastic deformation divided by the elastic limit deformation. A visual image of damage can be given by the residual deformation. The residual deformation in case of the elastic-perfectly plastic restoring force characteristics, δ_r is expressed by

$$\delta_r = \left| \delta_p^+ - \delta_p^- \right| = \left| \eta^+ - \eta^- \right| \delta_Y \tag{4.4}$$

Another important expression of damage is the maximum deformation or the apparent maximum plastic deformation as stated in 2.2.4.

The apparent maximum plastic deformation can be related to the cumulative plastic deformation. The nondimensionalised apparent maximum plastic deformation is expressed by μ as shown by Eqs.(2.30) and (2.31). Then, the estimate of η, $\left| \eta^+ - \eta^- \right|$ and μ is essential in the seismic design. Among them, η can express most directly the structural damage. $\left| \eta^+ - \eta^- \right|$ and μ can be subsequently qualified in relation to μ.

When W_p is dominant in Eq(2.10), $W_e + W_h$ can be neglected and η is quantified based on the following relationship.

$$W_p = E \tag{4.5}$$

When the system consists of single element, μ can be directly obtained from Eq(4.5).

When the system consists of plural elements, the total damage is clearly described by Eq(4.5). As for the distribution of W_{pi}, however, any possibilities can exist. Therefore, it is most important to clarify a law governing the distribution of W_{pi}.

4.2 Basic Damage Distribution Law Applied to Shear-Type Systems with Elastic-Perfectly Plastic Restoring-Force Characteristics

The yield strength of ith story is denoted by Q_{Yi} and the elastic limit deformation under Q_{Yi} is denoted by δ_{Yi}. Using the cumulative inelastic deformation ratio of ith story, η_i, W_{pi} is expressed as

$$W_{pi} = \eta_i Q_{Yi} \delta_{Yi} \tag{4.6}$$

The spring constant of ith story is denoted by k_i. Using the total mass, M and the fundamental natural period, T, the spring constant of an equivalent one-mass system, k_{eq} is define as

$$k_{eq} = \frac{4\pi^2 M}{T^2} \tag{4.7}$$

k_i is related to k_{eq} by the following expression

$$k_i = \kappa_i k_{eq} \tag{4.8}$$

Then, Eq(4.6) is rewritten as

$$W_{pi} = \frac{Mg^2 T^2}{4\pi^2} c_i \alpha_i^2 \eta_i \tag{4.9}$$

$$\text{where} \quad c_i = \left(\frac{\displaystyle\sum_{j=i}^{N} m_j}{M} \right)^2 \frac{1}{\kappa_i} \tag{4.10}$$

As a standard distribution of damage, the damage distribution under an uniform distribution of η_i is taken as follows

$$\frac{W_{pk}}{W_p} = \frac{c_k \alpha_k^2}{\displaystyle\sum_{i=1}^{N} c_i \alpha_i^2} \qquad (4.11)$$

The yield shear force coefficient distribution which realizes the standard damage distribution is defined to be the optimum yield shear force coefficient distribution and is expressed as

$$\bar{\alpha}_i = \alpha_i / \alpha_1$$

Eq(4.11) is also written as

$$\frac{W_{pk}}{W_p} = \frac{S_k}{\sum S_j} \qquad (4.12)$$

$$\text{where} \quad s_i = c_i \kappa_1 \bar{\alpha}_i^2 = \sum_{j=i}^{N} \left(\frac{m_j}{M}\right) \bar{\alpha}_i^2 \left(\frac{k_1}{k_i}\right) \qquad (4.13)$$

Eq(4.13) signifies that the standard damage distribution is given by the mass distribution, the spring constant distribution and the optimum yield shear force coefficient distribution.

4.3 Optimum Yield Shear Force Coefficient Distribution

If the optimum yield shear force coefficient distribution is unchanged, irrespective of the quantity of η_i, $\bar{\alpha}_i$ can be given by the shear force coefficient distribution of the elastic system which corresponds to the case of infinitesimally small constant damage.

The above-mentioned inference has been checked to be applied to practical cases, and an unified expression for $\bar{\alpha}_i$ has been already obtained [1].

In Fig.4.2, the optimum yield shear force coefficient distribution which was obtained by a trial-and-error approach applied to multi-mass systems with N of 3 to 9 is shown. The ordinate is taken to be $(i-1)/N$.

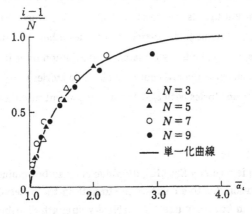

Fig.4.2   Optimum Yield Shear Force Coefficient Distribution

The sequence of the numerical analysis is as follows; First, a_1, and a_i/a_1 is set, and the response is obtained. A relatively weaker story suffers damage concentration.

Then, in the stories where η_k is larger than the averaged damage η_0, a_i is increased and in the stories where η_k is smaller than η_0, a_i is decreased.

This procedure is repeated several times until the condition of $\eta_k = \eta_0$ is almost satisfied. Used accelerogram is of El Centro (1940). The result of analysis can be expressed by an unified curve as shown in Fig.4.2.

The unified curve is expressed by

$$
\left.
\begin{aligned}
&\text{for} \quad x \geq 0.2, \\
&\bar{a}_i = 1 + 1.5927x - 11.8519x^2 + 42.5833x^3 - 59.48272 + 30.1586x^5 \\
&\text{for} \quad x < 0.2, \\
&\bar{a}_i = 1 + 0.5x
\end{aligned}
\right\}
\tag{4.14}
$$

$$
\text{where} \quad x = \frac{i-1}{N}
$$

4.4 Damage Distribution Law

4.4.1 Basic Expression

Under the optimum yield shear force coefficient the damage distribution is given by Eq(4.12) Based on numerical analyses for systems equipped with strength distribution different from $\bar{a}_i$, it was found that the damage distribution in general case can be expressed by a following expression continuous to Eq(4.12)

$$
\frac{W_{pi}}{W_p} = \frac{s_i p_i^{-n}}{\displaystyle\sum_{j=1}^{N} s_j p_j^{-n}} = \frac{1}{\gamma_i}
\tag{4.15}
$$

$$
\text{where} \quad p_j = \frac{a_j}{a_1 \bar{a}_j}
$$

n : damage concentration index

γ_1 : damage dispersion factor

p_j expresses the extent of discrepancy of a_j/a_1 from the optimum distribution.

When $p_j = 1$, it is obvious that Eq(4.15) is reduced to Eq(4.12). γ_i is termed damage dispersion factor, since γ_i indicates the extent of damage dispersion into stories other than ith story. n is a positive exponent. As n increases, the dependency of damage distribution on p_j increases.

In case of $n = 0$, the damage concentration does not take place. In stories with $p_j > 1$, damage is reduced as n increases. Reversely, in stories with $p_j < 1$, damage concentration is emphasized, as n increases.

4.4.2 Damage Concentration Index

When the damage distribution is given by Eq(4.15), the value of n can be obtained by observing the chance of the damage distribution due to a decrease of strength in the observed story as follows. First, the damage distribution in kth story under an arbitrary strength distribution is obtained as

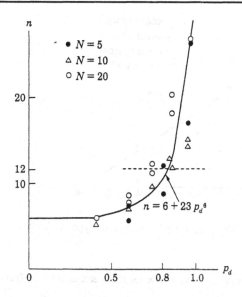

Fig.4.3   Value of n for Weak Column Type of Frames

follows.

$$a = \frac{W_{pk}}{W_p} = \frac{s_k p_k^{-n}}{\displaystyle\sum_{j=1}^{N} s_j p_j^{-n}} \qquad (4.16)$$

Next, the damage distribution under another strength distribution in which the strength in kth story is modified by multiplying p_d is obtained as follows.

$$b = \frac{W_{pk}}{W_p} = \frac{s_k p_k^{-n} p_d^{-n}}{\displaystyle\sum_{j \neq k} s_j p_j^{-n} + s_k p_k^{-n} p_d^{-n}} \qquad (4.17)$$

From Eqs.(4.16) and (4.17), the value of n is obtained as follows

$$n = -\ell_n \left\{ \frac{b(1-a)}{a(1-b)} \right\} \Big/ \ell_n p_d \qquad (4.18)$$

Herewith, the result of analysis is shown. Fig.4.3 shows the value of n for weak-column type of frames with elastic-perfectly plastic restoring force characteristics. It is obvious that n can be expressed by an increasing function of p_d. It is already ascertained that the value of n for the weak-column type of frames should be 12[1].

$n = 12$ corresponds to an upper bound value for $p_d = 0.8$. Then, it can be deduced that the value of n for practical use can be obtained by taking $p_d = 0.8$ in Eq(4.18)

In Fig.4.4, the value of n for weak-beam type of frames is shown, taking $p_d = 0.8$ and varying stiffness ratios between columns and beams, k_{cb}.

k_{cb} is the ratio of stiffness of column to stiffness of beam. A reasonable result that the damage concentration is mitigated by the increase of stiffness of column is clearly seen in decreasing tendency

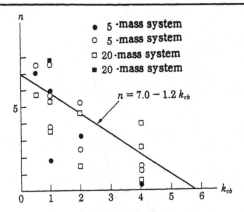

Fig.4.4 Value of n for Weak Beam Type of Frames

of n as k_{cb} increases.

As a representative and conservative value of n for the weak-beam type frames, $n = 6$ can be taken.

4.5 Relationship between Cumulative Inelastic Deformation and Maximum Deformation

A load-deformation relationship under a monotonic loading is indicated in Fig.4.5, in which Q_Y is the yield strength and δ_Y is the elastic deformation corresponding to Q_Y.

The inelastic strain energy under the monotonic load-deformation curve, W_{pm} is defined to be

$$W_{pm} = \int_{\delta_Y}^{\delta_m} Q \cdot d\delta \qquad (4.19)$$

Using the inelastic deformation ratio, μ,
δ_m is written as

$$\delta_m = (1+\mu)\delta_Y \qquad (4.20)$$

W_{pm} is an increasing function of μ.

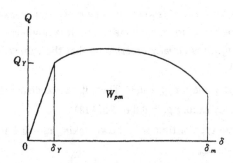

Fig.4.5 Load-Deformation Curve under Monotonic Loading

Under a seismic excitation, a structure develops maximum deformations in positive and negative directions, δ^+, δ^-. μ^+ and μ^- correspond to δ^+ and δ^-, and $\bar{\mu}$ is the average of μ^+ and μ^-.

The cumulative strain energy of a story can be formally expressed as follows, using W_{pm} given by Eq(4.19).

$$W_p = 2W_{pm}(\bar{\mu})a_p \tag{4.21}$$

a_p is a coefficient and is defined to be

$$a_p = \frac{W_p}{2W_{pm}(\bar{\mu})} \tag{4.22}$$

The cumulative inelastic deformation ratios, η^+ and η^- are defied to be

$$\eta^+ = \frac{W_p^+}{Q_Y\delta_Y}, \quad \eta^- = \frac{W_p^-}{Q_Y\delta_Y} \tag{4.23}$$

where W_p^+, W_p^- : cumulative inelastic strain energy in positive and negative loading domain

The sum of η^+ and η^- is the cumulative inelastic deformation ratio, η and the average of η^+ and η^- is the mean cumulative inelastic deformation ratio, $\bar{\eta}$.

A non-dimensional expression of W_{pm} is defined by

$$A_{pm}(\bar{\mu}) = \frac{W_{pm}(\bar{\mu})}{Q_Y\delta_Y} \tag{4.24}$$

Using these quantities and referring to Eq(4.21), the following equation is obtained.

$$\bar{\eta} = A_{pm}(\bar{\mu})a_p = \frac{A_{pm}(\bar{\mu})}{\bar{\mu}}a_p\bar{\mu} \tag{4.25}$$

Referring to Eq(4.25), the following expression is obtained.

$$\frac{\bar{\eta}}{\bar{\mu}} = \frac{A_{pm}(\bar{\mu})}{\bar{\mu}}a_p \tag{4.26}$$

The greater of η^+ and η^- is defined to be μ_m. Then, knowing $\eta = 2\bar{\eta}$, the next equation holds.

$$\frac{\eta}{\mu_w} = \frac{2A_{pm}(\bar{\mu})a_p}{\bar{\mu}}\cdot\frac{\bar{\mu}}{\mu_m} \tag{4.27}$$

$\bar{\mu}/\mu_m$ expresses the deviation of maximum deformations in one direction, being in the range of

$$0.5 \leq \frac{\bar{\mu}}{\mu_m} \leq 1.0 \tag{4.28}$$

Referring to Eq(4.27), the following relationship is obtained.

$$\frac{A_{pm}}{\bar{\mu}}a_p \leq \frac{\eta}{\mu_m} \leq 2.0\frac{A_{pm}}{\bar{\mu}}a_p \tag{4.29}$$

When the monotonic load-deformation curve is elastic-perfectly plastic type, $A_{pm}/\bar{\mu}$ is unify and

Eq(4.29) is reduced to

$$a_p < \frac{\eta}{\bar{\mu}_m} < 2.0\,a_p \qquad (4.30)$$

When a_p and $\bar{\mu}/\mu_m$ are quantified, the maximum deformation is related to the cumulative inelastic deformation. Since η is definitely related to the total energy input, the maximum deformation is related to the total energy input. In order to obtain a_p and $\bar{\mu}/\mu_m$, the numerical analysis is indispensable.

Using shear type of multi-story frames, $\eta - \mu_m$ relationship is quantified.

A general form of the shear-type frame can be expressed by the flexible-stiff mixed frame. The flexible-stiff mixed frame is characterized by the mixture of the flexible elastic element and the stiff elastic-plastic element in each story as shown in Fig.4.6.

Ordinary structures, which are not intentionally equipped with flexible elements, are identified to be structures singly equipped with stiff element (ordinary structures).

By introducing flexible elements, hysteretic behaviors tend to orient the point of origin.

As a result, the residual deformation is reduced and the maximum deformation is restrained under a certain amount of energy input. The flexible-stiff mixed structure is characterized by the rigidity ratio, r_k and the shear force ratio, r_q defined by

$$r_k = \frac{k_f}{k_s} \qquad (4.31)$$

$$r_q = \frac{_f\bar{Q}_m}{_sQ_Y} \qquad (4.32)$$

where k_f : spring constant of flexible element

k_s : spring constant of rigid element

$_f\bar{Q}_m$: average of maximum shear forces developed in the positive and negative loading domains in flexible element

$_sQ_Y$: yield shear force of rigid element

$\bar{\mu}$: average inelastic deformation ratio

r_q under a maximum deformation of $\bar{\mu}_s\delta_Y$ ($_s\delta_Y$: elastic limit deformation of stiff element) is expressed as follows, knowing

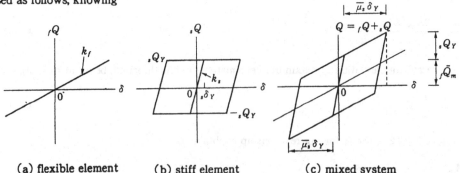

(a) flexible element   (b) stiff element   (c) mixed system

Fig.4.6   Flexible-Stiff Mixed Structure

$$_sQ_Y = k_s\delta_Y, \quad _f\bar{Q}_m = k_f(1+\mu)_s\delta_Y.$$

$$\tau_q = (1+\mu)\tau_k \tag{4.33}$$

When k_f and $_sQ_Y$ are kept constant values, it is anticipated that as k_s becomes greater, the stiff element absorbs energy more efficiently, thus the maximum deformation being more effectively restrained. However, as k_s becomes greater, τ_k is decreased. Therefore, τ_k can not be a major quantity effective to restrain the maximum deformation. On the other hand, τ_q expresses directly the degree of origin-orienting tendency, thus, can be a major quantity to restrain the maximum deformation. In this context, the results of numerical analysis are arranged by using τ_q. Used restoring force characteristics are elastic-perfectly plastic type.

In Fig.4.7, deformation responses are shown for three different seismic records.

a) Deviation of Maximum Deformation

In Fig.4.7(a), $\mu_m/\bar{\mu} - \tau_q$ relationship is shown for $T_f = 2.5$ sec These results are summarized as follows.

Upper bound value:

$$\left.\begin{array}{ll} \text{for } \tau_q \le 1.0, & \dfrac{\mu_m}{\bar{\mu}} = \dfrac{2+2\tau_q}{1+2\tau_q} \\[4mm] \text{for } \tau_q > 1.0, & \dfrac{\mu_m}{\bar{\mu}} = \dfrac{4}{3} \end{array}\right\} \tag{4.34}$$

Medium value:

$$\left.\begin{array}{ll} \text{for } \tau_q \le 1.0, & \dfrac{\mu_m}{\bar{\mu}} = \dfrac{3+\tau_q}{2+2\tau_q} \\[4mm] \text{for } \tau_q > 1.0, & \dfrac{\mu_m}{\bar{\mu}} = 1.0 \end{array}\right\} \tag{4.35}$$

b) **Deviation of Cumulative Inelastic deformation**

In Fig.4.7(b), $\eta_m/\bar{\eta} - \tau_q$ relationship is shown for $T_f = 2.5$ sec. As τ_q increases, $\eta_m/\bar{\eta}$ rapidly converges to unity. Referring to Eq.(4.4), it can be seen that residual deformation almost vanishes as τ_q reaches 0.2.

c) **Correspondence between $\bar{\eta}$ and $\bar{\mu}$**

In Fig.4.7(c), $\bar{\eta}/\bar{\mu} - \tau_q$ relationship is shown for $T_f = 2.5$ sec. Results are summarized as

Lower bound value:

$$\left.\begin{array}{ll} \text{for } \tau_q \le 1.0, & \dfrac{\bar{\eta}}{\bar{\mu}} = 2+2\tau_q \\[4mm] \text{for } \tau_q > 1.0, & \dfrac{\bar{\eta}}{\bar{\mu}} = 4.0 \end{array}\right\} \tag{4.36}$$

Design value:

$$\left.\begin{array}{ll} \text{for } \tau_q \le 1.0, & \dfrac{\bar{\eta}}{\bar{\mu}} = 3+\tau_q \\[4mm] \text{for } \tau_q > 1.0, & \dfrac{\bar{\eta}}{\bar{\mu}} = 4.0 \end{array}\right\} \tag{4.37}$$

The design value means the value which is proposed for practical use in design and is selected to be slightly larger than the lower bound value.

d) Correspondence between η and μ_m

In Fig.4.7(d), $\eta/\mu_m - r_q$ relationship is shown. Knowing $\bar{\eta}/\bar{\mu}$ and $\mu_m/\bar{\mu}$, η/μ_m is given by

$$\frac{\eta}{\mu_m} = \frac{2\bar{\eta}}{\mu_m} = 2\left(\frac{\bar{\eta}}{\bar{\mu}}\right)\left(\frac{\bar{\mu}}{\mu_m}\right) \tag{4.38}$$

Using the lower bound value of $\bar{\eta}/\bar{\mu}$ (Eq.(4.36)) and the upper bound value of $\mu_m/\bar{\mu}$ (Eq.(4.34)), the lower bound value of η/μ_m is obtained as follows.

$$\left.\begin{array}{ll} \text{for } r_q \leq 1.0, & \dfrac{\eta}{\mu_m} = 2 + 4r_q \\[3mm] \text{for } r_q > 1.0, & \dfrac{\eta}{\mu_m} = 6.0 \end{array}\right\} \tag{4.39}$$

To take the lower bound value of η/μ_m for practical design purposes seems to be over-conservative. Taking the design value of $\bar{\eta}/\bar{\mu}$ (Eq.(4.37)) and the medium value of $\bar{\mu}/\mu_m$ (Eq.(4.35)), the design value of η/μ_m is obtained as follows.

$$\left.\begin{array}{ll} \text{for } r_q \leq 1.0, & \dfrac{\eta}{\mu_m} = 4 + 4r_q \\[3mm] \text{for } r_q > 1.0, & \dfrac{\eta}{\mu_m} = 8.0 \end{array}\right\} \tag{4.40}$$

The obtained formulas can be applicable for different values of T_f than 2.5sec as shown in Fig.4.7.

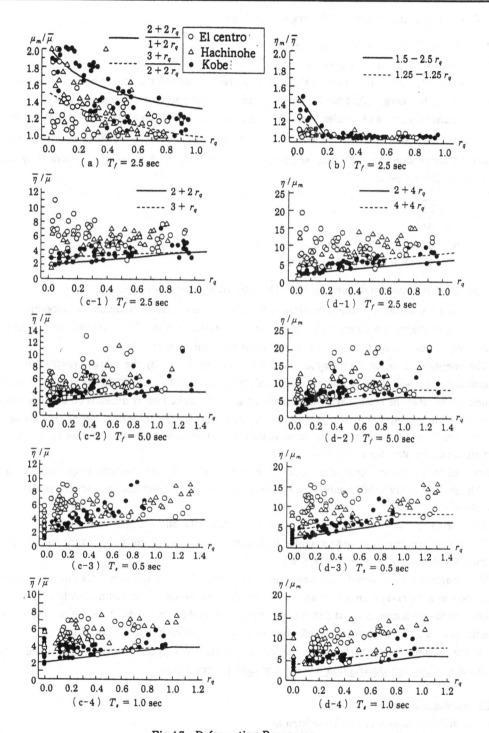

Fig.4.7 Deformation Responses

5. Practical application of Energy Approach

5.1 Ordinary Earthquake Resistant Structures

5.1.1 Loading Effect of Earthquakes

The loading effect of earthquakes on structures can be grasped most concisely by means of the energy input. The energy input can be expressed in a form of energy spectrum as follows [1].

- The Fourier spectrum of accelerogram of an earthquake motion coincides with the $V_E - T$ relationship of the same earthquake motion. V_E is the equivalent velocity obtained through the following conversion from the total energy input in the undamped one-mass vibrational system with the natural period of T.

$$V_E = \sqrt{\frac{2E}{M}} \qquad (5.1)$$

   where E : total energy input into the system
   M : total mass of the system

- The $V_E - T$ relationship defined by Eq.(5.1) is termed as the energy spectrum.

   The energy spectra for damped systems or inelastic systems can be obtained by smoothing (or averaging) the energy spectrum for the undamped elastic system. The extent of smoothing increases proportionally to the extent of nonlinearity of the system.

- The energy spectra for the inelastic systems can be represented by the energy spectrum for the elastic system with 10% of fraction of critical damping ($h = 0.1$). In this sense, it can be easily understood that the total energy input made by an earthquake mainly depends on the total mass and the fundamental natural period of the structure, and is scarcely influenced by the strength, strength distribution, stiffness distribution, mass distribution and type of restoring force characteristics of the structure.

Compared to the conventional acceleration response spectrum and the velocity response spectrum which can be directly applied only to elastic systems, the energy spectrum has decisive advantages as following.

- The energy spectrum can be directly applied both to elastic and inelastic systems.
- The energy spectrum can be represented by a single curve which corresponds to the energy spectrum for the elastic system with $h = 0.1$.
- Two major indices of structural damage are expressed in terms of the cumulative inelastic deformation and the maximum deformation. The cumulative inelastic deformation can be directly related to the total energy input and it is not difficult to find a relationship between the maximum deformation and the cumulative deformation through numerical response analyses.
- On the basis of energy spectrum, the earthquake resistant design can be clearly formulated to be Control of Damage Distribution under a Constant Energy Input.

5.1.2 Earthquake Resistant Design

The equilibrium of energy can be written as

$$W_e + W_h + W_p = E \qquad \text{where} \quad W_e : \text{elastic vibrational energy} \qquad (5.2)$$
$$W_h : \text{energy absorbed by damping}$$
$$W_p : \text{cumulative strain energy}$$

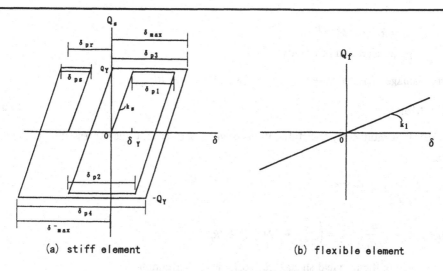

(a) stiff element   (b) flexible element

Fig.5.1 Restoring-Force Characteristics

The total energy input E is a very stable amount irrespective of structural behavior.

On the other hand, the distribution of energy over a structure depends on the structural type, and mechanical properties of structural components. The structural damage corresponds to W_p.

In order to know the distribution of energy, numerical response analyses are indispensable. By summarizing the results of numerical analyses, it is possible to construct a simplified and conceptual design method based on tие energy spectra.

W_p consists of cumulative inelastic strain energy in every story W_p. Thus,

$$W_p = \sum_{i=1}^{N} W_{pi} \tag{5.3}$$

where N : number of story.

W_p and W_{pi} can be regarded as structural damage. Each story of a shear type multi-story structures is considered to be composed of a stiff element and a flexible element. The flexible element has a small stiffness and remains elastic, whereas the stiff element has a large stiffness and behaves inelastically. The relation between the shear resistance and the story displacement is depicted in Fig.5.1, where the elastic-perfectly plastic restoring force characteristics of the stiff element is assumed. Assuming that the spring constant of the stiff element, k_s is sufficiently larger than that of the flexible element, k_f and the contribution of energy absorption of the flexible element can be neglected, then, the damage of the first story of a building is written as

$$W_{P1} = \frac{Mg^2T^2}{4\pi^2} \times \frac{2\alpha_1^2\bar{\eta}_1}{\kappa_1} \tag{5.4}$$

where α_1 : yield shear force coefficient of the first story ($= Q_{Y1}/Mg$)

   Q_{Y1} : yield shear force of the stiff element in the first story

   $\bar{\eta}_1$: averaged comulative inelastic deformation ratio of the stiff element in the first story (=cumulatiye inelastic deformation / two times of yield displacement)

   δ_Y : yield displacement (see Fig.5.1(a))

$$\kappa_1 = k_1/(4\pi^2 M/T^2)$$

g : acceleration of gravity

The total damage of a structure. W_p can be formally related to W_{p1} as

$$W_p = \gamma_1 W_{p1} \tag{5.5}$$

In the shear-type multi-story structures, it has been made clear that γ_1 is expressed by the following formula.

$$\gamma_1 = 1 + \sum_{j \neq 1} s_j (p_j/p_1)^{-n} \tag{5.6}$$

where $\quad p_j = \dfrac{\alpha_j}{\alpha_1 \bar{a}_j}, \quad s_j = \left(\sum_{i=j}^{N} m_i/M \right)^2 \bar{a}_j^2 (k_1/k_j)$

$\bar{a}_j$: optimum yield shear force coefficient distribution

α_j/α_1 : actual yield shear force coefficient distribution

m_i : mass of ith floor

k_i : spring constant of ith story

p_j means a deviation of the actual yield shear force distribution from the optimum yield shear force distribution under which the damage of every story $\bar{\eta}_j$ is equalized, and is termed the damage concentration factor. n is termed the damage concentration index. When n becomes sufficiently large, γ_1 becomes unity. It means that a sheer damage concentration takes place in the first story. When n is nullified, a most preferable damage distribution is realized. Practically, the value of n ranges between 2.0 and 12.0. Weak-column type of structures are very susceptible of damage concentration, and the n-value for them should be 12.0. In weak-beam structures, the damage concentration is considerably mitigated due to the elastic action of columns, and the n-value can be reduced to 6.0. In Fig.5.2 (c), a generalized form of weak-beam type structure is shown. The presence of a vertically extending elastic column is essential to this type of structure. The elastic column by itself is not required to withstand any seismic forces. While ordinary frames pin-connected to the elastic column absorbs inelastically seismic energy, the elastic column plays a role of damage distributor.

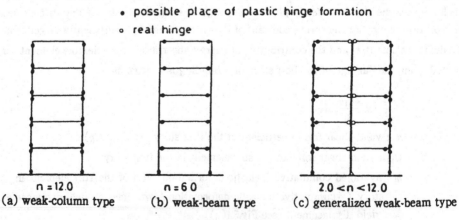

- possible place of plastic hinge formation
- real hinge

| n = 12.0 | n = 6 0 | 2.0 < n < 12.0 |
| (a) weak-column type | (b) weak-beam type | (c) generalized weak-beam type |

Fig.5.2 Various Structural Types

By applying this type of structure, the n-value can be reduced to 2.0.

The damage concentration is also governed by the value of p_j. To simply estimate the damage concentration in the first story, an unified value may be applied as p_j as follows.

$$p_1 = 1.0 \qquad\qquad p_{j \neq 1} = p_d \tag{5.7}$$

Eq.(5.7) signifies that the strength gap is assumed between the first story and the other stories. It is impossible to make the yield shear force distribution of an actual multi-story building agree completely with the optimum distribution. The reasons are easily found in the scatter in mechanical properties of material and the rearrangement of geometrical shapes of structural members for the purpose of simplification in fabricating process. Taking account of such a situation, the following value is proposed as a probable strength gap to be taken into account in the design procedure.

$$p_d = 1.185 - 0.0014N \qquad\qquad p_d \geq 1.1 \tag{5.8}$$

The elastic vibrational energy is expressed as follows.

$$W_e = \frac{Mg^2 T^2}{4\pi^2} \times \frac{\alpha_1^2}{2} \tag{5.9}$$

Taking account of damping, Eq.(5.2) is reduced to

$$W_p + W_e = E \times \frac{1}{\left(1 + 3h + 1.2\sqrt{h}\right)^2} \tag{5.10}$$

where h = damping constant

Substituting Eqs.(5.9) and (5.5) into Eq.(5.10) and using Eq.(5.4), the following formula is obtained.

$$\alpha_1 = \frac{\alpha_e}{\sqrt{1 + 4\frac{r_1 \bar{n}_1}{\kappa_1}}} \qquad\qquad \alpha_e = \frac{2\pi V_E}{Tg} \cdot \frac{1}{1 + 3h + 1.2\sqrt{h}} \tag{5.11}$$

Eq.(5.11) is rewritten as

$$\alpha_1(T) = D_s(\bar{n}_1)\alpha_e(T) \tag{5.12}$$

where $\alpha_e(T)$: required minimum yield shear force coefficient for the elastic system with the fundamental natural period T

$\alpha_1(T)$: required minimum yield force coefficient of the first story for the inelastic system with T

$D_s(\bar{n})$: reduction factor for the yield shear force coefficient, which depends on $\bar{n}_1$

The distribution of masses is assumed to be uniform. The yield deformation of every story is also assumed constant. Then, the stiffness distribution k_i/k_1 becomes equal to the strength distribution Q_{Yi}/Q_{Y1}.

The optimum yield shear force coefficient distribution $\bar{\alpha}_i$ is given by the following formula.

$$\bar{\alpha}_i = f\left(\frac{i-1}{N}\right) \tag{5.13}$$

for $x > 0.2$, $f(x) = 1 + 1.5927x - 11.8519x^2 + 42.583x^3 - 59.48x^4 + 30.16x^5$

for $x \leq 0.2$, $f(x) = 1 + 0.5x$

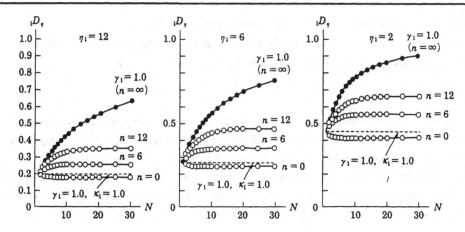

Fig.5.3 D_s-Values for Multi-Story Frames

Using the above-mentioned parameters, γ_1 in Eq.(5.6) and κ_1 are calculated and approximated by the following relations.

$$\gamma_1 = 1+0.64(N-1)p_d^{-n} \tag{5.14}$$
$$\kappa_1 = 0.48+0.52N \tag{5.15}$$

Then, the D_s-value is written as

$$D_s = \cfrac{1}{\sqrt{1+\cfrac{4\{1+0.64(N-1)p_d^{-n}\}\bar{\eta}_1}{0.48+0.52N}}} \tag{5.16}$$

Fig.5.3 shows the relationship between the D_s-value and the number of story N for specific values of $\bar{\eta}_1$.

The difference of D_s-values is caused by the difference of the damage concentration index n which governs the damage distribution in multi-story buildings. D_s-values inevitably increase with the increase N due to the effect of damage concentration.

The goal of earthquake resistant design can be summarized as follows.

1) to minimize α_1
2) to minimize the maximum story displacement δ_{max}
3) to minimize the residual story displacement δ_{pr} (see Fig.5.1(a))

To attain the first item, the following two measures are practicable.

1) to increase the deformation capacity $\bar{\eta}_1$
2) to reduce the damage concentration index n

The former is realized by applying mild steels to the stiff element. When the structural members are carefully selected so as to avoid structural instability such as local buckling and lateral buckling, it is not impossible to attain the value of $\bar{\eta}_1$ greater than 100. The later is realized by applying the weak-beam type structure or more general damage dispersing systems as shown in Fig.5.2(c).

High-strength steels can be most effectively used as a vertical damage distributor.

To discuss the maximum story displacement, the inelastic deformation ratio, $\bar{\mu}$ is introduced as follows.

$$\bar{\mu} = \left(\bar{\delta}_{max} - \delta_Y\right)/\delta_Y \tag{5.17}$$

where $\bar{\delta}_{max}$: average value of the maximum story displacement in the positive and negative
directions.

The residual story displacement, $\bar{\delta}_r$ is equal to the difference between the cumulative inelastic deformations of positive and negative directions as seen in Fig.5.1(a). To reduce $\bar{\delta}_r$ and $\bar{\mu}$, the most effective measure is the application of "the flexible-stiff mixed structure". Only slight participation of the flexible element enables to nullify $\bar{\delta}_r$ and to reduce $\bar{\mu}$ remarkably as is seen in the following empirical relations (see 4.5).

$$\text{for } r_q = 0, \ \bar{\mu} = \frac{\bar{\eta}}{2} \quad \text{for } r_q > 1.0, \ \bar{\mu} = \frac{\bar{\eta}}{4} \quad \text{to } \bar{\mu} = \frac{\bar{\eta}}{6} \tag{5.18}$$

The stiff elements are the source of energy absorption, whereas flexible elements restrain effectively the development of excessive deformations and one-sided deformations.

5.2 BASE-ISOLATED STRUCTURES

In the light of earthquake resistant design method described in the foregoing section, characteristics of base-isolation technique can be summarized as follows.

- Since, by applying flexible isolators at the base of structure, the superstructure can be assumed to be relatively rigid, the base-isolated structure can be assumed to be a single-degree of freedom system under horizontal ground motions
- Since the elastic energy absorption capacity of isolators is large enough to meet the total energy input due to an earthquake, the superstructure is liberated from several restrictions required for ordinary earthquake resistant structures to secure inelastic energy absorption capacity.
- By applying dampers, the horizontal displacement at the base is effectively reduced.

In turn, dampers must absorb almost all of the total energy input. When leads or steels are used for the material of dampers, the restoring-force characteristics at the base of the base-isolated structure take a shape as shown in Fig.5.4. Q denotes the total shear force and δ denotes the horizontal displacement at the base of base-isolated structure. α_s and α_f are defined as

$$\alpha_s = \frac{{}_sQ_Y}{Mg}, \ \alpha_f = \frac{{}_fQ_{max}}{Mg} \tag{5.19}$$

where ${}_sQ_Y$: yield strength of dampers

${}_fQ_{max}$: maximum shear force of isolators

M : total mass of superstructure

The results of numerical response analysis are summarized as follows.

- The maximum displacement δ_{max} takes place in almost same amount in the positive and negative directions.

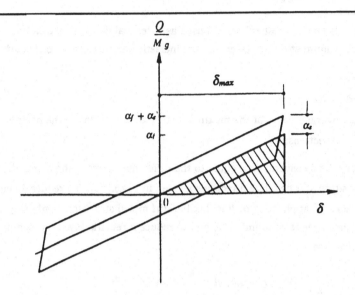

Fig.5.4. Restoring-Force Characteristics of the Base-Isolated Story

- The energy absorbed by dampers at the instant when the maximum displacement is reached
  can be approximately expressed by the area twice as large as the area covered by the closed
  loop in Fig.5.4.

The equilibrium of energy at the instant when the maximum displacement is reached can be expressed by

$$W_e + W_p = E(t_m)$$ (5.20)

where W_e : energy stored in isolators

W_p : energy stored in dampers

$E(t_m)$: total energy input at $t = t_m$

t_m : time when the maximum displacement is reached

The energy absorption due to damping of isolators is ignored. Referring the response characteristics, W_e and W_p are written as

$$W_e = \frac{Q_{max}\,\delta_{max}}{2}$$ (5.21)

$$W_p = 8\,_sQ_Y\,\delta_{max}$$ (5.22)

As the total energy input E is defined to be the energy input exerted by an earthquake during whole duration of time, E is generally greater than $E(t_m)$. Therefore, by applying E in place of $E(t_m)$ in Eq.(5.20), the maximum responses can be obtained with some errors of over-estimate. Then, the basic equation for the estimate of the maximum responses can be written as

$$\frac{Q_{max}\,\delta_{max}}{2} + 8\,_sQ_Y\,\delta_{max} = \frac{MV_E^2}{2}$$ (5.23)

δ_{max} can be expressed as

$$\delta_{max} = \frac{f Q_{max}}{k_f}$$ (5.24)

where k_f : spring constant of isolators

k_f is written as

$$k_f = \frac{4\pi^2 M}{T_f^2}$$ (5.25)

where T_f : period of base-isolated structure without dampers

Using Eqs(5.19), (5.24) and (5.25), Eq.(5.23) is reduced to

$$\frac{\alpha_f}{\alpha_0} = -a + \sqrt{a^2 + 1}$$ (5.26)

where $a = 8\left(\dfrac{\alpha_s}{\alpha_0}\right)$

$$\alpha_0 = \frac{2\pi V_E}{T_f g}$$

α_0 signifies the shear force coefficient for the base-isolated structure without dampers. The total shear force coefficient at the base of base-isolated structure α is obtained as

$$\alpha = \alpha_f + \alpha_s = \left(-\frac{7a}{8} + \sqrt{a^2 + 1}\right)\alpha_0$$ (5.27)

Then, the D_s-value for the base-isolated structure is obtained as

$$D_s = -\frac{7a}{8} + \sqrt{a^2 + 1}$$ (5.28)

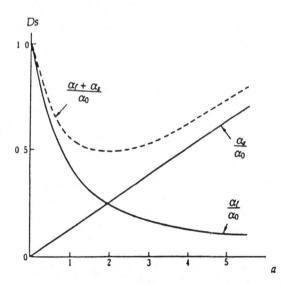

Fig.5.5 Ds-Values for Base-Isolated Structure

The maximum displacement is obtained as

$$\delta_{max} = \frac{Q_{max}}{k_f} = \frac{T_f^2 \alpha_f g}{4\pi^2} \tag{5.29}$$

The D_s-value given by Eq.(5.28) is shown in Fig.5.5. As is seen in the figure, the D_s-value for the base-isolated structure is hardly less than 0.5. The major advantage of base-isolated structure in reducing shear force response must be ascribed to the reduction of α_0. Since the V_g-value in the range of longer periods can be assumed to be constant, to make T_f large is of primary importance for the reduction of α_0. In order to make T_f large, the most effective measure is to reduce the number of isolators and this can be realized by properly evaluating huge compressive load-bearing capacity of isolators.

6. Prospect and Lessons

6.1 Advanced Design Method (Flexible-Stiff Mixed Structures)

6.1.1 Introduction

Previous seismic design methods have been developed with structural safety as the major consideration, while performance in other areas has been neglected, with the following results.

- During the severe earthquakes, structure are inevitably damaged to some extent and sometimes, repair is very expensive.
- Strengthening structures in order to reduce structural damage results in an increased acceleration response, which causes overturning of furniture and equipment. Thus, it causes an interruption of daily activities such as medical treatment and results in the loss of property.

On the other hand, the recently developed base-isolation technique has overcome the above-mentioned difficulties, without deterioration of the performance obtained by using conventional earthquake-resistant design methods.

In this section, an earthquake resistant design method for buildings which meets the requirements for both structural safety and reduction of the acceleration response is discussed.

The proposed design method is consistent to the method applied to base-isolated structures and is developed based on the balance between the seismic energy input and the energy absorption capacity of the structure. Structures in general are very complicated and prediction for exact behavior of them is, sometimes, very difficult. Therefore, in order to develop the performance-based design method, it is also necessary to exploit preferable structural types of which prediction of structural behavior can be explicitly made.

As a preferable structural type, the flexible-stiff mixed structure is introduced. The flexible-stiff mixed structure consists of the flexible elements which remain elastic even under severe earthquakes with a relatively low elastic rigidity and the stiff elements which behave mainly plastically with a relatively high elastic rigidity.

Conventional type of multi-story buildings can be remodeled to be a flexible-stiff mixed structure by definitely allotting a role of the flexible element or the stiff element to each structural element. Major damage indices such as the cumulative plastic deformation, the maximum deformation, the residual deformation and the maximum yield shear force coefficient are clearly related to the level of seismic input.

6.1.2 Flexible-Stiff Mixed Structure

The structure which is composed of the flexible elements remained elastic and the stiff elements with a high elastic rigidity and a high plastic deformation capacity is defined as the flexible-stiff mixed structure [1]. In flexible-stiff mixed structures, the yield strengths in positive and negative loading domains, $|Q_Y^+|$ and $|Q_Y^-|$, become different as the deformation develops. Generally, cumulative plastic deformations are liable to concentrate in the element with a relatively weak yield strength. Therefore, a further development of the plastic deformation in a loading domain where plastic deformations have developed with an increase of the yield strength is restrained autonomously in the flexible-stiff mixed structure, thus resulting in equalization of deformations in positive and negative loading domains. Main features in the response characteristics of the flexible-stiff mixed structures are summarized as follows.

1) The cumulative plastic defamations in positive and negative loading domains are nearly equal.

2) The maximum deformation in positive and negative loading domains are nearly equal.

3) Efficiency of the energy adsorption with respect to a maximum deformation is high.

4) The residual deformation can be made considerably small.

Referring to these characteristics, in comparison to the ordinary structures consisting of monotonous elastic-plastic elements, the flexible-stiff mixed structures are considered to be a preferable structural type of which performance in the seismic resistance can be clearly stated.

The cumulative plastic deformation, δ_p, is related to the maximum deformation, δ_m, by the following empirical equation in the flexible-stiff mixed structure, neglecting the elastic deformation of the rigid element.

$$\delta_p = 8\delta_m \qquad (6.1)$$

Also, the residual deformation in the flexible-stiff mixed structure, δ_r, is expressed empirically as

$$\delta_r = 0.2 \, {}_sQ_Y \left(\frac{1}{{}_fk} - \frac{1}{{}_sk} \right), \text{ and also } \delta_r < \delta_m \qquad (6.2)$$

where ${}_sQ_Y$: yield strength of the rigid element

   ${}_fk$: rigidity of the flexible element

   ${}_sk$: rigidity of the stiff element

6.1.3 Response of the Flexible-Stiff Mixed Structure

The form of the energy spectrum can be represented by a bilinear curve shown in Fig.6.1. That is, the V_E–T relationship is expressed by a line passing through the point of origin in the short-period range and takes a constant value in the long-period range as is expressed by

$$\text{for } T \le T_G, \quad V_E = \frac{V_{Em}}{T_G} T$$

$$\text{for } T > T_G, \quad V_E = V_{Em} \qquad (6.3)$$

where V_{Em} : maximum value of V_E

The energy input attributable to the structural damage, E_D, is also converted to the equivalent velocity, V_D, through the equation similar to Eq.(5.1).

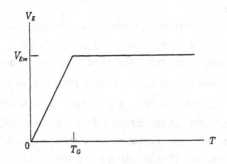

Fig.6.1 Energy Spectrum

V_D is related to V_E by the following empirical formula.

$$V_D = \frac{V_E}{1+3h+1.2\sqrt{h}}$$

(6.4)

The flexible-stiff mixed structure is assumed to be a shear type of multistory frame. The restoring force characteristics of the stiff element in each story is assumed to be elastic-perfectly plastic type. A hysteretic behavior of one story is shown in Fig.6.2. The rigidity of the stiff element is denoted by $_sk$ and the rigidity of the flexible element is denoted by $_fk$. $_s\delta_Y$ is the yield deformation of the stiff element. Under the maximum deformation, δ_m the instantaneous period of vibration of the system takes a value of T_m. The secant rigidity, k_s, associated with δ_m can be applied in order to predict T_m by using the following formula.

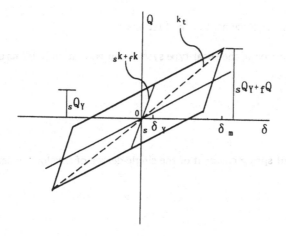

Fig.6.2 Hysteresis Loop

$$T_m = 2\pi\sqrt{\frac{M}{k_s}}$$

(6.5)

The energy attributable to the damage can be expressed in terms of potential energy under the gravity field i.e., by the equivalent height of the mass, h_E, according to the following equation.

$$h_E = \frac{E_D}{Mg} = \frac{V_D^2}{2g}$$

(6.6)

where g : acceleration of gravity

Generally, the energy input attributable to the damage is finally absorbed by structural skeletons of a structure in a form of cumulative plastic deformation. Therefore, the following equation holds.

$$E_D = \sum_{i=1}^{N} W_{pi}$$

(6.7)

where W_{pi} : cumulative plastic strain energy in each story

   N : number of the story

Eq.(6.7) can be written in respect to the damage of the first story as

$$E_D = W_{p1}\gamma_1 \qquad (6.8)$$

where γ_1 : the ratio of the total damage to W_{p1}, given by Eq.(5.14)

The damage concentration index, n is taken to be 6.0 for the flexible-stiff mixed structure. By dividing Eq.(6.8) with Mg, the following equation is obtained.

$$h_E = {}_s\alpha_{Y1}\delta_{p1}\gamma_1 \qquad (6.9)$$

where ${}_s\alpha_{Y1} = {}_sQ_{Y1}/Mg$: yield shear force coefficient in the first story

Using Eqs.(6.1) and (6.9), ${}_s\alpha_{Y1}$ is determined as

$${}_s\alpha_{Y1} = \frac{h_E}{8\gamma_1\delta_{m1}} \qquad (6.10)$$

where δ_{m1} : maximum displacement of the first story

The fundamental natural period of the shear-type system can be written in terms of the spring constant of the first story, k_1, as

$$T = 2\pi\sqrt{\frac{M\kappa_1}{k_1}} \qquad (6.11)$$

where $\kappa_1 = k_1/k_{eq}$

$\quad\quad k_{eq}$: equivalent spring constant of the single-degree of freedom system with M and T

κ_1 is given by Eq.(5.15).

$$\kappa_1 = 0.48 + 0.52N \qquad (6.12)$$

While in the long-period range, the energy input is given regardless of the value of the period, the energy input in the short-period range depends on the period. Therefore, the period must be precisely estimated.

The substantial period for the sysytem of which the period of vibration changes between T_0 and T_m is calculated for the short-period range as

$$T_s = \sqrt{\frac{T_0^2 + T_0 T_m + T_m^2}{3}} \qquad (6.13)$$

where T_s : substantial period of the sysytem

$\quad\quad T_0$: period in the elastic range

Referring to Fig.6.2, T_0 and T_m are written as

$$T_0 = 2\pi\sqrt{\frac{\kappa_1}{g\left(\dfrac{{}_s\alpha_1}{{}_s\delta_{Y1}} + \dfrac{{}_f\alpha_1}{\delta_{m1}}\right)}} \quad , \quad T_m = 2\pi\sqrt{\frac{\kappa_1}{g\left(\dfrac{{}_s\alpha_1 + {}_f\alpha_1}{\delta_{m1}}\right)}} \qquad (6.14)$$

where ${}_f\alpha_1 = \dfrac{{}_fk_1\delta_{m1}}{Mg}$: shear force cocfficient of the flexible element in the first story

6.1.4 Illustrative Example

As an illustrative example, the sysytem in which $_sk$ is sufficiently greater than $_fk$ and $_s\delta_Y$ is negligible small, is taken.

Applying these assumptions, T_0 becomes nullified, and T_e is reduced to

$$T_e = \frac{2\pi}{\sqrt{3}} \sqrt{\frac{\kappa_1 \delta_{m1}}{(1+f)_s a_{Y1}}}$$

(6.15)

where $f = {}_f a_1 / {}_s a_{Y1}$

Denoting h_E at $T = T_G$ by h_{Em}, h_E in the short-period range characterized by the linear $V_D - T$ relationship is written as

$$h_E = h_{Em} \left(\frac{T_e}{T_G} \right)^2$$

(6.16)

Using Eqs.(6.10)(6.15) and (6.16), $_s a_{Y1}$ for the energy input in the short-period range is determined as follows, regardless of δ_{m1}.

$$_s a_{Y1} = \frac{\pi}{\sqrt{6}\, T_G} \sqrt{\frac{h_{Em}\kappa_1}{(1+f)g\gamma_1}}$$

(6.17)

On the other hand. in the long-period range, since $h_E = h_{Em}$, $_s a_{Y1}$ is obtained as

$$a_{Y1} = \frac{h_{Em}}{8\gamma_1\delta_{m1}}$$

(6.18)

As an illustrative example, the maximum level of the ground motion which competes with the Hyogoken-nanbu earthquake, 1995 is applied, i.e., T_G and V_{Em} in the energy spectrum shown in Fig.6.1 are selected to be

$$V_{Em} = 400cm$$

(6.19)

$$T_G = 1.0sec$$

The damping of $h = 0.02$ is assumed. Then, the maximum value of V_D, V_{Dn}, and h_E become as

$$V_{Dm} = \frac{V_{Em}}{1+3h+1.2\sqrt{h}} = 325\ cm/sec$$

(6.20)

$$h_{Em} = 54\ cm$$

As a structural performance, the maximum story dislacement in the first story is assumed to be

$$\delta_{m1} = 5cm,\ 6.67cm,\ 10cm$$

(6.21)

A weak-beam type of structure is assumed, that is, in estimating the damage distribution, $n = 6.0$ is taken.

$_s a_{Y1}$ obtained by Eq.(6.17) is denoted by $(_s a_{Y1})_I$ and $_s a_{Y1}$ obtained by Eq.(6.18) is denoted by $(_s a_Y)_{II}$. The smaller of those becomes the real value of $_s a_{Y1}$ which corresponds to the given energy spectrum. The larger value of those corresponds to the extended lines of the two line segments of the energy spectrum. The $_s a_{Y1} - N$ relationships for $f = 1.0$ are shown in Fig.6.3 in which the solid lines are valid due to the above-mentioned reason.

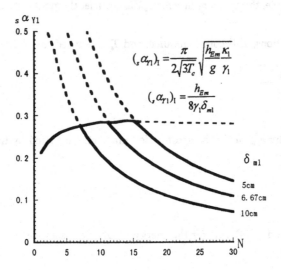

Fig.6.3 $_s\alpha_{Y_1}-N$ Relationship

Table 6.1 $_s\alpha_{Y_1}-N$ Relationship

N	γ_1	κ_1	$(_s\alpha_{Y1})_I$	$\bar{\delta}_{m1}(cm)$	$(_s\alpha_{Y1})_{II}$		
					$\delta_{m1}(cm)$		
					10.0	6.67	5.0
1	1.00	1.00	0.213	31.7	0.675	1.011	1.350
2	1.23	1.52	0.237	23.2	0.548	0.822	1.096
3	1.47	2.04	0.251	18.3	0.459	0.688	0.918
4	1.71	2.56	0.261	15.1	0.394	0.591	0.788
5	1.96	3.08	0.267	12.9	0.344	0.516	0.688
6	2.21	3.60	0.272	11.2	0.305	0.458	0.610
8	2.71	4.64	0.279	8.9	0.249	0.373	0.498
10	3.23	5.68	0.285	7.4	0.209	0.313	0.418
12	3.77	6.72	0.284	6.3	0.179	0.268	0.358
14	4.32	7.76	0.289	5.5	0.156	0.234	0.312
16	4.89	8.80	0.286	4.8	0.138	0.207	0.276
18	5.47	9.84	0.286	4.3	0.123	0.185	0.247
20	6.07	10.88	0.285	3.9	0.111	0.167	0.222
22	6.68	11.92	0.283	3.6	0.101	0.151	0.202
24	7.32	12.96	0.283	3.25	0.092	0.138	0.184
26	7.97	14.00	0.282	3.0	0.085	0.127	0.169
28	8.64	15.04	0.281	2.8	0.078	0.117	0.156
30	9.32	16.08	0.280	2.6	0.072	0.109	0.145

Also, in Table 6.1, the values of $_s\alpha_{Y1}$ for $f = 1.0$ are shown. The values of $(_s\alpha_{Y1})_{II}$ listed above the horizontal line are larger than $(_s\alpha_{Y1})_{I}$. Accordingly, those value are not valid. In the short-period range, $_s\alpha_{Y1}$ takes a constant value irrespective of δ_{m1}. In this case, however, δ_{m1} is limited by the condition that T_e does not exceed T_G. Denoting δ_{m1} which corresponds to T_G by $\bar{\delta}_{m1}$, $\bar{\delta}_{m1}$ is written as

$$\bar{\delta}_{m1} = \frac{3g_s\alpha_{Y1}T_G^2}{(1+f)\pi^2\kappa_1} = \frac{\sqrt{3}\,T_G}{2(1+f)\pi}\sqrt{\frac{gh_{Em}}{\kappa_1\gamma_1}} \tag{6.22}$$

In the short period range, an arbitrary value of δ_{m1} can be taken under a constant value of $_s\alpha_{Y1}$. However, under the given condition $_f\alpha_{Y1} = f_s\alpha_{Y1}$, $_fk_1$ must be

$$_fk_1 = \frac{f_s\alpha_{Y1}}{\delta_{m1}} \tag{6.23}$$

Actually, the restraining condition of Eq.(6.23) can be mitigated, that is, the rigidity higher than that given by Eq.(6.23) can be allowed on the reason that a higher rigidity makes T_e smaller than the prescribed value due to Eq.(6.23), resulting in a decrease of the energy input in the short-period range. In the region of $\delta_{m1} > \bar{\delta}_{m1}$, $_s\alpha_{Y1}$ is given by $(_s\alpha_{Y1})_{II}$ and the rigidity is secured also by Eq.(6.23). Assuming that $_fk/_sk$ is negligibly small and substituting $_sQ_Y = _fQ_Y/f$ into Eq.(6.2), the residual plastic deformation in the first story is obtained as

$$\delta_{r1} < \frac{0.2\,\delta_{m1}}{f} \tag{6.24}$$

6.2 Lessons Learnt from Hyogoken-Nanbu Earthquake

6.2.1 Typical Type of Damage

A common phenomenon found in both the Northridge earthquake (Jan. 17. 1994) and the Hyogoken-nanbu earthquake (Jan. 17. 1995) was unexpected fractural type of failure of steel moment frames [5], [6]. Steel structural members are composed of plate elements. It has been believed that the maximum strength of the plate elements is generally limited by the local buckling in the plastic range with rare exception in which the maximum strength is determined by the tensile strength of the material. And also it has been believed that the deformation capacity of the plate element limited by breaking under tension is greater than that limited by local buckling under compression.

Typical examples of fracture in the Northridge earthquake are found around the field-weld joint between heavy H-shape columns and deep H-shape beam (Fig.6.4). The flange of beam is welded to the flange of column together with the use of the bolted connection in the web of beam. In such a connection, the bending moment at the web of beam is hardly transmitted to the column, resulting in the stress concentration in flanges at the end of beam. Defects in materials of the heavy column must be an incentive for the propagation of brittle cracks on the side of the column.

Typical examples of fracture in the Hyogoken-nanbu earthquake are found around the weld connection between rectangular-hollow section columns and H-shape beams. While the stress transmission between the beam-flange and the column is made completely through the diaphragm plate, the bending moment in the web of the beam can't be transmitted completely to the column, since the stress transmission is made through the out-of-plane bending of the flange plate of the column. Therefore the stress concentration can occur in the flange of beam. The diaphragm plate is usually

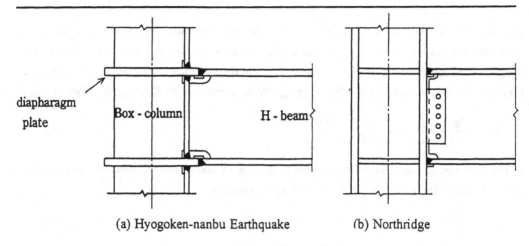

(a) Hyogoken-nanbu Earthquake   (b) Northridge

Fig.6.4 Beam-to-Column Connection

thicker and wider than the flange of beam. Then, the fracture develops on the side of beam. As a yield-mechanism, the weak-beam type is preferred both in Japan and USA. This situation characterizes the fracture at the end of beams. In order to secure the deformation capacity of beams, compact beam section with small width to thickness ratios were selected and the stress concentration on flanges at the end of beam hindered the extension of the inelastic region along the beam-axis. These must have brought avoidance of local buckling and revelation of fractural mode of failure in beams. Moreover, an extremely high intensity of the ground motion must be also responsible for such a wide range of damage. In this paper, referring to the current seismic design method in Japan, the following points relevant to the Hyogoken-nanbu earthquake are made clear.

Under what condition could occur the fractural mode of failure of the steel moment frames?

What measure will be effective to prevent the fractural mode of failure?

6.2.2 Earthquake Resistant Design Method in Japan

The earthquake resistant design method shown in the Japanese building code which revised in 1981 is summarized as follows.

a)   The buildings should be proportioned on the basis of allowable stress design method under the seismic input which corresponds to the intensity level of $C_0 = 0.2$.

b)   The building should be equipped with the energy absorption capacity for the seismic input which corresponds to $C_0 = 1.0$.

C_0 is a coefficient which indicates the level of seismic input, and the simplified design spectra for $C_0 = 1.0$ are shown in Fig.6.5, specifying ground conditions by I, II and III.

Observing the giving and receiving of the energy input, the basic relationship between the strength for a structure and the seismic input is obtained as shown in Eq.(5.11).

Eq.(5.11) can be rewritten as

$$V_D = \frac{gТа_1}{2\pi}\sqrt{\frac{1+2\eta_1\gamma_1}{\kappa_1}} \tag{6.25}$$

where $V_D = \dfrac{V_E}{1+3h+1.2\sqrt{h}}$

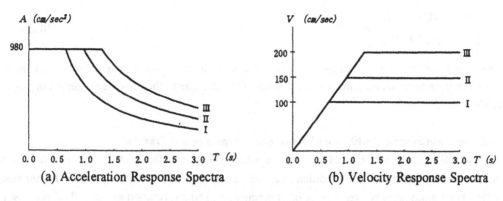

(a) Acceleration Response Spectra   (b) Velocity Response Spectra

Fig.6.5 Design Spectra for $C_0 = 1.0$

By applying Eq.(6.25) to existing buildings, we can evaluate the level of seismic input to which the existing building could resist.

6.2.3 Condition for Eliminating Brittle Fracture

In 1930s, brittle fracture became most prevailing mode of failure in welded steel ships. The extensive investigation arrived at a conclusion that Charpy impact test can provide a reliable basis to establish a criterion in order to eliminate brittle fracture [7].

Under a certain temperature, the fracture surface of V-notch specimen of Charpy impact test changes into perfectly crystallized one and indicates no trace of ductility. This temperature is defined to be the nilductility temperature, NDT.

As for the criteria to eliminate brittle or unstable fracture, the following conditions have been established for mild steels.

- Under the temperature higher than, NDT+15°C, the unstable fracture does not occur under the stress level less than $\sigma_Y/2$, where σ_Y is the yield strength of materials.
- Under the temperature higher than NDT+30°C, the unstable fracture does not occur under the stress level less than σ_Y.
- Under the temperature higher than NDT+60°C, the unstable fracture does not occur under the stress level less than σ_B, where σ_B is the tensile strength of materials.
- NDT+60°C is the temperature limit to eliminate brittle fracture in the plastic range and is defined to be the fracture-transition plastic, FTP.
- NDT+30°C is the temperature limit to eliminate the brittle fracture in the elastic range and is defined to be the fracture-transition elastic, FTE.

In order to develop a sufficient inelastic deformation, the condition for used temperature greater than FTP must be satisfied. Aforementioned criteria for FTP was established for ships. The problem is; what is the criteria for FTP in the seismic design of buildings? In order to grapple with the problem, a full scale shaking table tests become inevitably necessary, since the real situation of stress concentration, strain rate and heat generation accompanied with plastic deformation can be adequately simulated only in the full scale shaking table tests.

The results of full scale shaking tests are summarized as follows [8].

$$FTP = NDT + 40°C \atop FTE = NDT + 30°C \Bigg\} \qquad (6.26)$$

The reason for the reduction of the temperature of FTP in the shaking table tests by 20°C is as-cribed to the rise of temperature around the plastified zone due to the rapid development of plastic strains.

6.2.4 Ultimate Seismic Resistance of Weak Beam Type Moment Frames

The inelastic deformation capacity for frames which collapse in the fractural mode of failure in beams is estimated under the condition that beams are used under the temperature higher than FTP. The relationship between the applied moment, M, and the rotation at the end of beam sub-jected to the asymmetric moment distribution, θ becomes such as shown in Fig.6.6. When the maxi-mum strength is limited by local buckling or lateral buckling, the beam can continue to absorb energy in the range of strength deterioration. On the other hand, when the fracture takes place, the energy absorption capacity is limited at the maximum strength point. When buckling does not takes place, the full-plastic moment M_p, and the fractural moment, M_B, are expressed by

$$M_p = \sigma_Y Z_p \atop M_B = \sigma_B Z_p \Bigg\} \qquad (6.27)$$

where Z_p : plastic section modulus

The flexural rigidity in the strain-hardening range of the $M-\theta$ relationship, D_{st} is approximately re-lated to the rigidity in the elastic range, D as follows.

$$D_{st} = 0.03D \qquad (6.28)$$

Approximating the $M-\theta$ relationship with the solid line in Fig.6.6, the cumulative inelastic defor-mation ratio under monotonic loading, $\bar{\eta}_B$, is expressed by

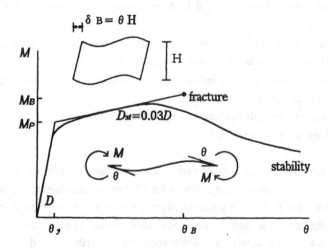

Fig.6.6 $M-\theta$ Relationship of Beam

$$\bar{\eta}_B = \frac{(M_B/M_p)^2-1}{2D_{st}/D} = \frac{(\sigma_B/\sigma_Y)^2-1}{0.06} \tag{6.29}$$

$\bar{\eta}_B$ can be related to η_1 in Eq.(6.25) as follows

$$\eta_1 = \frac{2\delta_{BY}}{\delta_{Y1}} \cdot a_B \cdot a_p \cdot \bar{\eta}_B \tag{6.30}$$

where a_B : amplification factor due to Bauschinger effect

a_p : amplification factor due to the plastification of structural components other than

beams

δ_{BY} : yield deformation of the first story calculated on the assumption that members

other than beams are rigid (see Fig.6.6 in which H is the height of story)

The factor of 2 in Eq.(6.30) corresponds to the assumption that the inelastic deformation takes place with equal amount both in positive and negative directions. Assuming $\delta_{BY}/\delta_{Y1} = 1/3, a_B = 2.0$ and $a_p = 1.5$, Eq.(6.30) is reduced to

$$\eta_1 = 2\bar{\eta}_B \tag{6.31}$$

In beam-to-column connections as is shown in Fig.6.4, the transmission of bending moment through the web plate of beam is incomplete, and an effective section modulus, $Z_{pe} = rZ_p$ must be introduced to estimate the maximum strength, M_B, as follows.

$$Z_{pe} = rZ_p \tag{6.32}$$

$$r = \frac{A_f + r_w A_w/4}{A_f + A_w/4} \tag{6.33}$$

where r : reduction factor of the section

r_w : reduction factor of the web

A_f : area of the flange (one side)

A_w : area of the web

In such a case, $\bar{\eta}_B$ becomes $\bar{\eta}_B = \dfrac{(r\sigma_B/\sigma_Y)^2-1}{0.06}$ \hfill (6.34)

Beams are generally connected to concrete slabs with stud bolts. In such a composite beams, since the plastification of upper flange of beams is not likely to occur under the bending moment which produces a compressive stress in the concrete slab, the energy absorption capacity of beams can be reduced to three fourths of the original. Considering such a situation, the ultimate seismic resistibility of the moment frames is evaluated on the basis of Eq.(6.25). As a practical example, the following conditions are taken.

structures : $\delta_Y = \dfrac{H}{150}, H = 400cm$; $n = 6$ (weak beam type)

material : $\sigma_Y = 1.2\sigma_{Y0}, \sigma_B = 1.1\sigma_B$

$$\left(\sigma_Y = 3.3t/cm^2, \sigma_{B0} = 5.0t/cm^2 \atop \text{specificated values for SM490 steels} \right)$$

others : $h = 0.02$

From Fig.6.5, the shear force coefficient used for the allowable stress design, α_e, is read as follows.

For $T \leq 1.28$sec, $\alpha_e = 0.2$

For $T > 1.28$sec, $\alpha_e = \dfrac{0.256}{T}$ (6.35)

Since the skeletons are designed on the basis of the elastic analysis, the strength which corresponds to α_e is the elastic limit strength, Q_{e1}. The yield strength, Q_{Y1}, and Q_{e1} can be roughly related to be

$$Q_{Y1} \geq 1.5 Q_{e1}$$ (6.36)

Therefore, considering also the increase of yield point stress by 20 percents, α_1 can be assumed to be

$$\alpha_1 = 1.5 \times 1.2 \alpha_e = 1.8 \alpha_e$$ (6.37)

The fundamental natural period T is written as

$$T = 2\pi \sqrt{\frac{M}{k_{eq}}} = 2\pi \sqrt{\frac{M\kappa_1}{k_1}}$$ (6.38)

Knowing $k_1 = Q_{y1} / \delta_{y1} = \alpha_1 Mg / \delta_{Y1}$, T is reduced to

$$T = 2\pi \sqrt{\frac{\kappa_1 \delta_{Y1}}{\alpha_1 g}}$$ (6.39)

Applying Eqs.(6.29) and (6.31) for the given condition,

$$\eta_1 = 31.0$$ (6.40)

For reference's sake, two other values of η_1 are taken, i.e. $\eta_1 = 20.0$ and $\eta_1 = 10.0$. Values of r and r_w which corresponds to the selected values of η_1 are obtained as follows by applying Eqs.(6.32) and (6.33) and practical values of A_f / A_w ranging from 1.0 to 2.0.

For $\eta_1 = 20.0$, $r = 0.91$, $r_w = 0.55 \sim 0.73$

For $\eta_1 = 10.0$, $r = 0.82$, $r_w = 0 \sim 0.46$

Considering the influence of composite-beam action, η_1 factored by 0.75 is also applied. V_D-values obtained by Eq.(6.25) are shown in Fig.6.7. V_D-spectra along the fault line of the Hyogoken-nanbu earthquake are indicated by three bi-linear curves according to the classification of ground and the V_D-spectrum for the record at Kobe Meteorological Observatory by Japan Meteorological Agency (JMA) is also shown. The bi-linear relationship shown by broken line is the V_D-spectrum on the soft ground prescribed in Japanese Building Code.

Comparing the intensity of the seismic input and the capacity of frames, the following facts can be seen form Fig.6.7.

1) The maximum intensity of the seismic input in the Hyogoken-nanbu Earthquake was one point five to two times as large as the intensity prescribed in the current Japanese Building Code.

2) The deficiency of the moment-transmission through the web plate at the beam-to-column connection governs the deformation capacity of the beam.

3) The allowable stress design under the seismic input of $C_0 = 0.2$ is very effective to secure a minimum required level of strength of frames.

4) Under the condition of $r_w > 0.5$, η_1 reaches the level of 20.0. Therefore, it can be said that the fractural mode of failure could be avoided by applying the current design method as far as the condition of $r_w > 0.5$ is kept.

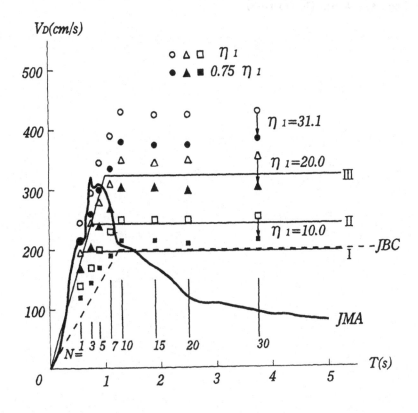

Fig.6.7 $V_D - T$ Relationship

References

1) Akiyama, H.: Earthquake-Resistant Limit-State Design for Buildings, University of Tokyo Press, 1985

2) Housner, G. W.: Limit Design of Structures to Resist Earthquakes, Proc. of 1st WCEE, 1956

3) housner, G. W.: Behavior of Structures during Earthquakes, ASCE, EM4, Oct. 1959.

4) Akiyama, H.: Earthquake-Resistant Design Method for Buildings Based on Energy Balance, Gihodoshuppan, 1999. (In Japanese)

5) Akiyama, H. and Yamada, S.: Seismic Input and Damage of Steel Moment Frames, Stessa '97, Kyoto, Japan, 3-8 August, 1997

6) Engelhardt, M. D. and Sabol, T. A.: Testing of Welded Steel Moment Connections, in Response to the Northridge Earthquake, Progress Report to the AISC advisory Subcommittee on Spécial Moment Resisting Frame Research.

7) Tetelman, A. S. and McEvily, A. J.: Fracture of Structural Materials, John Wiley & Sonns, Inc. 1967

8) Akiyama, H., Yamada, S, Minowa, C., Teramoto, T., Otake, F. and Yabe, Y.: Experimental Method of the Full Scale Shaking Table Test Using the Inertial Loading Equipment, J. Struct. Constr. Eng., AIJ, No.515. (In Japanese)

CHAPTER 4

DESIGN OF MOMENT RESISTING FRAMES

F.M. Mazzolani
University "Federico II" of Naples, Naples, Italy

1 General Features of MRF

1.1 MRF Practice

Moment-resisting frames (MRF) are structures with a satisfactory behaviour under severe earthquakes. As it has been already noted (Chapter 1, Section 4), they can provide a large number of dissipative zones, where plastic hinges form with potentially high dissipation capacity. In order to maximise the energy dissipation capacity, MRFs have to fail with a mechanism of global type. As a consequence proper design criteria have been conceived to fulfil this condition (Mazzolani and Piluso, 1996a). According to the assumed design approach (see Section 1.2), moment resisting frames can provide different level of strength and ductility.

It must be noted that MRFs present a weak point in their lateral flexibility. In fact this structural typology is not able to provide sufficient stiffness for reducing sway deflections as far as the height of the building increases, also under moderate earthquakes. Then, for high rise buildings the fulfilment of the requirements, which are necessary to guarantee the check against the serviceability limit state, can be very severe and consequently MRFs become uneconomical in developing the design stiffness required by the drift control (Bruneau et al., 1998).

The necessity to obtain rigid beam-to-column connections creates constructional problems, because usually field-welds are less reliable than shop-welds, due to the presence of welding defects. In fact, during the earthquakes of Northridge (1994) and Kobe (1995) some steel MRF buildings exhibited many failures located at the beam-to-column connections, where dissipative zones were developed. In particular brittle fractures occurred in weldments (Bertero et al., 1994; Akiyama and Yamada, 1997).

A suitable constructional procedure for MRFs is based on shop-welded/field-bolted connections, which guarantee a more reliable behaviour, by means of the so-called "column-tree" technique (see Figure 1a,b,c,d,e).

1.2 Code Provisions

According to code provisions and practice, three design levels can be recognised (Mazzolani, 1999).

The first level consists of the dimensioning of the member sections according to the internal forces computed by means of a simple elastic analysis. It is clear that in such a way there is not any control of the collapse mechanism and, therefore, plastic hinges can be located either in beams or in columns and in panel zones.

Figure 1. The "column-tree" constructional technique for MRF: (a) The shop-welded column-tree is ready for erection; (b)The column-tree is lifted at the relevant floor; (c) The column-tree is field-bolted; (d) The frame structure is quite completed; (e) A detail of the structural joints for principal and secondary beams, completed with the trapezoidal sheeting.

The second level is based on the same procedure as previously explained, but it is integrated by the member hierarchy criterion. This criterion, adopted in most of the modern international seismic codes would lead to a member sizing, according to which beams represent the weaker element, compared to columns and panel zones. To this purpose it states that at any beam-to-column joint the sum of the plastic moments of the columns has to be greater than the sum of the plastic moments of the beams. Even though this provision aims at the control of the failure mode, many numerical analyses have pointed out that it is able to avoid failure modes having very unsatisfactory ductility, such as story mechanism, but it is not sufficient to lead to a collapse mechanism of the global type.

On these bases the American seismic codes (UBC and AISC, 1997) introduced a classification of MRF in three groups, namely OMF (Ordinary Moment Frames), IMF (Intermediate Moment Frames) and SMF (Special Moment Frames). OMF are sized according to the first level design, SMF are sized according to in the second level design, while IMF represent a compromise between the previous ones. These three types are characterised by different energy dissipation capacity and ductility levels, therefore they have to be designed assuming correspondent reduction factors of the elastic design spectrum. The American rules recommend to use a reduction factor R equal to 8 for SMF, 6 for IMF, 4 for OMF.

On the contrary the modern European seismic codes (ECCS and EC8) give provisions corresponding to the second design level only, being forbidden the use of OMF even in case of low seismicity areas. According to Eurocode 8 and to ECCS Recommendations, the elastic design spectrum reduction factor, so-called behaviour factor or simply q-factor, for rigid moment-resisting frames is equal to 5 α_u/α_y, being α_u and α_y respectively the maximum and the first yielding values of the horizontal forces multiplier versus top sway displacement of the monotonic behavioural curve, obtained by push-over analyses.

The results of many numerical analyses on the inelastic response of steel frames demonstrated that the codified design methods are often inadequate to guarantee that frames fail in a global mode (Landolfo and Mazzolani, 1990).

The third design level is represented by the use of more sophisticated design procedures based on elasto-plastic analyses under a given distribution of seismic horizontal forces. They allow the direct control of the failure mode and consequently the design of frames presenting a collapse mechanism of global type under the design distribution of seismic forces. For these reason the frames designed according to this third level can be defined as Global Moment-Resisting Frames (GMF, Mazzolani, and Piluso, 1995a). In order to design GMF a new method has been recently proposed (Mazzolani and Piluso, 1997a), based on the application of the kinematic theorem of the plastic collapse. Other proposals are under study to evaluate their efficiency and possibility to be introduced in seismic codes by means of simple rules (see Section 3).

From the codification point of view, it can be interesting to compare the results coming from the application to MRFs of the outstanding codes which polarise three main geographical areas (Mazzolani, 1999):

- Europe (ECCS Recommendations, 1988, and EC8, 1994);
- America (AISC, 1997 and UBC, 1997);
- Asia (AIJ - Japanese code, 1990).

An extensive comparison among these codes is not easy in general, but a consistent effort has been done within the European RECOS project (Mazzolani, 2000).

In Table 1 the comparison is referred to the simple case of the evaluation of the seismic actions on a moment-resisting frame (Mazzolani, 1995a), in order to emphasise an important aspect of the design conception. For a given moment resisting frame of ductile type, like SMF, the following common assumptions are considered:

- building for normal use;
- high sismicity zone;
- firm soil condition;
- plateau zone of the spectrum;
- 5% damping.

Table 1

Code	V/W	S/g	q
ECCS (1988)	0.131	1.050	8
EC 8 (1994)	0.146	0.875	6
AISC (1997)	0.138	1.100	8
UBC (1997)	0.129	1.100	8,5
AIJ (1990)	0.250	1.000	4

Notations are defined as:

V = design base shear for ultimate limit state;
S = maximum elastic spectral acceleration,
q = behaviour factor for ductile MR-frames;
W = dead and partial live load.

It can be observed that the difference among the maximum values of the dimensionless acceleration S/g is slightly influenced by the difference among the elastic response spectra for the severest design earthquake. On the contrary the main reason of difference for the design base shear ratio V/W is due to the values assumed for the behaviour factor q.

Hence, as the spectral accelerations provided by the Japanese code are comparable with the others, the greater magnitude of the design seismic forces is mainly due to the very restrictive value adopted for the behaviour factor (q = 4).

This remark should be considered in the light of the unquestionable fact that the value of the behaviour factor has to be derived not only by engineering judgement, but also by cost-benefit considerations. In fact, seismic-resistant steel structures in Japan look actually very heavy as compared to the American and European ones and it corresponds to higher safety against earthquakes (Figure 2).

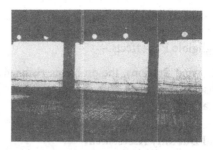

Figure 2. Seismic-resistant steel structures in Japan.

Despite these differences in the code values, during either the Northridge or Hyogoken-Nanbu earthquakes steel structures suffered severe unexpected damages, leading to the conclusion that in both cases the code provisions were inadequate to prevent serious shortcoming and to guarantee a reasonable safety level (Mazzolani, 1998).
Both in U.S. and in Japan, a very intensive activity has been devoted to understand the mistakes and to change the corresponding provisions. New innovative activities are also going on in Europe during the convertion phase of EC8 from ENV to EN, aiming at setting-up updated and revised design rules for seismic resistant structures (see Section 7 of Chapter 1).

1.3 Global Ductility Assessment

As it has been already pointed out, the ability of a structural system to resist seismic actions in forfaitary post-elastic range is represented by one parameter, the so-called q-factor, which synthetically and approximately for a MDOF system takes into account the energy dissipation capacity of a ductile response (Mazzolani, 1988, Mazzolani and Piluso, 1996a)

As it has been shown before, in case of dissipative steel structures q-factor ranges from 2 to 8, according to the degree of ductile behaviour of the structure typology, from the worst (concentric V-bracing) to the best (ductile moment resisting frames).

The values of q-factor given by each code represent the design values, which are valid under some important prerequisites such as (Mazzolani, 1995b):

– structural regularity;
– ductile global failure mode;

-   absence of local buckling;
-   ductile behaviour of connections;
-   negligible P-Δ effects.

In the following Sections, the problems rising from the absence of one or more prerequisites will be discussed and the corresponding influences on the evaluation of the q-factor will be also emphasised.

1.4 Local Ductility Assessment

The assessment of the member local ductility in the European codes is based on a limitation of the width-thickness ratio b/t in the compressed part of the cross-section. In dissipative zones, three ductility classes are defined as in Table 2 (similar to those given in Eurocode 3, 1993) and to each one a limit value of q-factor is associated (Mazzolani, 1995a).

Table 2

Ductility class	b/t_f	h/t_w		q
		Beam	Beam-column	
1	10ε	72ε	33ε	$q > 4$
2	11ε	83ε	38ε	$2 < q \leq 4$
3	15ε	124ε	42ε	$q \leq 2$

$$\varepsilon = \sqrt{235 / f_y}$$

This approach given in the Eurocode 8 (1994) seems to be rather incomplete, because it depends exclusively on the geometrical dimensions of single plates forming the cross-section and on the material strength.

The Japanese code (AIJ, 1990), for example, defines three ductility classes depending on given values of the rotation capacity demand ($\eta = 0, 2, 4$), which corresponds to a cross-section classification based on interaction formulae for the width-to-thickness ratio limitation. Each class is also associated to a limit value of the behaviour factor D_s (equivalent to 1/q, Kato, 1995, see Table 3).

Table 3

η	D_s	$1/D_s$
$4 <$	0.25	4.00
$2 <$	0.35	2.86
$0 <$	0.45	2.22

An exhaustive classification criterion is needed in codification in order to substitute the present cross-section behavioural classes, by introducing the concept of member behavioural classes (Mazzolani and Piluso, 1993a). It should take into account the plastic rotation capacity of the member which can be evaluated either according to the Mazzolani – Piluso proposal (1996), as a function of the non-dimensional buckling stress s and the non-dimensional axial load ρ or by means of the relationship proposed by Gioncu (Gioncu and Mazzolani, 1995), on the basis of the method of plastic mechanism (Gioncu, 1997).

As far as the connection ductility is concerned, the European seismic codes require that joints in dissipative zones have to possess sufficient overstrength to allow for the yielding of the ends of connected members (Mazzolani, 1998).

It is deemed that the above design condition is satisfied in case of welded connections with full penetration welds, while partial penetration welds, when overstressed in bending and/or direct tension, are susceptible to brittle fracture (Krawinkler, 1995). On the contrary, in case of fillet weld connections and in case of bolted connections the design resistance of the joint has to be at least 1.20 times the plastic resistance of the connected members. This means that the use of full-strength joints is requested.

Partial strength connections and their contribution in dissipating the earthquake input energy are not forbidden in principle. Notwithstanding, their use is strongly limited because the experimental control of the effectiveness of such connections in dissipating energy is requested.

On the other hand, the joints commonly used in practice even for MRF, when submitted to cyclic loads, give an experimental answer which is rather far from the perfectly rigid one, so they mostly have to be classified as semi-rigid joints (see Section 5).

The above overstrength level (1.20) should be properly related to the width-to-thickness (b/t) ratios of the beam section. In fact, the maximum flexural strength, that the beams are able to withstand, is developed at the occurrence of local buckling.

As far as the overstrength factor is concerned, recent analyses have shown that the value 1,20 is not enough to cover the random variability of material and geometry. A value at least equal to 1.3, as it is considered in the Japanese code, should be more adequate (Mazzolani et al., 1998).

Either for full-strength or for partial-strength semi-rigid frames, an approximated approach for evaluating the behaviour factor to be used in design has been recently proposed for both American (Astaneh and Nader, 1992) and European codes(Mazzolani and Piluso, 1995b), but not yet codified.

The results of theoretical and experimental research on steel connections (see Sections 5) can provide up-to-now enough data to justify the introduction of more quantitative provisions on the required behaviour of connections in seismic resistant structures. This need is very urgent, because after the experience of Northridge and Kobe it seems very clear that the existing provisions are not sufficient to guarantee against the unexpected brittle behaviour of connections (Mazzolani, 1998).

1.5 Influence of structural configuration

One of the lessons learned from the last earthquakes, which has not been widely investigated, is the importance of the design configuration of the space structure considered as a whole

(Mazzolani, 1998).

With reference to moment-resisting frames, different solutions for the configuration of building structure are available, leading to different distributions of inertia forces and to a significant variation in the number of dissipative zones (Mazzolani and Piluso, 1997b).

Within the general typologies of moment-resisting frames, a classification can be done looking to the space distribution and the type of seismic resistant substructures:

- space moment-resisting frames (S-MRF);
- perimeter moment-resisting frames (P-MRF);
- few bays perimeter moment-resisting frames (FBP-MRF), also called "dual structure".
- hybrid configuration.

In order to optimise lateral stiffness and energy dissipation capacity, the traditional space moment-resisting frames would present all the beam-to-column connections as moment-resisting and full-strength type. On the other side, it is common in nowadays practice to design steel moment resisting frames as perimeter frames and/or dual rigid-pinned frames (De Matteis et al., 1999). The main advantage of this configuration is both to reduce the number of moment resisting connections, what is convenient from the practical and economical points of view, and to reduce (in case of FBP eliminate) the number of weak axis connections (beam framing into the column web). But the reduction of moment connections gives rise to a significant increase of the damage index, as it clearly appears from the comparison between S-MRF and FBP-MRF (Figure 3).

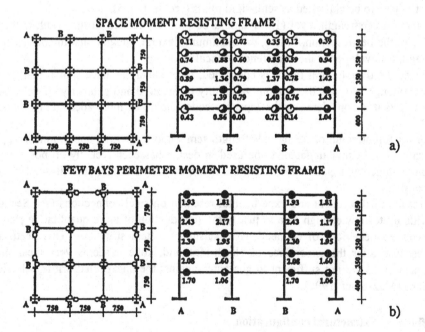

Figure 3. Damage index in beam-to-column connections for two limit configurations:
S-MRF (a) and FBP-MRF (b).

As it was foreseeable, most of the buildings which experienced severe damage during Northridge earthquake had the seismic-resistant scheme realised according to the few bays perimeter frame system. The review of experimental test results carried out on welded beam-to-column connections starting from the 60's, as well as the analysis of recent experimental results obtained within the SAC Steel Project developed in U.S.A. following the Northridge earthquake, have shown that the rotational capacity of welded connections adopted in usual perimeter frames can often be lower than the rotations required during strong earthquakes. This is because the redundancy of steel structures adopting perimeter frames is reduced, so that the number of dissipative zones is also reduced (Astaneh, 1995a). The consequence is a ductility demand concentration at beam-to-column connections.

A possible way to mitigate this problem is to use semirigid and partial resistant connections instead of complete pinned ones (De Matteis et al., 1999). In such a solution, duality must be intended in terms of joint mechanical properties .This structural configuration has became almost popular in United States and Japan (Shen, 1996, Kishi et al., 1996).

Another structural typology can be characterized by semirigid joints everywhere in the structure (Figure 4). But, in this case, the difficulty to yield the serviceability conditions imposed by code provisions become more and more important, because of the reduction of lateral stiffness and the increase of the period due to the different soil-to-structure interaction (Astaneh and Nader, 1992; Astaneh, 1995b). On the other side, the latter effect is beneficial due to the decreasing of the earthquake equivalent loading on the structure.

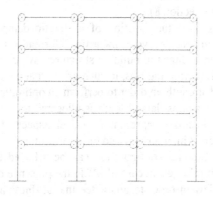

Figure 4. Homogeneous structural configuration with semirigid joints

The variation of joint mechanical properties throughout building frames could be considered as an additional source of structural irregularity.

The last typology is the hybrid configuration, in which moment resisting frames are associated in resisting the horizontal loads with other structural components, like reinforced concrete core or walls (Figure 5a), or infilled cladding (Figure 5b) or also steel walls (Figure 5c). The fundamental differences among the three solutions can be synthesised as follows.

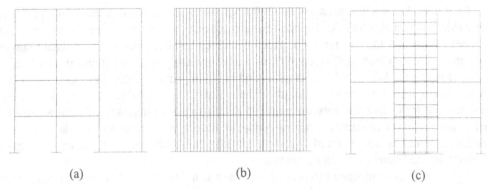

(a)   (b)   (c)

Figure 5. The hybrid structural typology

In the first case the reinforced concrete core or wall represents a very rigid element, generally located around staircases and elevators. A repartition of lateral loads takes place according to the lateral stiffness of both the resistant substructures. Critical points are the connections between concrete and steel members and the base section of the wall, where the only plastic hinge of the reinforced concrete wall can develop.

In the second case the infilled panel, which usually represents a non structural component, is called to stiffen the structure against lateral loads, conferring to MRF an additional source of energy dissipation. Problems arise to conceive the connections between the bare steel structure and the panel in a proper way (see Section 8).

In the third case steel walls have the function of hysteretic dampers, which are able to dissipate the seismic energy by means of cyclic inelastic deformations of the panel. This last solution has been recently applied in Japan in building structures, where panels are characterised by a low-yield strength steel with reinforcing ribs, in such a way that panel buckling is prevented before the attainment of the yield strength, in order to perform an optimum ductile behaviour.

Generally when the resistance against lateral loads is allocated to few elements of the space structure, the corresponding foundations represent a critical aspect of the structural design, because they have to be dimensioned in order to avoid the overturning.

In all these typologies the structural configuration can be affected by plan and/or vertical irregularities. In the modern seismic codes, in case of MRF, the presence of "set-backs" is limited from the geometrical point of view in order to guarantee that stiffness and mass properties are approximately uniform along the building height (Mazzolani, 1999).

The design values of q-factor given by the Eurocode 8 are valid for high regular buildings; if the building is not regular in elevation the listed values must be reduced of 20 %.

It has to be pointed out that such provisions are not able to account for all the factors affecting the inelastic behaviour of structures and as a consequence the penalisation of some geometrical configurations is not justified (Guerra et al., 1990a). It can be concluded that the so-defined "geometrical irregularity" does not necessarily lead to "structural irregularity" (Calderoni et al., 1996). For these reasons, new provisions have to be set-up by introducing more general irregularity parameters accounting for mass, stiffness and strength distributions, which allow for a correct evaluation of the structural irregularity, together with the definition of the range in which a worsening of the seismic behaviour is actually produced (Mazzolani and Piluso, 1997c).

At least for some specific cases, such as symmetric schemes with set-back, the introduction of simplified procedures for the evaluation of the q-factor could be recommended (see Section 6).

2   Evaluation of q-factor

2.1 Code provisions

Within the general methodology of limit states, severe earthquakes have to be examined in the safety check against the ultimate limit state, which is associated with the collapse conditions of the structures.

It has commonly accepted that for a strong earthquake with a long return period the elastic limit may be exceeded in order to permit the dissipation of a part of the seismic energy by means of cycles of alternating plastic deformation. In this context the modern seismic codes generally base the design of constructions on a linear elastic analysis of the structure under seismic horizontal forces, which statically model the ground motion actions, combined with a check of the structure inelastic deformation capacity. The design seismic horizontal forces are defined by reducing the base shear required in the structure in order to remain in the elastic range during the severest design earthquake, or in other words by reducing the linear elastic design response spectrum (Figure 6), by means of a coefficient, namely in Europe the q-factor

This parameter represents a "magic" number accounting for the energy dissipation capacity and the ductility of a structure, as a function of the structural typology.

According to Eurocode 8, Part 1.1 (EC8, 1994), "the behavior factor q" is an approximation of the ratio of the seismic forces, that the structure would experience if its response was completely elastic with 5% viscous damping, to the minimum seismic forces that may be used in design – with a conventional linéar model – still ensuring a satisfactory response of the structure.

$$F_d = \frac{a_u R(T) M}{q}$$

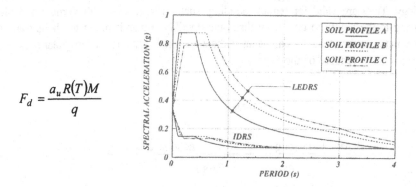

Figure 6. Elastic and inelastic design response spectra

Considering the inelastic deformation and energy dissipation capacities, structural typologies can be classified in two groups (Mazzolani and Piluso, 1996a):

1. Non dissipative structures, which are not able to dissipate earthquake input energy by means of cyclic inelastic behaviour and for which the q-factor is equal to 1.
2. Dissipative structures, which are able to withstand severe earthquakes, thanks to the ductile hysteretic behaviour of some of their components, the so-called dissipative zones, and for which the q-factor is always greater than 1 ($1 < q < 8$). In this case non-dissipative parts have to be designed in order to remain in the elastic range.

Table 4

Structural System	q-factor
Moment resisting frames	$5\,\alpha_u/\alpha_y \leq 8$
Concentric braced frames	
diagonal bracings	4
V-bracings	2
K-bracings	1
Eccentric braced frames	$5\,\alpha_u/\alpha_y \leq 8$
Cantilever structures	2
Structures with concrete cores or concrete walls	See recommendations for r.c. of EC8
Mixed structures (MRF with r.c. or masonry infills)	2

The values of the behaviour factor suggested in EC8 are synthesised in Table 4 as a function of the structural typology. They are valid for regular building, such as structures with vertical and plan regularity, referring to the ability to uniformly distribute ductility demands along the height and to vibrate separately in two orthogonal vertical planes without torsional coupling. For irregular buildings, the q-factor values must be reduced of about 20%.

Table 5

Behaviour factor q	Cross sectional classes
$4 < q$	Class 1
$2 < q \leq 4$	Class 2
$q \leq 2$	Class 3

In Table 4, α_y and α_u represent the multipliers of the horizontal design seismic action which respectively correspond to the first yielding, such as to the attainment of the plastic resistance in the most strained cross section, and to the collapse, such as to the attainment of the plastic moment resistance of a number of cross section sufficient to the development of the overall structure collapse.

The α_u/α_y ratio can be defined as redistribution factor (see Figure 7). According to ECCS Recommendations (1988) for moment resisting frames the α_u/α_y ratio can be approximately assumed equal to 1.2 without any evaluation, but it cannot exceed the value 1.6 (EC8, 1994).

Apart from the structure regularity, the design values of q-factor are reliable if the following conditions are satisfied:

a. The collapse mechanism is global;
b. Connections are ductile;
c. Rotation capacity is provided both by members and connections;
d. Geometrical non-linearity is negligible.

As a consequence, reduction of the suggested design q-factor values has to be applied if the following situations are of concern:

a. Random variability of f_y;
b. Vertical irregularity;
c. Partial mechanism;
d. Semi-rigid (partial strength) connections;
e. P-Δ effects.

As it can be deduced, there is a strict correlation between the value of the q-factor and the ductility of the structure, at local and global levels, the ductility being the ability of a member or a structure to sustains deformations in inelastic range without significant loss of strength.

Locally, dissipative zones, such as plastic hinges, must be more ductile as far as the assumed design q-factor value is higher. EC8 indicates cross-section classes to use according to the assumed q-factor, as it is shown in Table 5, where cross-sections are classified as function of the b/t ratio and of the structural material as indicated in EC3.

2.2 General Definitions

The global ductility defined as the ratio between the top sway displacement of the frame evaluated at the ultimate and the yield conditions ($\mu=\delta_u/\delta_y$) by means of a monotonic load condition (Figure 7), has been largely used in the past to extend to MDOF systems the knowledge acquired for the inelastic behaviour of SDOF systems.

The most important behavioural parameters affecting the inelastic performance of a structure, whose response is evaluated under monotonic horizontal loads, can be summarised as follows:

- The plastic redistribution parameter α_u/α_y;
- The slope parameter of the softening branch γ;
- The member rotation capacity;
- The type of collapse mechanism.

The critical point of this extension is due to the number of parameters characterising the pattern of yielding of MDOF structures. In fact it is clear that SDOF system inelastic behaviour can be characterised by means of a simple single parameter such as the global ductility, while in case of a MDOF system different patterns of yielding could correspond to the same inelastic displacement, so the control of local ductility is also necessary.

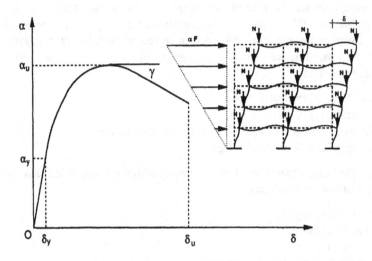

Figure 7. Behavioural α -δ curve

A first problem concerns the choice of the design seismic horizontal forces distribution, because various distribution can be adopted depending upon the number of predominant vibration modes of the structure. Once it has been done, the ductility ratio is not univocally determined, because of different possibility to define the ultimate and yield states (Figure 8).

Moreover it has been noted that this way to define the global ductility does not represent a sufficient mean to estimate the structural damage, because it does not take into account the number of yield excursions and reversals occurring during a severe earthquake. Therefore a more complete characterisation of the inelastic response of structures requires new damage parameters, such as the cyclic ductility, the hysteretic ductility, the Park and Ang damage index, the low cycle fatigue index (Mazzolani and Piluso, 1996a).

The situation becomes more complicated if one attempts to determine the structural response for a MDOF system, when a dynamic time-history analysis is used instead of a static monotonic analysis.

Among the steel structures researchers, the most popular definition of q-factor based on dynamic analysis seems is given by the ratio between the maximum ground acceleration leading to ultimate limit state (a_u) and that corresponding to the first yield (a_y):

$$q = \frac{a_u}{a_y}$$ (1)

In such form the q-factor can be evaluated performing dynamic inelastic analyses repeated for a sufficiently great number of different ground motions. As a consequence, the corresponding computational effort is particularly cumbersome and, therefore, simplified methods have been proposed by many researchers (Mazzolani and Piluso, 1996a).

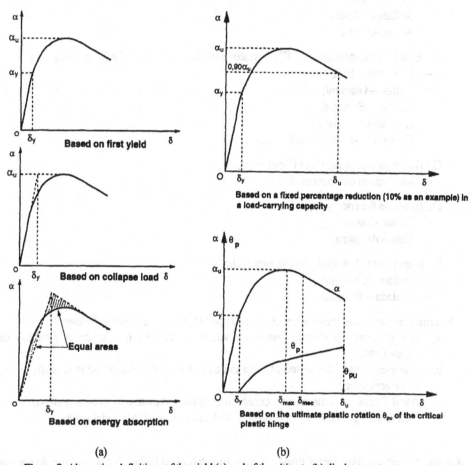

(a)   (b)

Figure 8. Alternative definitions of the yield (a) and of the ultimate (b) displacement.

The above definition yields results not always similar to those derived from the definition based on the ratio of shear forces coming from the static analysis. Furthermore, the q-factors obtained by dynamic analysis are highly sensitive to the characteristics of the seismic input, showing a large scatter for the same structure in the values obtained with different ground motions. Differences are due to dissimilar time-histories and dissimilar spectral shapes.

It should be also noted that this definition is not consistent with the EC8 definition, since the reference level is the first yield and not the design level, which usually are different due to overstrength.

2.3 Simplified methods

They can be classified into five groups:

A. Based on the inelastic response of a MDOF system:
- Ballio – Setti;
- Sedlacek – Kuck;
- Aribert-Grecea.

B. Based on the extension to MDOF systems of the response of SDOF systems:
- Newmark – Hall;
- Giuffrè – Giannini;
- Palazzo – Fraternali;
- Krawinkler – Nassar;
- Cosenza - De Luca - Faella – Piluso.

C. Based on the application of limit design:
- Mechanism curve method.

D. Based on the energy approach.
- Como – Lanni;
- Kato – Akiyama.

E. Based on the low cycle fatigue approach:
- Ballio – Castiglioni,
- Calado – Azevedo.

The critical points of the above approaches, A, B and C, can be gathered as follows:
a. It is necessary to identify parameters which characterise the pattern of yieldings of multi-storey frames;
b. The increasing of the period of vibration and participation factor due to inelastic deformations is not considered;
c. Axial forces in columns can exceed the values predicted by modal analysis leading to a reduction of plastic moment capacities and an increase of required ductility.

Group A. The ductility factor theory is used to establish relations between the q-factor and some simple parameters characterising the inelastic behaviour of steel frames. It is also used to interpret the results of inelastic dynamic analysis.

The ductility factor theory is reliable for steel structures with T>0.5s. It states that the q-factor is coincident with the global ductility, provided that there are no limitations due to local ductility (Figure 9). If the inelastic response is lower than the indefinitely elastic one, the design based on an elastic spectrum will be on the safe side. In the opposite case the design will be on the unsafe side.

The hypotheses of structural regularity and global type mechanism are required for the following methods.

Ballio (1985) and Setti (1985) proposed a method to identify this condition, by plotting the curve δ/δ_d versus a/a_d (Figure 10a), where δ_d and a_d are respectively the design maximum displacement and acceleration evaluated by means of a first order elastic analysis and δ are the maximum displacements obtained by performing a series of dynamic inelastic analysis in which the peak ground acceleration a is increased step by step.

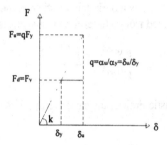

Figure 9. The ductility factor theory.

The bisectrix of the axes δ/δ_d and a/a_d represents the indefinitely elastic response. The maximum value which can be assigned to the q-factor is given by the intersection of the curve δ/δ_d-a/a_d coming from the dynamic analysis and the bisectrix, because at that level the structure is near the state of global dynamic instability, which means that the reserve of structural resistance is exhausted and a range of large deflection arises.

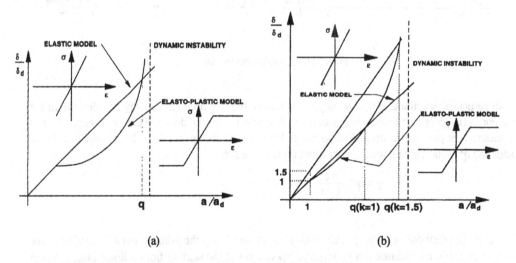

(a)   (b)

Figure 10. (a) Ballio and Setti method; (b) Sedlacek and Kuck method.

Sedlacek and Kuck (1993) introduced a correction of this procedure, because the intersection point definition is not always accurate due to an overvaluation of second order effects and consequently the corresponding q-factor values are too conservative. Therefore, they proposed to increase the slope of the linear response by means of a coefficient k, which is a conventional value usually assumed equal to

1.5, in order to evaluate the corrected q-factor at the intersection between the dynamic curve $\delta/\delta_d - a/a_d$ and the k-dependent linear curve, as shown in Figure 10b.

Aribert and Grecea (1999) have defined the q-factor as the ratio between the base shear force at collapse $(V^{e,th})$ of the structure considered as indefinitely elastic and the maximum inelastic shear force at the structure base (V^{inel}), both obtained from a time-history analysis (see Figure 11a):

$$q = \frac{V^{(e,th)}}{V^{(inel)}} = \frac{V^{(e)}}{V^{(inel)}} \cdot \frac{\lambda_u}{\lambda_e} \tag{2}$$

where $V^{(e)}$ is the base shear at the elastic limit of the structure and λ is coincident with the acceleration a in the previous methods.

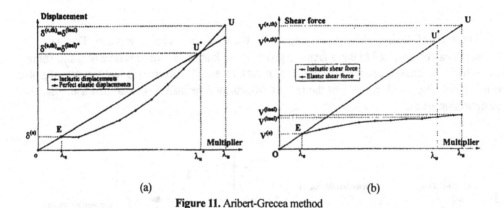

(a)   (b)

Figure 11. Aribert-Grecea method

In particular, according to the ductility factor theory, the displacement equality criterion (point U* in Figure 11a) gives a q* value which can be used in practice for the determination of q-factor. In fact by means of a parametric investigation the following relationship has been found between the q-factor and q* values, giving results very often on the safe side:

$$q = q^* \cdot \frac{\lambda_u}{\lambda_u^*} \tag{3}$$

Group B. The methods belonging to this group aim at correlating the q-factor and the ductility factor in order to obtain the inelastic design response spectrum (IDRS) starting from a linear elastic design response spectrum (LEDRS).

A synthetic overview of the corresponding relationships is reported in the following (for details see Mazzolani and Piluso, 1996a, Chapter 7).

Newmark and Hall (1973) proposed the formulas shown in Figure 12 for steel structures with T>0.5s and with T>0.5s.

Giuffrè and Giannini (1982) found out an expression of q in function of the global ductility μ and the period of vibration T, by using ductility response spectra of stiffness degrading SDOF systems.

Palazzo and Fraternali (1987a, b) used a similar procedure to take into account the second order effects due to vertical loads, by introducing the slope of the softening branch of the forces versus displacement curve. They assumed the stiffness degradation as a function of displacement and energy dissipation in previous cycles.

Krawinkler and Nassar (1992) proposed a relation which renders by means of mathematics some fundamental properties, such as the elastic behaviour (μ=1, q=1), the infinitely rigid behaviour (T=0, q=1), the infinitely flexible behaviour (T→∞, q=μ), considering an elastic strain hardening SDOF model.

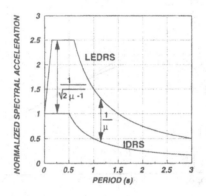

Figure 12. Newmark and Hall method

All these methods have been formulated for a SDOF system. Their extension to MDOF systems can be assessed by means of the definition of an equivalent SDOF system, having the same period and the same ductility of the actual structure, starting from the behavioural curve. The hypothesis of structural regularity and global type mechanism are strongly required. Considering a constant value for viscous damping, usually 5%, the q-factor can be expressed as a function of the global ductility μ, the period T, the slope coefficient γ and the plastic redistribution factor α_u/α_y (CDFP method, see Cosenza et al., 1988):

$$q = q\left\{\mu, T, \gamma, \frac{\alpha_u}{\alpha_y}\right\} \tag{4}$$

Group C. The mechanism curve method (Horne and Morris, 1981; Mazzolani and Piluso, 1996a) was originally devoted to the approximate estimation of the ultimate multiplier, but it can be usefully adopted also for evaluating the structural ductility and hence the q-factor values, either for new and existing buildings. It can be considered as a simplified procedure, which can substitute the more cumbersome static inelastic analysis. The method consists on the plotting of a trilinear diagram, composed by the linear elastic versus top displacement relationship (curve OA in Figure 13), which corresponds to the initial tangent of the behavioural curve of the frame, the mechanism curve (curve BD in Figure 13), which represents the slope of the softening branch of the behavioural curve corresponding to the assumed collapse mechanism, the horizontal curve identified by the ultimate multiplier α_u of the horizontal forces.

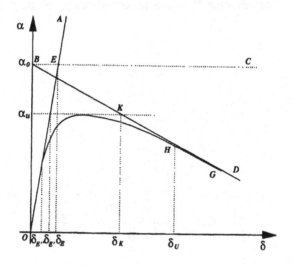

Figure 13. The mechanism curve method

The mechanism curve can be obtained by means of the cinematic theorem of limit design evaluating the minimum kinematically admissible multiplier of the horizontal forces related to the ultimate conditions. In order to determine α_u the K point (see Figure 13) has to be evaluated considering δ_k as equal to 2.5 δ_e, because this condition corresponds to a sufficient accuracy in the estimate of the ultimate multiplier, as it has been shown studying a great number of frames (Horne and Morris, 1973). Furthermore by determining the ultimate and the yielding displacements, using one of the above definition also the global ductility μ can be evaluated and, therefore, the q-factor can be estimated as a function of α_u/α_y, T, μ, γ.

Group D. The method of Como and Lanni (1983) is based on a simplified model of the energy exchanges occurring during the earthquake. The seismic motion of a structure has been divided in a series of simplified cycles. In each of them in a first phase the kinetic energy is stored into the structures and in a second phase the energy cumulated is transformed in elastic-plastic work (energy dissipation).

The q-factor has been defined as follows:

$$q = \sqrt{\frac{W_u}{W_y}}$$ (5)

where W_u is the total energy stored and dissipated up to failure and W_y is the elastic strain energy stored by the system in the yield state. Energy can be calculated by means of static inelastic analysis, calculating the work done by a distribution of statically equivalent horizontal forces.

The method proposed by Kato and Akiyama (1982; see also Chapter 3) attempts to account for the damage distribution. It represents the background of the Japanese seismic code.

The principle is to assess the safety of a structure against a design destructive earthquake by comparing the energy dissipation capacity of the structure with the earthquake input energy. The hysteretic energy dissipation of a steel frame in one direction is assumed equal to the plastic energy absorption under monotonic loading with a displacement amplitude equal to the cumulative plastic deformation (Figure 14), which accounts for all plastic excursions in both positive and negative ranges.

The cumulated ductility ratio is defined as .

$$\eta_i^c = \sum \frac{\delta_{pi}}{\delta_{yi}}$$ (6)

where the factors are shown in Figure 14.

The distribution of plastic work along the height of the structure is affected by the distribution of the yield shear coefficient defined as the ratio between the shear at the ith floor and the total mass:

$$\alpha_i = \frac{Q_{y,i}}{\sum_{i=1}^{n_i} m_j g}$$ (7)

The distribution of α_i, which produces a uniform distribution of cumulated plastic ductility is defined as the optimum distribution. When the shear coefficient differs from the optimal value, the structure is poorly sized and the input energy will concentrate into a particular story.

As a function of the ratio between α_i and $\alpha_{i,opt}$ a damage partition coefficient is obtained, which represents the ratio between the plastic work done by the first story and the plastic work done by the whole structure. Therefore, imposing the energy balance and expressing the energy quantities by means of the above defined parameters, the base shear can be derived and consequently an expression of q-factor as a function of the same parameters can be deduced.

This method is valid for shear type systems (weak column-strong beam frame), considering that the weak column structures are the most common. The purpose is to determine the design q-factor, so that the structure can be designed in such a way that the weak story is other than the first one. The q-factor is evaluated at each story and the design one for the structure is the minimum value.

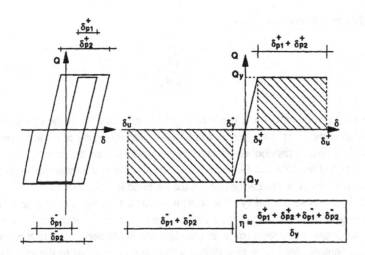

Figure 14. Cumulated ductility ratio

Group E. In the above mentioned methods, the structural damage has been approximately identified: the collapse conditions have been defined mainly through the maximum displacement criterion or sometimes through the energy criterion. The first criterion gives a good approximation for deformation histories characterised by only one excursion with a large plastic engage. The second one gives good approximation for deformation histories characterised by many cycles with large plastic engage, where the accumulated damage mainly depends on the dissipated energy.

The actual process of damage cumulation is a compromise between the two eventualities. The fatigue theory in the low cycle range and the linear damage cumulation assumption according to the Miner's rule have been recognised as a useful tool.

Ballio and Castiglioni (1993) defined the q-factor as the one that corresponds to equality of the damage index ID, estimated with linear and non linear dynamic analyses repeated for increasing peak ground acceleration until the mentioned condition has been attained. The method in its original formulation can be applied to single story buildings or to shear type structure, where the columns can be modelled as cantilevers under cyclic bending.

A similar approach has been applied to cantilever columns by Calado and Azevedo (1989).

This approach are requires a great computational effort due to the evaluation of the non linear damage index for each plastic hinge and also the high sensitivity of fatigue curve to variation of axial load. But for structures failing in global mode, where plastic hinges forms only in beams, the prediction is very accurate.

2.1 Comparisons

It seems useful to summarize the results of some comparison made among the above presented simplified methods (Mazzolani and Piluso, 1996a).

In Figure 15 the methods based on the response of SDOF systems are compared. It appear evident that the values of q-factor obtained by means of the Krawinkler and Nassar formulation, referred to an elastic perfectly plastic model, are greater. It could be explained by the fact that the energy dissipation capacity of the bilinear model, for a given value of the available ductility μ, is greater than for the stiffness degrading model.

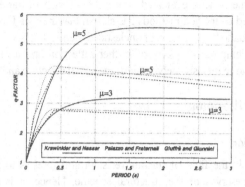

Figure 15. Comparison between the different formulations for SDOF systems

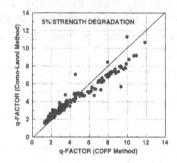

Figure 16. Comparison between the q-factor values computed by means of the Como-Lanni method and the CDFP method.

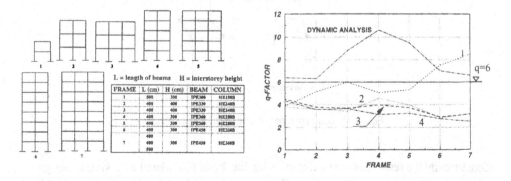

Figure 17. Comparison between the q-factor values computed by means of different simplified methods for the selected frames: 1. Ballio and Setti; 2. Sedlacek and Kuck; 3. Como and Lanni; 4. Kato and Akiyama.

In Figure 16 the q-factor values calculated by means of methods belonging to group C (Como and Lanni) and B (Cosenza et al.) have been compared. They present a good agreement even if the two methods are based on completely different theories.

A comparison between the q-factor evaluated on various schemes of frame structures by means of some of the above mentioned simplified methods has shown that they lead to cautelative results with respect to the "exact" evaluation by means of inelastic dynamic analysis, but they are also very disperse (Figure 17), evidencing the necessity of further reflection and investigation mainly on the correlation between the dynamic inelastic behaviour of the actual structure and the q-factor definition.

3   Ductile Design Methodology

3.1 The Hierarchy Criterion

The goal of design criteria in the field of seismic resistant structures is to guarantee the complete exploitation of the plastic reserves of the structural scheme. Hence, it is requested to avoid the formation of partial story mechanisms and to try to obtain global mechanism at collapse (Figure 18).

It has been shown that as far as the column dimensions increase, a major number of plastic hinges forms mainly in beams, involving more extensively the whole structures. This observation pointed out the convenience to dimension frame structures in such a way that beams are the weak structural elements where inelastic deformations develop first.

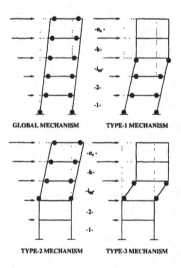

Figure 18. Different types of frame collapse mechanism.

Consequently the resistance hierarchy criterion has been formulated as a simple design tool aimed at the collapse mechanism control. The general principle is that increasing applied loads, the most stressed section of beams have to be the first to deform inelastically, while columns are still elastic. The hierarchy condition can be imposed stating that at each beam-to-column joint the sum of column plastic moments ($M_{pl,c,i}$) must be greater than the sum of the beam plastic moments ($M_{pl,b,j}$):

$$\sum M_{pl,c,i} > \sum M_{pl,b,j} \tag{8}$$

with $M_{pl,c,i} > M_{S,c,i}$,where $M_{S,c,i}$ is the elastic bending moment in the ith column, due to design loads (Figure 19).

It is clear that the cross section plastic moment has to be reduced to take into account the degrading of resistance due to the presence of axial loads.

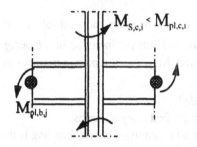

Figure 19. The hierarchy condition at a structural joint.

The application methodologies existing in literature can be distinguished in three groups on the base of a different interpretation of the hierarchy criterion:

A. The hierarchy condition can be obtained designing the columns converging in a node to resist the elastic bending moment amplified by a coefficient α, which guarantees the respect of equation (8). The main feature of the methods belonging to the first group is the definition of a design bending moment for each column of the structure.

B. The equation (8) is checked a posteriori.

C. More refined approaches are based on the application of the limit design theory.

The design methods belonging to each of the above groups are listed in the following:

Group A:
-   CNR-GNDT method (1984),
-   ECCS method (1988),
-   Revised amplification factor method (1999).

Group B:
-   EC8 provisions (1996),
-   UBC and AISC provisions (1997),
-   Lee's method (1996).

Group C:
-   Mazzolani-Piluso method,
-   Ghersi-Neri method.

3.2 Simplified design methods

CNR-GNDT method (1984). The respect of equation (8) is expressed as follows:

$$\alpha \cdot \sum M_{S,c,i} = \sum M_{pl,bJ} \tag{9}$$

The coefficient α is the amplification factor of the total design column moment, which can be evaluated by means of the following condition:

$$\alpha \cdot = \frac{\sum M_{pl,bJ}}{\sum M_{S,c,i}} \tag{10}$$

and: $M_{S,c,i} = M_{S,c,i,v} + M_{S,c,i,s}$ is the bending moment of the ith column converging in a structural node, due to both vertical loads (v) included in the design seismic load combination and design seismic actions (s) and $M_{pl,b,j}$ is the plastic moment of the jth beam converging in the same node.

Moreover it has to be provided:
–   a 20% increase of α for the first story columns;
–   a 30% increase of α for all columns, if biaxial bending is not considered.
Finally the column design bending moment is obtained from:

$$M_{c,d,i} = \alpha \cdot \left(M_{S,c,i,v} + M_{S,c,i,s} \right) \tag{11}$$

ECCS method (1988). The equation (8) is expressed in the following form, where the vertical (v) and lateral (s) load contributions to bending moment in the column have been separated:

$$\sum M_{S,c,i,v} + \alpha \cdot \sum M_{S,c,i,s} = \sum M_{pl,bj} \tag{12}$$

The coefficient α is the amplification factor of the design column moment due to seismic actions only (s), by keeping constant the vertical loads (v).

It can be evaluated by means of the following condition:

$$\alpha \cdot = \frac{\sum M_{pl,bJ} - \sum M_{S,c,i,v}}{\sum M_{S,c,i,s}} \tag{13}$$

Moreover it has to be provided:
– $\alpha=1$ for first story columns;
– $\alpha=1$ for top story columns.
The column design bending moment is obtained from:

$$M_{c,d,i} = M_{S,c,i,v} + \alpha \cdot M_{S,c,i,s} \tag{14}$$

Lee's method (1996). It represents a correction to the methods used by the Eurocode 8 (1994) and the US seismic codes (UBC and AISC, 1997), belonging to group B.

From the examination of the bending moment distribution near the collapse, obtained by means of a static push over analysis on a six floor-two bays steel frame, it has been observed that after the plastic hinges formation in beams, at each node the bending moment of the superior column drastically reduces, contrary the bending moment in the inferior column rapidly increases, being about 3 the ratio between inferior and superior column bending moments (Figure 20). It means that the 3/4 of the sum of plastic bending moments in beams converging in the node are transmitted to the inferior column.

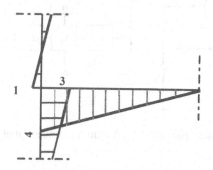

Figure 20. The bending moment distribution near collapse at a beam-to-column joint.

As a consequence the equation (8) become:

$$M_{pl,c,inf} \geq 0.75\gamma \sum M_{pl,b} \tag{15}$$

where γ is a strain-hardening factor assumed equal to 1.1.

Revised amplification factor method (Faggiano and Mazzolani, 1999). This proposal consists on a different evaluation of the amplification factor. In fact it is defined as the amplification factor of the bending moment in the beam due to design seismic loads, leading to the inelastic strain-hardening deformation until reaching the local instability in the most stressed section of the beam.

By considering a given column between node (i) and node (j) (Figure 21), at each beam end the following condition is valid:

$$M_{S,b,v,i,I} + \alpha_{i,I} \cdot M_{S,b,s,i,I} = s_I \cdot M_{pl,b,i,I}$$

which gives:

$$\alpha_{i,I} = \frac{s_I \cdot M_{pl,b,i,I} - M_{S,b,v,i,I}}{M_{S,b,s,i,I}} \tag{16}$$

At each node (i) or (j)

$$\alpha_i = \max{(\alpha_{i,1}, \alpha_{i,2})}; \qquad\qquad \alpha_j = \max{(\alpha_{j,1}, \alpha_{j,2})};$$

Then, for the column (i,j)

$$\alpha = \max{(\alpha_i, \alpha_j)}$$

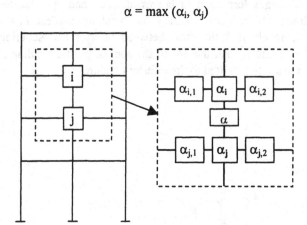

Figure 21. The amplification coefficient α characterisation.

The parameter s_i in equation (16) is defined as the ratio between the stress leading to the local instability (f_u) and the yielding stress of the steel (f_y). The more is $f_u > f_y$, the more is the available local ductility (Mazzolani and Piluso, 1996a).

At the same time a member ductility classification is proposed, as alternative to the cross section classification, depending on the value of the parameter s (Table 6), as a function of the following parameters:

- the slenderness of cross-section flanges and web;
- the material properties;
- the bending moment distribution along the axis.

Table 6 . Member classification

s >1.2	ductile
1< s <1.2	plastic
s <1	slender

This suggested classification is more general than the EC3 one.

The column design internal force and moment are then evaluated as follows:

$$M_{c,d,i} \geq \chi \cdot \left(M_{S,c,v,i} + \alpha \cdot M_{S,c,s,i} \right) \qquad\qquad (17')$$

$$N_{c,d,i} = \left(N_{S,c,v,i} + \alpha \cdot N_{S,c,s,i} \right) \tag{17''}$$

where the factor $\chi = \dfrac{1}{1-2COV}$ accounts for the overstrength due to random material variability (see Section 7).

Mazzolani-Piluso method (1993b, 1995c, 1997a). It can also be called "the global mechanism method". It applies the kinematic theorem of limit design theory, stating that the kinematically admissible multiplier of the horizontal forces corresponding to the attainment of a global mechanism is determined as the lesser than those corresponding to the other 3ns-1 kinematically admissible mechanisms, being n the number of stories. Therefore the actual collapse mechanism is of the global type.

The procedure consists on the application of the energy balance that explicits the kinematic theorem according to each of the 3n mechanism types. The beams are dimensioned to resist vertical loads and, as a consequence, the unknowns of the design problem are the column sections. The influence of axial loads on the plastic resistance of column sections is taken into account. The initial version of the method (1993b) has been successively refined, by introducing the P-Δ effects (1995c) and the influence of distributed loads acting on the beams (1997a).

Ghersi-Neri method (1999). Considering a triangular distribution of the static seismic forces along the frame height, firstly the evaluation of the collapse multiplier of seismic actions α_c is performed by equating the energy dissipated by the plastic hinges to the work of horizontal forces owed to a rigid displacement according to the global mechanism.

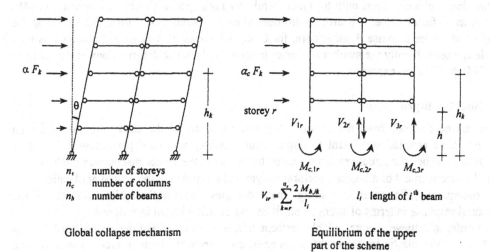

Figure 22. The Ghersi-Neri method.

Secondly the bending moments at each column of a story are evaluated by imposing the rotation equilibrium of the upper part of the frame and assuming an equal repartition of bending moment among the column of the story (Figure 22). Beam cross sections are dimensioned to resist vertical loads.

3.3 Comparison among methods.

The codified design methods of EC8, ECCS and CNR-GNDT have been extensively applied to a great number of different frame typologies (Landolfo and Mazzolani, 1990). The study frames differ for the number of storeys, the design q-factor, the orientation of floor structures, the total vertical loads and the ground acceleration values. The analysis showed that results are very dispersed and no trend in q-factor values or collapse mechanism types can be identified. But it can be observed that:

- the global mechanism is not a sufficient condition to achieve q-factor values greater than the design values;
- the tapering of columns can prevent the formation of global mechanism;
- the global interpretation of the examined cases showed a linear relationship between q-factor and structural weight.

More recently (Mazzolani and Piluso, 1995a) different structural typologies, classified as OMF and SMF (according to the U.S. codes) and GMF (designed by means of the Mazzolani - Piluso method), have been analysed to point out the differences in terms of structural weight, q-factor and collapse mechanism. The number of stories were variable from 1 to 8, the number of bays from 1 to 6.

It has been observed that the structural weight increases from the OMF to the GMF mainly as the number of stories increases. Results from the seismic analysis present q-factor values higher than the design ones only for GMF, while for both special (SMF) and ordinary (OMF) frames the q-factor values are strongly influenced by the number of stories, decreasing as the number of stories increase. Furthermore, for OMF the values of q-factor are always less than the design ones. Finally the number of stories involved in inelastic deformations increases from the OMF to GMF, as expected.

3.4 Ductility Inprovement

As it can be deduced from the previous subsections, many of the above mentioned design methods have resulted inefficient to obtain good ductility structural performance. The best results are reached for structures dimensioned by means of the global mechanism methods, but with the inconvenient of a higher structural weight and a certain complexity of application.

Consequently, the research on design methodologies aimed at the optimisation of seismic structural response in terms of strength, stiffness and ductility is still in progress.

In order to achieve a more reliable performance, it was underlined the necessity of forcing the formation of plastic hinges in the beam ends, away from the column face. Proposals for a ductile design of frame, which may lead to the attainment of this scope are based on the introduction in frame structures of two opposite types of interventions: the so-called "Strengthening" and "Weakening" the beam ends (Figure 23). The first is aimed at moving the plastic hinge formation from the column by suitably strengthening the beam ends which could be damaged. The second solution exploits the weakening of a given beam section in a relevant position, far from the node, where inelastic strains have to be concentrated. This technique is also well-known under the name of "dog-bone". In both cases the stress state near the beam-to-column connections is conveniently reduced, because, at the beam-to-column interface, in the former the beam resistance is increased, in the latter the bending moment near the column is decreased (SAC 96-03, 1997; Bruneau et al., 1998).

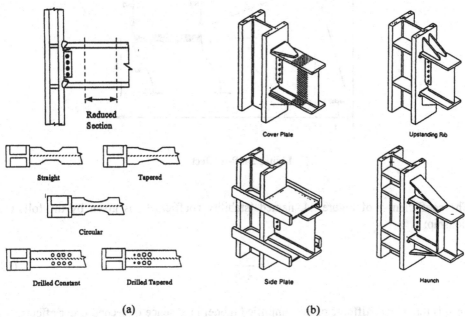

(a)   (b)

Figure 23. The weakening (a) and strengthening (b) strategies.

The complete implementation of these solutions in the structural design practice is not yet achieved, being at the moment the use of strengthened sections or weakened sections only a suggestion, which are proposed in design guidelines for seismic structures.

4   P-Δ effect

4.1   General

The locution 'P-Δ effect' refers to the influence of vertical loading on the behaviour of a multi-storey frame, by producing a reduction of lateral stiffness and strength and the appearance of an unstable branch in the lateral restoring force characteristics at the story level. Figure 24 shows the modification of the monotonic restoring force characteristics of a generic story of a shear-type frame, due to geometric non-linearity. An elastic-perfectly plastic material model is assumed. This behaviour can be simply derived by writing two equilibrium conditions for two behavioural phases, the elastic phase (increasing branch) and the plastic phase (decreasing branch). In writing these equations a first-order bending moment diagram is assumed for members.

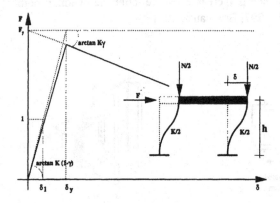

Figure 24. P-Δ effect.

The coefficient γ of Figure 24, named 'stability coefficient', is given by the following relationship:

$$\gamma = \frac{N}{k_1 h}$$

(18)

where k_1 is the lateral stiffness of the simplified model in absence of second order effects, N is the total vertical load acting above the examined story and h is the story height.

4.2 P-Δ Effects on Seismic Response of SDoF Systems

SDoF models are extensively used to obtain reduced design values of seismic actions compatible with a predetermined level of structural damage. Several definitions of damage can be introduced. The simplest parameter, which could be adopted as damage index, is the kinematic ductility $\mu = x_{max}/x_y$, x_{max} and x_y being the maximum displacement and the yielding displacement, respectively. It is well known that it does not take into account the effect of plastic fatigue, related to the repetition of inelastic deformations.

The influence of P-Δ effects on the behaviour factor q can be highlighted by the following relationship:

$$q(T,\mu_1,\gamma) = \frac{q(T,\mu_1)}{\varphi_\Delta}$$

(19)

where T is the natural period of vibration, μ_i is the damage index to be kept constant, γ is the already defined stability coefficient and φ_Δ is a reduction factor. The latter satisfies the following limitation:

$$\varphi_\Delta \geq 1 \tag{20}$$

because geometric non-linearity must lead to an increase of damage if strength is kept constant and, therefore, to an increase of strength required to keep damage constant.

Cosenza et al. (1989) computed the φ_Δ spectra, assuming as damage index the kinematic ductility μ. They considered ten artificial accelerograms generated from the elastic design response spectrum, which corresponds to the one given in EC8 with the numerical parameters of the Italian code GNDT (CNR-GNDT, 1984). Thirty different values of the design level ($\eta = F_y/MA$, where A is the peak ground acceleration) were considered and the ductility demand was computed by varying the period of vibration from 0.1 to 2.0 s . Analyses were repeated for values of the stability coefficient equal to 0.0, 0.025, 0.050, 0.075, 0.1, 0.15, 0.20 . Figure 25 shows the obtained reduction factor spectra. It can be seen that the reduction factor is quite constant over the considered period range.

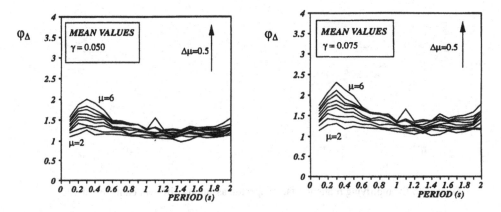

Figure 25. Reduction factor spectra.

Faella et al. (1993) carried out another investigation on this subject, considering 8 acceleration records of historical earthquakes. All the records were scaled to the same peak ground acceleration of 0.35g. Periods of vibration and stability coefficients were varied in the same previous ranges. Design levels η ranging from 0.02 to 2.50 with a very small step ($\Delta\eta = 0.02$) were considered. q-factor spectra were computed for both $\gamma=0$ and $\gamma\neq0$ and the reduction factors were obtained as ratios between spectra for $\gamma = 0$ and spectra for $\gamma\neq0$. Results, showed in Figure 26 for the case $\gamma = 0.20$, agree with the previous observation that the reduction factor is substantially independent of the period, which gives rise only to random scatters. Therefore, the influence of T can be considered only from the statistical point of view.

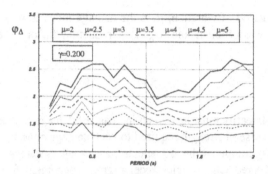

Figure 26. Exemplifying reduction factor spectrum, assuming the kinematic ductility as damage index.

This conclusion is confirmed when considering different damage criteria. For instance, Figure 27 and 28 show the mean values of the reduction factors computed with reference to four different damage criteria: the cyclic ductility, the hysteretic ductility, the Park and Ang and, finally, the low-cycle fatigue (Mazzolani and Piluso, 1996*b*).

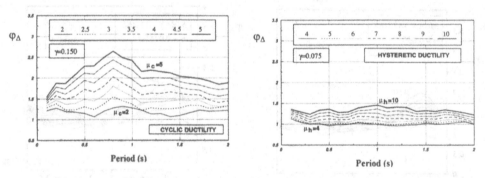

Figure 27. Exemplifying reduction factor spectra, assuming different damage criteria.

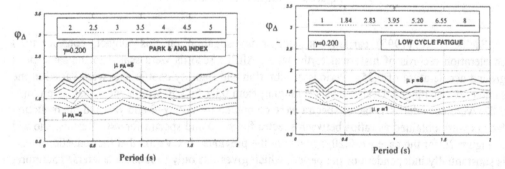

Figure 28. Exemplifying reduction factor spectra, assuming different damage criteria.

The analysis of results by numerical regression has led to the following formula:

$$\varphi_\Delta = \psi_3 + \frac{(1-\psi_3)\left[1+\psi_1\left(\mu^* - k\right)^{\psi_2}\gamma\right]}{1-\gamma} \qquad (21)$$

where μ^* is the assumed damage index. With reference to the mean values of the reduction factor φ_Δ, the values of the parameters ψ_1, ψ_2, ψ_3 and k, computed by means of a linear-logarithmic mixed regression, are given in Table 7 for the different damage criteria. Besides, the characteristic values of these parameters, which take into account the random influence of both the period of vibration and the seismic input, are given in Table 8.

Table 7. Mean values of parameters needed for the evaluation of the reduction factor φ_Δ.

Damage parameter	k	ψ_1	ψ_2	ψ_3
Kinematic ductility	1	1.4276	1.0415	0.2790
Cyclic ductility	0	0.2887	1.8273	0.0935
Hysteretic ductility in one direction	1	0.3045	1.3791	-0.4500
Total hysteretic ductility	1	0.1171	1.3788	-0.4687
Park and Ang parameter	1	0.9944	0.9542	0.2912
Low-cycle fatigue parameter	0	0.4010	0.5858	0.2101

Table 8. Characteristic values of parameters needed for the evaluation of the reduction factor φ_Δ.

Damage parameter	k	ψ_1	ψ_2	ψ_3
Kinematic ductility	1	4.3891	0.8282	0.3082
Cyclic ductility	0	1.4604	1.4751	0.2440
Hysteretic ductility in one direction	1	1.4155	1.2410	0.0065
Total hysteretic ductility	1	0.5990	1.2407	0.0074
Park and Ang parameter	1	3.2132	0.6659	0.2938
Low-cycle fatigue parameter	0	1.8164	0.3967	0.1962

4.3 P-Δ Effects on Seismic Response of Multistorey Frames

The possibility of extending the results obtained for the SDoF system to multi-storey frames has been investigated (Guerra et al., 1990b; Mazzolani and Piluso, 1993c; 1996b). Dynamic inelastic analyses, carried out by means of the computer program DRAIN 2D, were performed in order to analyse the inelastic seismic response of the frames shown in Figure 29.

Three artificial accelerograms were considered. The peak ground accelerations leading to the frame yielding and those leading to the frame collapse were computed with and without P-Δ effects. Correspondingly, q-factor was computed with and without P-Δ effects, and the ratio between these two quantities provided the computed reduction factors φ_c. These values were compared to those obtained by the application of results referring to SDoF system.

Figure 30 shows this comparison and illustrates the very good agreement between the numerical simulation and the simplified procedure based on the SDoF system.

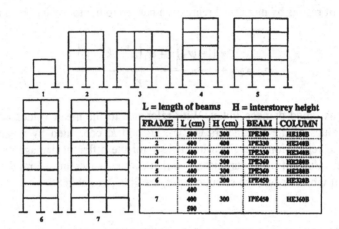

FRAME	L (cm)	H (cm)	BEAM	COLUMN
1	500	300	IPE300	HE180B
2	400	400	IPE330	HE240B
3	400	400	IPE330	HE240B
4	400	300	IPE360	HE280B
5	400	300	IPE360	HE280B
6	400 400	300	IPE450	HE320B
7	400 500	300	IPE450	HE360B

Figure 29. Analysed frames.

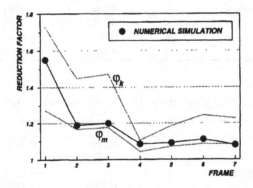

Figure 30. Comparison between the values the reduction factors obtained by numerical analyses and by the proposed approximated formulation (φ_m and φ_k are mean and characteristic values, respectively).

5 Influence of Connections

5.1 General

With reference to a MRF with usual values of vertical loads and beam spans, looking at the bending moment diagram illustrated in Figure 31, it is clear that dissipative zones in such type of structural scheme are located near beam-to-column joints, at beam and/or column ends.

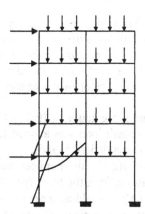

Figure 31. Qualitative shape of the bending moment diagram for usual framed structures

Moreover, according to Section 2 (Ductile Design Methodology) the "weak beam - strong column" design methodology should be implemented for increasing the energy absorption capacity of the frame and, contemporary, decreasing the frame susceptibility to damage concentration, thus obtaining the best structural performance.

Hence, beam ends are individuated as the locations where plastic deformations should develop and, consequently, beam-to-column connections directly or indirectly influence the dissipative behaviour of the frame. It is now worthy to remember that also the column panel zone may be involved in the plastic mechanism, if it is not adequately designed to remain elastic. Therefore, the joint mechanical behaviour must be concerned when designing a MR frame.

According to modern design standards (ECCS, 1988 and EC8, 1994) "joints in dissipative zones should possess sufficient over-strength to allow for yielding of the ends of the connected members" and "the experimental control of the effectiveness in dissipating energy is requested" when designing partial strength joints. Therefore, according to European aseismic codes full strength joints are more reliable than partial strength joints, whose application in seismic zones is not explicitly denied but in practice strongly limited with the requisite of experimental verification of their ductility. Rigid - full strength joint is the ideal for seismic applications. When the stiffness or the resistance are not adequate, a semi-rigid and/or partial strength joint is produced and it is possible to look at this situation as a deviation from the ideal condition, that is like a frame imperfection. This imperfection modifies the value of the behaviour factor to be applied because of a decrease of capacity and/or an increase of demand in terms of frame ductility. The following sections aim to clarify the effect of connection typology on the behaviour factor.

5.2 Experimental Behaviour of Connections

In the 80s years a joint research project has been developed by the Universities of Naples and Milan (Ballio et al., 1986; 1987). A broad experimental campaign has been carried out for investigating the cyclic behaviour of the most common connection typologies. 14 specimens were designed in order to cover a wide range of possible solutions for connecting I shaped beams to I shaped columns, with different degrees of rigidity varying from fully welded joints to bolted angle connections. Joint details were designed to obtain the same flexural strength of the connected beam. The 14 specimens can be grouped into four categories, indicated by letters A, B, C and D. For each category a 'base joint' was assumed as the least rigid and was indicated by number 1. Other joints of the same category were obtained by introducing different local stiffening elements to the base joint.

In all cases both beams and columns are double-T IPE 300 shapes, made of Fe360 steel. An axial load of about 0.30 times the squash load was applied to the column. Cyclic tests were conducted according to ECCS Recommendations (1986).

Type A1 is a cover plate joint, obtained by three splices welded to the column and bolted to flanges and web of the beam. Variations from the basic type A1 are obtained by introducing a diagonal stiffener in the panel zone (A2), by adding inner reinforcing cover plates for the beam flanges (A3), or by means of both these variations. Figure 32 shows the obtained results. The progressive reduction of the hysteretic area, because of slippage in the plate joint due to bolt - hole clearance, can be observed.

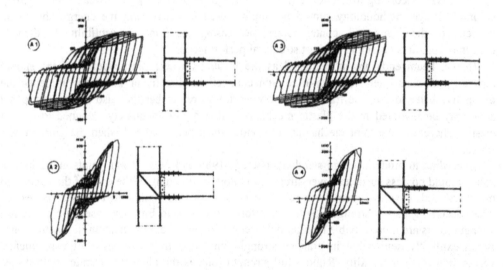

Figure 32. Type A specimens.

Type B1 is a top and seat angle with web angle connection. Variations from the basic type are obtained by stiffening the column at the end of the leg of top and bottom angles (B2), by reinforcing the angles on beam flanges with transversal triangular plates (B3), or by adopting both these interventions (B4). Figure 33 illustrates the hysteretic behaviour of this typology. It can be observed that the same degradation effects of type A are present, but more pronounced because of angle permanent deformations.

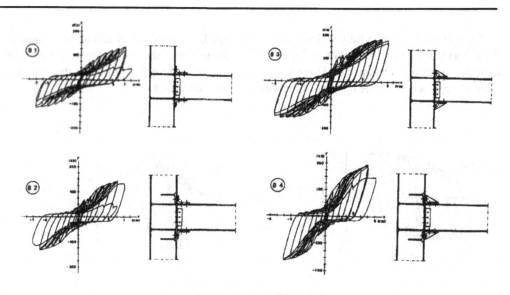

Figure 33. Type B specimens.

Type C1 is an extended end-plate connection with rigid column stub. Variations from the basic type are obtained by introducing full-depth stiffeners (C2 and C4) or partial-depth stiffeners (C3) in the beam web, or also by increasing the thickness of the end plate (C3 and C4). Results are given in Figure 34. They show a better performance with respect to previous typologies because of the absence of significant pinching effects in the hysteretic cycle shape.

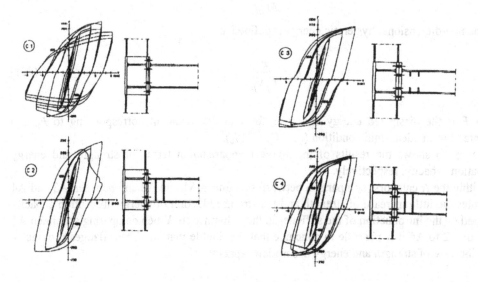

Figure 34. Type C specimens.

Type D1 is a common fully welded joint, while type D2 is obtained by adding two plates on the column web. Results, shown in Figure 35, illustrate round-shaped hysteretic cycles with high energy absorption capacity. However, in D2 case, there are quite significant deterioration effects when increasing the number of inelastic deformation cycles.

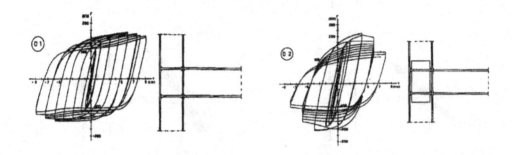

Figure 35. Type D specimens.

The comparison of the cyclic performance of the examined joints is quantitatively made by means of non-dimensional strength and non-dimensional dissipated energy. The non-dimensional strength is defined as the ratio between the maximum value F_{max} of the force recorded at the beam end during the whole deformation history and the conventional plastic resistance F_p, which is obtained by dividing the fully plastic moment M_{pl} of the beam cross-section by the beam length L:

$$\overline{F} = \frac{F_{max}L}{M_{pl}} \tag{22}$$

The non-dimensional hysteretic energy is defined as:

$$\overline{E} = \frac{E}{F_p v_p} \tag{23}$$

where E is the dissipated energy and v_p is the end displacement corresponding to F_p and evaluated for the ideal rigid condition ($v_p = F_p L^3/3EI_b$).

Figure 36 shows the results of the above comparison in terms of strength and energy dissipation capacity, respectively.

Within the A category, comparison between specimens A1 and A2 and between A3 and A4 illustrates the little increase of strength and the strong reduction of energy absorption capacity provided by the introduction of the stiffener in the column web. When comparing specimen A1 to A3 or A2 to A4 it is possible to conclude that the double plate on beam flanges provides a slight increase of strength and energy absorption capacity.

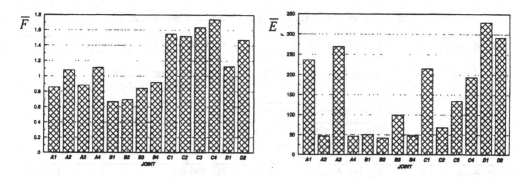

Figure 36. Comparison between specimens.

With reference to category B, it can be observed a slight increase of strength and a decrease of energy absorption capacity when going from B1 to B2 or from B3 to B4. If we compare joint B1 to B3 or B2 to B4, we can observe an increase of strength and energy dissipation capacity.

Comparison between C1 and C2 shows a strong reduction of energy dissipation capacity due to the stiffening of the beam web. When increasing the thickness of the end-plate (C3), there is an increase of energy absorption capacity (compare to C2), even if not so strong to compensate the loss due to partial-depth ribs on the beam web. A further stiffening of the beam web (full-depth ribs) gives rise to an increase of energy dissipation capacity, but it remains lesser than that of the base joint.

In the case of welded joints we can observe the increase of strength and the decrease of energy dissipation capacity due to the reinforcing of the column web panel zone.

On the basis of the previous explained results the following conclusions can be drawn:

1. If stiffeners are added to those parts of the connection which are mainly responsible for its deformability, then the amount of energy absorption is decreased, but the strength is increased.
2. If additional elements are introduced to increase the local strength of connection elements, but they do not substantially modify the evolution of the plastic mechanism, then there will be an increase of both strength and energy dissipation capacity, provided that the failure mode is ductile.

Ballio and Chen (1993*a*, *b*) carried out an experimental investigation on the cyclic behaviour of full-strength rigid and semi-rigid connections. They investigated the behaviour of the fully welded and the flange welded - web bolted connections among the rigid types and the extended end-plate connections among the semi-rigid types.

Table 9 gives a summary of the main features of the tested joints together with the outcome of this investigation, which is synthesised in the values of the plastic rotational capacity (*PRC*).

Table 9. Experimental values of plastic rotation capacity (*PRC*).

Joint Typology	Beam Characteristics	Column Characteristics	PRC
E1	HE 260 A $f_y = 318$ N/mm^2 (flanges)	HE 300 B $f_y = 318$ N/mm^2 (flanges) $f_y = 307$ N/mm^2 (web)	0.08
	Other Specific Characteristics		
	Continuity plates: 12.5 mm Full penetration welds		
F1	HE 260 A $f_y = 318$ N/mm^2 (flanges)	HE 300 B $f_y = 318$ N/mm^2 (flanges) $f_y = 307$ N/mm^2 (web)	0.093
	Other Specific Characteristics		
	Continuity plates: 12.5 mm Full penetration welds Reduced beam section ("dog bone")		
I1	HE 260 A $f_y = 302$ N/mm^2 (flanges)	HE 300 B $f_y = 282$ N/mm^2 (flanges) $f_y = 307$ N/mm^2 (web)	0.084
	Other Specific Characteristics		
	Continuity plates: 12.5 mm Shear tab: 12 mm Full penetration welds for flanges Web bolts: 2 bolts M27-10.9 $f_y = 960$ N/mm^2 ; $f_u = 1040$ N/mm^2		
D1	HE 260 A $f_y = 318$ N/mm^2 (flanges)	HE 300 B $f_y = 318$ N/mm^2 (flanges) $f_y = 307$ N/mm^2 (web)	0.067
	Other Specific Characteristics		
	End plate thickness: 44 mm Continuity plates: 12.5 mm Bolts: 8 bolts M24 $f_y = 960$ N/mm^2 ; $f_u = 1040$ N/mm^2 ; $\varepsilon_u = 12\%$		
D6	HE 260 A $f_y = 318$ N/mm^2 (flanges)	HE 300 B $f_y = 318$ N/mm^2 (flanges) $f_y = 307$ N/mm^2 (web)	0.045
	Other Specific Characteristics		
	End plate thickness: 26 mm Continuity plates: 12.5 mm Bolts: 4 bolts M30 $f_y = 960$ N/mm^2 ; $f_u = 1040$ N/mm^2 ; $\varepsilon_u = 12\%$		

Table 9. Continue.

Joint Typology	Beam Characteristics	Column Characteristics	*PRC*
J1	HE 260 A $f_y = 318$ N/mm^2 (flanges)	HE 300 B $f_y = 318$ N/mm^2 (flanges) $f_y = 307$ N/mm^2 (web)	0.082
	Other Specific Characteristics		
	Column web stiffeners: 12.5 mm End plate thickness: 50 mm Bolts: 4 bolts M30-10.9 $\quad$ $f_y = 960$ N/mm^2 ; $f_u = 1040$ N/mm^2 ; $\varepsilon_u = 12\%$		
k1	HE 260 A $f_y = 318$ N/mm^2 (flanges)	HE 300 B $f_y = 318$ N/mm^2 (flanges) $f_y = 307$ N/mm^2 (web)	0.082
	Other Specific Characteristics		
	Column web stiffeners: 12.5 mm End plate thickness: 50 mm Bolts: 4 bolts M30-10.9 $\quad$ $f_y = 960$ N/mm^2 ; $f_u = 1040$ N/mm^2 ; $\varepsilon_u = 12\%$ Reduced beam section ("dog bone")		
J4	HE 260 A $f_y = 318$ N/mm^2 (flanges)	HE 300 B $f_y = 318$ N/mm^2 (flanges) $f_y = 307$ N/mm^2 (web)	0.045
	Other Specific Characteristics		
	End plate thickness: 50 mm Bolts: 4 bolts M30-10.9 10 mm thick end plates		

5.3 Sensitivity to details

The results of an experimental investigation by Matsui and Sakai (1992) can highlight the strong sensitivity of the cyclic inelastic response to constructional details of beam-to-column joints. Test parameters of the recalled study were the collapse modes of steel frames and the presence of web copes at beam ends. Figure 37a shows the detail of the tested beam-to-column connections. In the case of the specimen without web cope, it was necessary to design an appropriate groove face for the beam ends, as shown in Figure 37b.

Figure 38 illustrates the cyclic inelastic behaviour of six tested beam-to-column joints. In this Figure the letter 'M' stands for Mode (of collapse); the second letter (after the 'M') indicates the type of collapse mode: 'P' when the plastic hinge is in the column web panel, 'B' when it is at the beam end, 'S' when both the column web panel and the beam end are in the plastic range; finally, the last letter indicates the presence (letter 'S') or the absence (letter 'N') of the web cope. From Figure 38 it can be observed that the presence of web copes produces different effects in function of the type of collapse mode. When the beam end is elastic, then there is no influence, as it is expected. In the other two cases the presence of the web cope produces a worsening of the ductility.

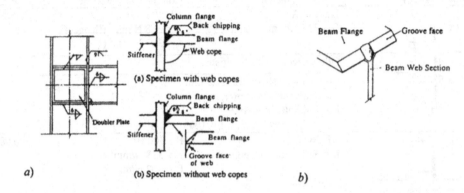

Figure 37. The tested beam-to-column connections (a) and the detail of the groove face of the beam end without web copes (b).

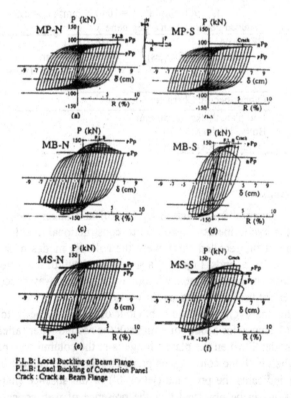

F.L.B: Local Buckling of Beam Flange
P.L.B: Local Buckling of Connection Panel
Crack : Crack at Beam Flange

Figure 38. Cyclic inelastic response of the beam-to-column steel joints tested by Matsui and Sakai (1992).

5.3 Analytical Prediction of Connection Influence

The simplified model (Mazzolani and Piluso, 1996a). To obtain closed-form relationships and, therefore, to clearly understand the effect of semi-rigidity and partial strength of connections on seismic behaviour of MRFs, it is effective to study the simplified model shown in Figure 39, which is deemed to adequately represent the global response of regular frames (see Section 6 for the concept of regularity).

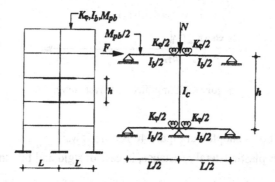

Figure 39. The simplified model.

The following three non-dimensional parameters govern the behaviour of the simplified model:

$$\zeta = \frac{EI_b/L}{EI_c/h}; \qquad \bar{k} = \frac{k_\varphi L}{EI_b}; \qquad \bar{m} = \frac{M_{uc}}{M_{pb}} \qquad (24)$$

where:
k_φ and M_{uc} are the rotational stiffness and the ultimate moment of connections;
I_b, M_{pb} and L are the inertia moment, the fully plastic moment and the span of the connected beam;
I_c and h are the inertia moment and the length of the column;
E is the Young modulus of the base material.

Connection influence on frame behaviour under lateral increasing monotonic loads. In case of regular frames, whose dynamic elasto-plastic behaviour is fundamentally governed by the first vibration mode, an adequate evaluation of frame performance can be made by means of a static push-over analysis. With reference to the above defined simplified model, by assuming an elastic-perfectly plastic behaviour of the members and connections of the frame, it is possible to obtain a simplified relationship between lateral load and top displacement, as illustrated in Figure 40.

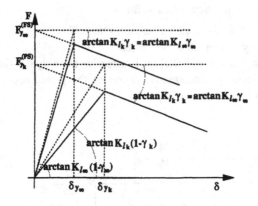

Figure 40. Simplified pushover curve.

The meaning of symbols adopted in Figure 40 is as follows:

$F_{y\infty}^{(FS)}$ is the first-order plastic lateral resistance, in case of rigid and full-strength connections;

$F_{yk}^{(PS)}$ is the first-order plastic lateral resistance, in case of semi-rigid and partial-strength connections;

$K_{l\infty}$ and γ_{∞} are the lateral stiffness and the stability coefficient (see Section 4), in case of rigid connections;

K_{lk} and γ_k are the lateral stiffness and the stability coefficient (see Section 4), in case of semi-rigid connections;

The lateral stiffness of the semi-rigid frame is related to the analogous parameter of the rigid one by the following equation:

$$\frac{K_{lk}}{K_{l\infty}} = \frac{\bar{k}(1+\zeta)}{\bar{k}(1+\zeta)+6} \tag{25}$$

which, remembering the definition of the stability coefficient ($\gamma = \dfrac{N}{K_l h}$), becomes:

$$\frac{\gamma_k}{\gamma_\infty} = \frac{\bar{k}(1+\zeta)+6}{\bar{k}(1+\zeta)} \tag{26}$$

As plotted in Figure 41, it shows that the decrease of connection stiffness produces the increase of the stability coefficient, which means the increase of frame sensitivity to second order effects. This effect is qualitatively present in case of both weak beam- strong column ($\zeta < 1$) and strong beam-weak column ($\zeta > 1$).

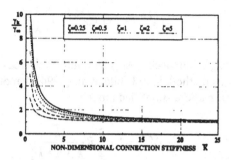

Figure 41. Connection influence on the stability coefficient.

The first-order plastic resistance is not influenced by the semi-rigidity of connections. Therefore, the following simple relationship is valid:

$$\frac{F_{yk}^{(PS)}}{F_{y\infty}^{(FS)}} = \overline{m} = \frac{M_{uc}}{M_{pb}} \tag{27}$$

But the second-order lateral resistance is influenced also by lateral stiffness. Therefore, the actual yield strength of the semi-rigid partial strength frame is obtained by combining the effect of partial-strength with the effect of semi-rigidity, as follows:

$$\frac{F_{yk}^{II(PS)}}{F_{y\infty}^{(FS)}} = \overline{m}(1 - \gamma_k) \tag{28}$$

The inelastic performance of the frame is strictly related to its deformation capacity, which in turn is related to the deformation capacity of members and connections. In case of full-strength connections the problem is governed by the plastic rotational capacity of beams. On the contrary, the plastic rotational capacity of connections is of concern in case of partial strength. By defining δ_y and θ_p as the frame yield displacement and the connection plastic deformation respectively, the global ductility of the frame is as follows:

$$\mu = 1 + \frac{\delta_p}{\delta_y} = 1 + \frac{\theta_p h}{\delta_y} \tag{29}$$

From this relationship it is apparent that both the stiffness and the ductility of connections influence the frame global ductility. Making reference to a full strength connection it is possible to highlight the influence of semi-rigidity alone. Taking in mind that semi-rigidity modifies only the yielding displacement, while the plastic portion of the ultimate displacement is unchanged, the following relationship can be set up:

$$\mu_k^{(FS)} = 1 + \left(\mu_\infty^{(FS)} - 1 \right) \frac{K_{lk}}{K_{l\infty}} \qquad (30)$$

Therefore, the decreasing of connection stiffness implies decreasing values of frame global ductility. By developing this methodology Faella et al. (1994) demonstrated the validity of the following relationship in case of the simplified model:

$$\mu_k^{(FS)} = 1 + \frac{3}{2} R \frac{\bar{k}}{\bar{k} + 6 + \bar{k}\zeta} \qquad (31)$$

which for $\bar{k} \to \infty$ provides also $\mu_\infty^{(FS)}$.

In Figure 42 the ratio $\mu_k^{(FS)} / \mu_k^{(PS)}$ is plotted as a function of the connection non-dimensional rotational stiffness for two values of the beam plastic rotation capacity ($R=2$ and $R=6$).

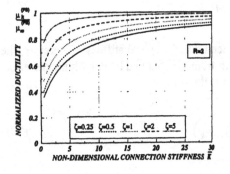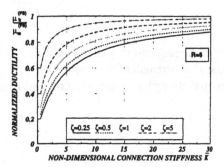

Figure 42. Connection influence on global ductility for full-strength connections.

The behaviour of partial-strength connections is quite different because the global ductility is directly influenced by an additional parameter:

$$\rho = \frac{R_c}{R_b} \qquad (32)$$

that is the ratio between the plastic rotational capacity of connections (R_c) and the plastic rotational capacity of the connected beams (R_b). For common connection typologies, as those considered in the previous Section 5.2, is $\rho \leq 1$. Therefore, the global ductility is further decreased, due to partial-strength of connections.

The plastic rotational capacity of connections is not a simple matter. It is nowadays subject of extensive research to find a good methodology for the prediction of their available local ductility. According to Bjorhovde et al. (1990), the ultimate rotation of connections can be estimated as a function of its non-dimensional strength ($\overline{m}$). Therefore, a functional relationship of the following type should be implemented:

$$\mu_k^{(PS)} = f(\overline{k}, \overline{m}, \alpha_i)$$
(33)

where α_i is a generic non-dimensional parameter considering other frame characteristics.

Following this direction Faella et al. (1994) have obtained the approximate expression:

$$\mu_k^{(PS)} = 1 + \left(\frac{5d_b\overline{k}}{L}\frac{5.4 - 3\overline{m}}{2\overline{m}} - 1\right)\frac{6}{\overline{k} + 6 + \overline{k}\zeta}$$
(34)

On the other hand, strength and stiffness for connections of usual type may be correlated, by analysing experimental data collected in literature. For example, the same Authors have obtained the following relationship referring to top and seat angle with double web angle connections by examining the experimental results by Aziziniamini and Radziminski (1988):

$$\overline{m} = 0.7474\left(\frac{L}{d_b\overline{k}}\right)^{-0.4120}$$
(35)

This relationship allows also for evaluating a limit value $\overline{k}^*$ of the connection rotational non-dimensional stiffness limit value (for $\overline{m} = 1$), which marks the boundary between the range of partial strength connections and that of full strength connections.

Figure 43 shows the application of the above relationships to a given special case, but the results are of general validity. From this Figure it clearly appears that full-strength connections always allow greater global ductility to be attained, when a very ductile beam ($R = 6$) is adopted (first-class sections according to EC3). The increase of connection deformation capacity due to decrease of its stiffness implies a more ductile global behaviour only when low ductile beams ($R = 2$) are assumed.

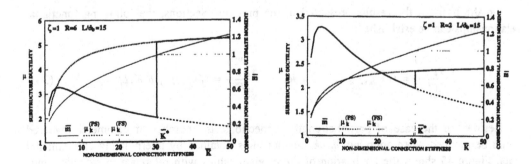

Figure 43. Connection influence on global ductility.

Connection influence on frame behaviour under seismic loads. It is well known that seismic action is far from the monotonic one, because it involves several cycles in inelastic range under alternate loading. As demonstrated in Section 5.2, the cyclic behaviour of a top and seat angle connection is quite different from that of an extended end-plate or a welded connection, because of pinching effects, which are significant in the former case and negligible or totally absent in the latter cases. Consequently, top and seat angle connections have a lower energy dissipation capacity than welded connections. To take into account the shape of hysteretic cycle by means of closed-form relationship is quite difficult. On the other hand, monotonic ductility is an useful index of plastic deformation capacity of members and connections also under cyclic loads. Therefore, a simplified approach could be the evaluation of the behaviour factor on the basis of a limit value of kinematic ductility μ, possibly reduced to take into account cyclic action.

The behaviour factor takes into account not only the capacity of frame, but also the seismic demand. The natural period of vibration is a fundamental parameter governing the seismic demand. The influence of semirigidity on the natural period of vibration for the assumed simplified regular frame model is very simple to derive on the basis of the previously recalled relationship between the lateral stiffness of the model with semi-rigid joints and that of the ideal rigid condition, as it is shown in Figure 44.

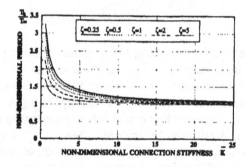

Figure 44. Connection influence on the period of vibration.

On the basis of the results presented in the previous Sections, the following functional relationships can be established:

$$\frac{q_k^{(FS)}}{q_\infty^{(FS)}} = f(\bar{k}, \zeta, \gamma_\infty, R) \qquad and \qquad \frac{q_k^{(PS)}}{q_\infty^{(FS)}} = f(\bar{k}, \zeta, \gamma_\infty, R, \bar{m}, L/d_b) \tag{36}$$

Besides, in the case of partial-strength connections, the number of parameter involved could be reduced when adopting $\bar{m} - \bar{k}$ relationships obtained by interpolating experimental data. Figure 45 shows the application of these relationships with reference to extended end-plate connections. It shows that semi-rigidity has a very slight influence on the q-factor in the range of full strength connections, where it is very close to 1. On the contrary, there are strong variations in the range of partial strength connections, with strong reduction of the q-factor. Only for very low values of the beam rotational capacity R ($R = 2$) it seems possible to attain adequate values of the q-factor, but this happens in an unstable zone of the relationship, where little variation of stiffness can produce a very sharp change of q.

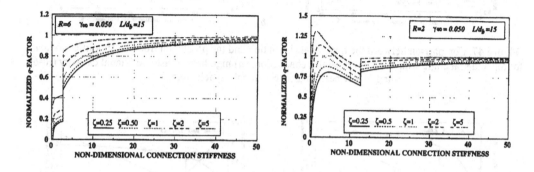

Figure 45. Connection influence on q-factor.

In order to evaluate the ability of the simplified model to interpret the influence of connection characteristics on the mechanical behaviour of the actual multi-degree of freedom frames, the results of static inelastic analyses were compared with the model prediction (Mazzolani and Piluso, 1996a). The analysed frame and the equivalent sub-structure are shown in Figure 46.

Figures 47 and 48 illustrate that the proposed simplified model is able to interpret, with a reasonable approximation, the complex influence of beam-to-column connections on the elastic and inelastic response of steel frames.

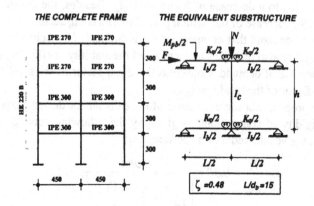

Figure 46. Analysed frame.

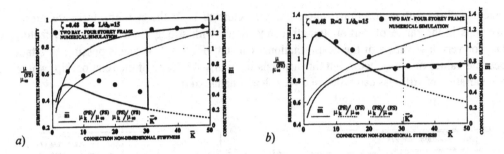

Figure 47. Comparison between the numerical simulation and the prediction from the simplified model, for top and seat angle with double web angles connections; *a*) high beam plastic rotational capacity (*R*=6) and *b*) low beam plastic rotational capacity (*R*=2).

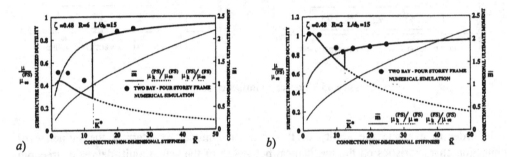

Figure 48. Comparison between the numerical simulation and the prediction from the simplified model, for extended end-plate connections with unstiffened columns; *a*) high beam plastic rotational capacity (*R*=6) and *b*) low beam plastic rotational capacity (*R*=2).

5.4 Codification proposals

At the light of previous results, it is apparent that both semi-rigidity and partial strength of connections always leads to a decrease of frame ductility. Besides, the advantageous effect of semi-rigidity, that is the increase of the natural period of vibration, seems to be not sufficient to produce a reduction of demand that compensate the reduction of capacity. It is worthy to take in mind that semi-rigidity also produces an increase of the stability coefficient, thus leading, for this aspect, to an increase of demand. Consequently, the *q*-factor of the semi-rigid frame q_k is ever less than the *q*-factor of the rigid one (q_∞).

An approximate proposal for the evaluation of the *q*-factor for design purposes, taking into account the semi-rigidity of connections, is given by the following equation (Mazzolani and Piluso, 1995*b*), which gives the reduction coefficient η:

$$\eta = \frac{q_k}{q_\infty} = 0.30 + \frac{0.08\bar{k}}{\left[1+\left(\dfrac{\bar{k}}{10}\right)^{1.58}\right]^{1/1.58}} \leq 1 \tag{37}$$

This proposal is compared in Figure 49 with that of Astaneh and Nader (1992). In this Figure the relationship η vs $\bar{k}$ is diagrammed, showing a continuous behaviour until $\bar{k}$ = 25, what is physically more advisable than the bi-linear proposal.

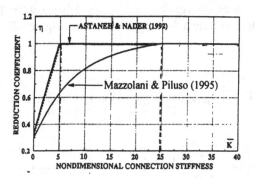

Figure 49. Proposals for the reduction coefficient η taking into account the semirigidity of connections.

Equation (37) is valid for full-strength connections only, as partial strength connections are usually not allowed in seismic resistant structures.

6   Geometric Regularity

6.1   General

Although the physical meaning of the structural regularity concept is quite intuitive, its quantitative definition is particularly difficult.

The concept of regularity has been introduced into modern seismic codes to allow for the extension of results obtained for single degree of freedom systems to actual multi-degree of freedom structures. In general, this extension is not allowed, because it is impossible to characterise multi-degree of freedom structure damage by means of one single parameter. However, the extension became possible if damage is uniformly distributed in the structure. On the other hand, the structural regularity is universally recognised as a fundamental requisite to avoid local damage concentration. Therefore, structural regularity could be considered as synonymous of uniform damage distribution.

Unfortunately, modern seismic codes oversimplify the problem of structural regularity, identifying it with the concept of geometrical regularity. The next section is devoted to illustrate that this codified assumption can lead to a misunderstanding of the problem and that it is not a sufficient approximation. The problem will be discussed with reference to plan frames, in relation to the damage distribution along the building height; obviously, it is believed that the analysed frames are extracted from buildings which are regular in plan.

6.2 Behaviour of Frames with Set-Back

In order to evaluate the possible worsening of the dissipative capacity of steel MR frames in the presence of set-back, the seismic inelastic response of three series of geometrically irregular frames was analysed. The first series derives from a 5-storey 3-bay regular frame (Figure 50), the second series derives from a 6-storey 5-bay regular frame (Figure 51) and, finally, the third series derives from a 5-storey 6-bay frame (Figure 52).

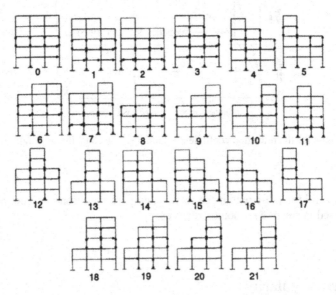

Figure 50. The first series of examined frames.

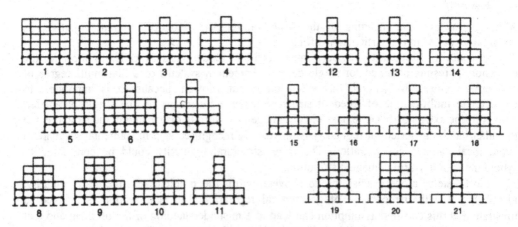

Figure 51. The second series of examined frames.

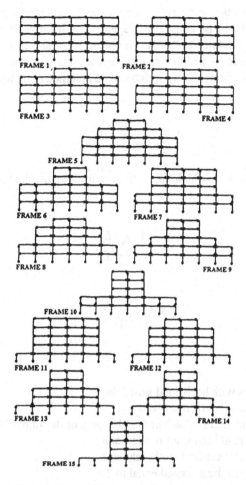

Figure 52. The third series of examined frames.

Member sections of the three frame series were designed according to three different methodologies. In the first methodology (first series) no account is taken for the inelastic behaviour and the structure was designed on the basis of a dynamic elastic analysis with reference to the first three modes of vibration. Moreover, frame column sections were varied along the building height. In the second methodology a simplified hierarchy criterion, similar to that proposed in EC8, was applied in order to improve the global ductility of frames. In this case, frame column sections were assumed constant along the height. Unfortunately these simplified design rules very often do not lead to the achievement of the desired results, as it is shown from the obtained results. Therefore, a more refined design procedure has been developed, which force the frame to collapse following a global mechanism (Mazzolani and Piluso, 1993*b*). This last methodology was applied when designing the third series of frames (Mazzolani and Piluso, 1996*c*). Looking to these three series at the light of the more recent classification, the frames of the first series can be defined "ordinary" moment resisting frames, those of the second series can be considered "special" moment resisting frames and, finally, those of the third series are defined as "global" moment resisting frames.

A quantitative evaluation of geometrical irregularity due to set-back (see Figure 53) was performed by means of the parameter Φ, which can be assumed as an irregularity index for set-back, given by:

$$\Phi = \frac{1}{1+K}(C + KV) \tag{38}$$

where the coefficients C and V, which take into account horizontal and vertical set-backs, are provided by

$$C = \frac{1}{N_p} \sum_{i=1}^{N_p} \frac{L_i}{L} \tag{39}$$

$$V = \frac{1}{N_c} \sum_{i=1}^{N_c} \frac{H_i}{H} \tag{40}$$

where:

N_p is the number of storeys with horizontal set-backs;
N_c is the number of bays with vertical set-backs;
L_i is the length of the horizontal set-back at the ith storey or the sum of the lengths of the set-backs, if they are present at both sides of the frame;
H_j is the height of the vertical set-back at the jth bay;
K is a constant value that can be assumed equal to 2.

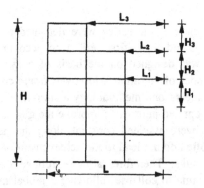

Figure 53. Definition of the geometrical irregularity index.

In order to obtain a forecast of the pattern of the yielding in the structure and, at the same time, to obtain a quick evaluation of its seismic behaviour, both static and dynamic inelastic analyses were performed. In particular, the former were used for determining the α-δ monotonic relationships, where α is the multiplier of horizontal forces and δ is the top sway displacement, as well as for evaluating the q-factor according to the energy method proposed by Como and Lanni (1983). The q-factor value is given by:

$$q = \left(\frac{W_u}{W_y}\right)^{\frac{1}{2}} \tag{41}$$

where W_y is the frame elastic strain energy at the yielding state and W_u is the total absorbed strain energy at the frame ultimate state.

Because of the variation in the fundamental period of vibration, the design q-factors (q_d) of the examined frames of the first series are different one from another. Therefore, the ratio $q^* = q/q_d$ was considered in order to examine the influence of the irregularity index Φ on the frame seismic response. In Figure 54 the ratio q^*/q^*_0 is plotted, where q^*_0 is the value of q^* when $\Phi = 0$. In this Figure the points evidenced by open circles are representative of the first series of frames (5-storey, 3-bay frames); the points denoted by solid squares represent the second series of frames (6-storey, 5-bay frames); the full circles correspond to the third series of frames (5-storey, 6-bay frames). From Figure 54 it can be observed that the design criteria used seem to have more significant effects than those related to the geometrical configuration. Moreover, the same Figure shows that for frames of the second and third series is always $q^* > q^*_0$ (with two exceptions only). On the contrary for frames belonging to the first series the q^* reduction can reach 20% for high irregularity index values ($\Phi \geq 0.6$). According to the above Figure the influence of the geometrical irregularity, in the case of frames designed without any control of the failure mechanism, i.e. ordinary MRF, could be accounted for by reducing the design value of the q-factor through the following coefficient:

$$\eta = 0.80 + \frac{4}{7}(0.60 - \Phi) \leq 1 \tag{42}$$

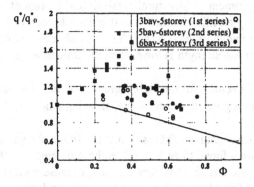

Figure 54. Normalised values of the q-factor computed for the three series of frames.

6.3 Remarks

On the basis of the previous numerical analyses, it can be stated that the code provisions unfairly penalise structural configurations characterised by the presence of set-backs. It seems to be essential to introduce in the definition of structural irregularity all the possible causes for non-uniform damage distribution (Mazzolani and Piluso, 1997c). In fact, geometrical irregularity does not always lead to structural irregularity, because the seismic inelastic behaviour is strictly dependent on the adopted design criteria rather than on the structural configuration. A worsening of the inelastic performances of structures can occur only in the case of ordinary moment resisting frames and in the case of special moment resisting frames characterised by a geometrical irregularity index Φ greater than 0.5. On the contrary, in the case of global moment resisting frames (i.e. frames designed to ensure a collapse mechanism of global type) the influence of set-back configurations can be practically neglected.

7   Random Material Variability

7.1 General

The fulfilment of design objectives in earthquake resistant limit state design is strictly related to material strength. In fact, distribution of dissipative zones through the structure depends upon the relative strength of structural components. Therefore, it is of primary importance to take into account the actual values of yield stress when performing an inelastic frame analysis aiming at identifying the dissipative behaviour under extreme earthquake loads.

7.2 Randomness of Material Properties

To assure global type collapse mechanism, columns must be stronger than beams. This leads to column flange thickness greater than that of beam flanges. Therefore, it is interesting to find the relationship between material strength and the thickness of the plate.

Figures 55 to 57 illustrate the variations of yield and ultimate stress with the thickness of tested plates for Fe360, Fe430 and Fe510 structural steels. Experimental data were collected in the last decades by the Department of Structural Analysis and Design of Naples University as part of its testing and quality certification activity. It is well known that the yield point of a plate decreases as far as its thickness increases. It has been observed that the influence of the thickness on the ultimate tensile strength is lower than the influence on the yield stress and in some cases it is negligible.

In Figure 58 the previous results for the Fe360 grade steel are summarised by means of a normal probability graph for a given thickness value, showing a normal distribution of data in the 5% confidence interval.

From the complete statistical analysis (Mazzolani et al., 1990a, b), values for the coefficient of variation of yield stress were found to be larger than those obtained for the ultimate strength. From the seismic point of view this result is very warning, because it means that an increase of the coefficient of variation of yield strength of members could induce the structural system to collapse in local failure modes, even if designed to collapse in a global mechanism. In practice the coefficient of variation of the yield stress is very often greater than 0.05, while an improvement of the controls in steel production could provide lower values, such as 0.025 following the proposal of Kuwamura and Kato (1989). In seismic design recommendations material specifications regarding the upper bound of the yield stress are usually absent and the lower limits are only specified; this gap must be eliminated and an appropriate coefficient of variation for the design stress should be introduced.

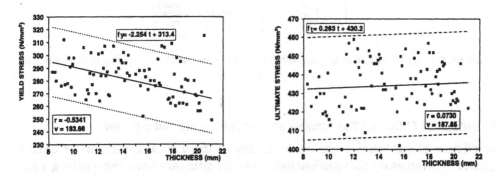

Figure 55. Yield stress (*a*) and ultimate stress (*b*) versus thickness regression, for Fe360 steel.

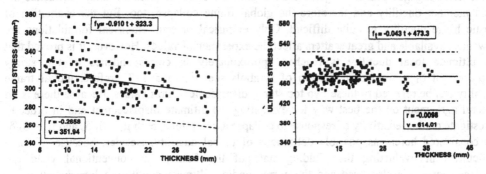

Figure 56. Yield stress (*a*) and ultimate stress (*b*) versus thickness regression, for Fe430 steel.

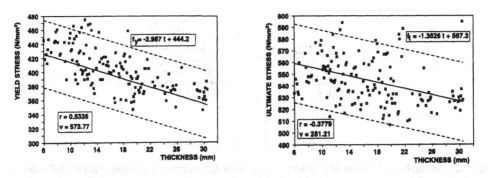

Figure 57. Yield stress (*a*) and ultimate stress (*b*) versus thickness regression, for Fe510 steel.

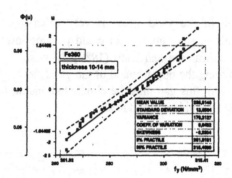

Figure 58. Data representation on normal probability graph paper and corresponding to 5% confidence interval.

7.3 Influence of Material Randomness on Global Ductility and *q*-Factor

The influence of yield strength random variability on frame inelastic performance could be evaluated by investigating the variations induced on the main parameters characterising a static pushover curve of the analysed frame.

Usually frame collapse is assumed to correspond to the attainment of a limit plastic rotation in the most engaged plastic hinge. Yield and ultimate stress random variations influence the local member ductility and therefore the global frame collapse, too. But the evaluation of plastic hinge ductility is quite difficult. Only empirical or semi-empirical formulations are nowadays available and great scatters affect the experimental values. However, it is proved that the ultimate local ductility is reached approximately in correspondence of top storey displacement δ_1 or δ_2 (for definition of symbols see Figure 59). Therefore, global frame ductility can be referred to these two top storey displacement limit values. Because there is no general agreement on the best way for evaluating the ultimate displacement and in order to investigate on the sensitivity of response to collapse assumptions, both $\mu_1 = \delta_1/\delta_y$ and $\mu_2 = \delta_2/\delta_y$ are considered herein as possible definitions of global ductility. Besides, because of some difficulties in evaluating the yielding state of the frame, the conventional yield top displacement δ_y^* is also used and the corresponding ultimate ductility is introduced: $\mu_2^* = \delta_2/\delta_y^*$.

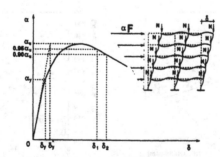

Figure 59. Multiplier of horizontal forces versus top sway displacement behavioural curve: definition of the examined parameters.

An approximate evaluation of the q-factor has been done on the basis of the pushover analysis, by means of the Como-Lanni method (see Section 2.3). As far as the computation of q is concerned, the following observations can be made.

When designing the frame, member sections have to be chosen among standard shapes and therefore an over-strength results. Consequently, first yielding is reached under a total shear force equal to $\alpha_y F_d$ greater than the design shear force F_d (= $MAR_e(T)/q_d$). The deterministic value of the frame behaviour factor must be evaluated with reference to actual deterministic frame strength. Therefore, the following relationship applies:

$$q^* = \alpha_y q_d = \left(\frac{W_u}{W_1}\right)^{\frac{1}{2}}$$

(43)

where W_1 is the elastic strain energy corresponding to design forces.

A statistical analysis of frame inelastic behaviour under random distributions of yielding stresses was performed (Mazzolani et al., 1991). The examined frame is illustrated in Figure 60. It was designed to collapse in a global mechanism. In fact, for this type of failure mode the plastic hinge formation process has the maximum influence on the ultimate behaviour.

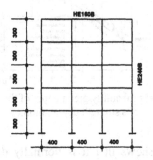

Figure 60. The examined frame.

A hybrid Monte Carlo simulation method was adopted for the analyses. In fact, in a first step random yield strength distributions were generated, neglecting the geographical distribution over the generic cross-section and adopting mean values for a given flange thickness (the yield strength of the I shaped sections was assumed to be coincident with that of the flanges only). As the examined frame is made of HE160B beams and HE240B columns, two different normal distribution laws were used, by considering both thicknesses of beam and column flanges. 100 sets of 35 yield stress random values were generated and assigned independently of beam and column, thus obtaining 100 geometrically identical but mechanically different frames. The section plastic modulus random variation was neglected because of its very small coefficient of variation when compared with that of the yield stress.

After each generation of a random frame, the inelastic static pushover analysis was performed and the previously defined behavioural parameters were computed. The set of 100 analysed frames was considered to be a statistical sample of the structural performance.

From the design point of view, the more interesting results are the 5% fractiles, which have to be compared with the deterministic values of the examined behavioural parameters, obtained on the basis of the nominal yield strength.

Figure 61 illustrates on a normal probability graph paper the distribution of global ductility μ_1, computed according to 5% strength degradation ultimate displacement (δ_1).

It can be seen that global ductility is not normally distributed, as confirmed by the 5% confidence interval and also by the skewness coefficient test. Therefore, different distribution laws have to be investigated. However, it can be assumed that the use of distribution laws different from the normal one does not produce significant variation of the results from the practical point of view. Hence, considering that the deterministic value of the global ductility was 9.59, it can be stated that the global ductility of the actual structure can be less than that computed on the basis of the nominal yield strength with a probability of almost 50%, which is non-negligible.

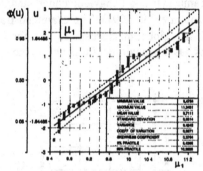

Figure 61. Representation on normal probability graph paper of the global ductility corresponding to a 5% strength degradation.

For the q-factor, Figure 62 shows that it is normally distributed. Considering that the deterministic value of the q-factor was equal to 7.94, it can be observed that the probability to have lesser values is practically zero. Therefore the deterministic analysis underestimates the q-factor.

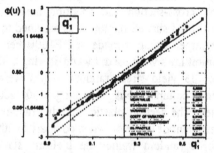

Figure 62. Representation on normal probability graph paper of the q-factor corresponding to a 5% strength degradation.

Moreover, by examining the collapse mechanisms of the 100 generated frames it can be observed that only in 10 cases a failure mode different from the global one was obtained, although very similar to it. Local failure modes have not been produced by random material variation so that the slopes of the softening branch of the α-δ curves are quite constant.

7.4 Effect of *COV*

A further research activity (Mazzolani et al., 1992) was devoted to analyse the effect of the coefficient of variation (*COV*) of the yield stress. It was assumed equal to 0.025, 0.050, 0.075, 0.100, 0.150, thus covering the whole variability range for structural steels. Contemporary, the dependence of the yield stress on the thickness of frame member flanges, the influence of column-to-beam strength ratio and the influence of the structural scheme were considered.

The yield stress-flange thickness relationship was assumed to be linear:

$$f_{y,m}(t_f) = -\alpha t_f + \beta \tag{44}$$

where $f_{y,m}$ is the average value of the yield stress and t_f is the flange thickness. The values of α and β were chosen as those corresponding to Fe510, because for this steel grade the influence of thickness is the greatest.

The parameter ρ, which is defined as the ratio between column and beam plastic moduli, influences the type of collapse mechanism. Two values of ρ (1 and 1.64) were considered by assuming everywhere the same HE200B shape for beams and two different shapes (HE200B and HE240B) for columns.

Finally, two different types of frame were considered. They are shown in Figure 63, together with the corresponding nominal failure modes, i.e. the failure modes determined by assuming a constant value of the member yield strength equal to the nominal value ($f_y = 355$ N/mm^2).

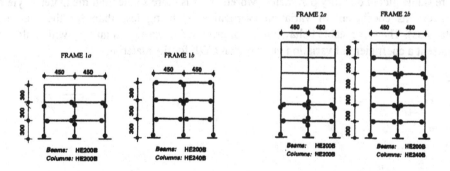

Figure 63. Analysed frames and corresponding nominal failure mode.

The deterministic global ductility was determined by an inelastic pushover analysis with a *constant yield stress* expressed as:

$$f_y(t_f) = f_{y,k}(t_f) = f_{y,m}(t_f) \, (1\text{-}2COV) \tag{45}$$

In the case of the six-storey frame with HE200B columns (frame 2a of Figure 63), Figure 64 shows that the deterministic global ductility is always greater than the corresponding 5% fractile computed by the probabilistic approach, independently of the adopted definition (μ_1, μ_2, μ_2^*). Besides, it should be noted that the scatter between deterministic and probabilistic values of global ductility is increasing with the value of COV. Analogous results were obtained in the other cases, clearly showing that the deterministic approach is not on the safe side and, on the other hand, the COV value should have to be reduced in order to mitigate the scatter between the deterministic behaviour and the probabilistic one, which is more close to reality.

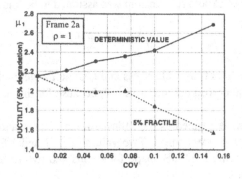

Figure 64. Comparison between probabilistic analysis and deterministic analysis for the global ductility corresponding to a 5 % strength degradation.

It is interesting to compare, by means of Figure 65, the stability, from a statistical point of view, of the different definitions of global ductility. It can be observed that μ_2^* based upon δ_y^* is the more stable global ductility parameter. Moreover, it is more stable than the material yield strength itself, its coefficient of variation (vertical axis) being less than for the material (horizontal axis). The same stable behaviour is not provided in case of μ_1 and μ_2, which almost always present a coefficient of variation greater than COV for the material.

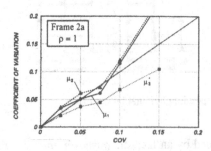 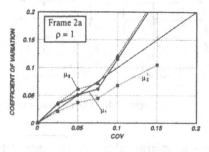

Figure 65. Statistical stability of definitions of global ductility for Frame 2.

These results have confirmed that the effects of randomness of material yield strength increase as the *COV* increases. The increase of the coefficient of variation of material yield strength (*COV*) can produce a significant reduction of the available global ductility, which is not negligible from the seismic design point of view. Besides, they have shown that the probabilistic approach can be an useful tool for defining stable parameters, which characterise the inelastic structural response of seismic resistant steel frames.

8   Influence of Claddings

8.1   General

Claddings are usually considered as non-structural elements. It is essentially due to the difficulties arising in the evaluation of claddings mechanical properties, but the experimental evidence shows that they may significantly contribute to enhance the structural performance under low to moderate earthquake motions and also to improve the structure energy dissipation capacity. On the other hand, interacting with the framing structure, claddings may suffer significant damage, whose correct evaluation can be made only by explicitly considering their actual mechanical behaviour. Moreover, it can be observed that the use of panels as collaborating structural components can allow for a simplification of the structural scheme, because simpler constructional details for the joints between frame members (i.e. pinned connections) can be adopted, leading to more economical structural solutions (see Figure 66).

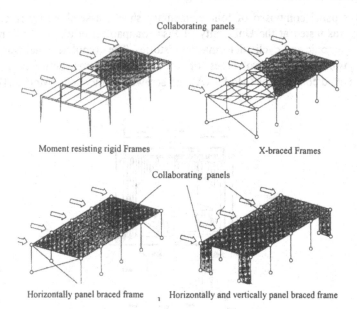

Figure 66. The use of collaborating panels in different structural schemes (MR frames, totally or partially braced frames, pinned frames).

The influence of claddings, acting as diaphragms, on structure mechanical behaviour is usually referred to as "skin-effect", giving rise to the so-called "stressed skin design", which has been dealt with for normal steel buildings in the field of activity of *ECCS* (1977).

From the seismic point of view, the degree of interaction between cladding panels and the main structure can be used as a behavioural parameter in order to classify the type of diaphragm action (Mazzolani and Sylos Labini, 1984):

a. Panels are designed to improve the structure performance under frequent earthquake motions, and therefore they are considered as collaborating in the serviceability limit state, but not in the ultimate limit state checking.

b. Panels are designed to contribute to ultimate structure seismic resistance, interacting with the framing structure.

c. Panels are designed to solely resist earthquake loads, without interacting with the framing structure which carries vertical loads only.

Beam-to-column connections are to be chosen according to the chosen design criteria and to the degree of co-operation with the panels in carrying earthquake loads.

The seismic behaviour of pin-jointed structures with bracing panels was examined (Mazzolani and Piluso, 1990), as an economical solution for buildings in low-seismicity zones. The important influence of the connecting system has also been pointed out (Mazzolani and Piluso, 1991), by considering that the transfer of shear forces from the structure to the panels can be guaranteed by means of different technological connecting systems which use bolts, rivets, screws or welds.

8.2 Some Experimental Data on Sheetings

A single layer panel composed of four elementary sheets, assembled by rivets, is shown in Figure 67. It was tested at the University of Pisa (Sanpaolesi et al., 1983; Sanpaolesi, 1984) under both monotonic and cyclic deformation histories. The complete panel was screwed to the upper and lower chords. Two types of connecting systems between sub-panels were considered: riveted connections (Model 1A) and screwed connections (Model 1B).

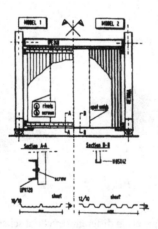

Figure 67. Tested panels.

Figure 68 shows the results obtained when testing Model 1A. It can be seen the strong non-linear behaviour, which is apparent also for very low values of the shear force. Besides, the cyclic behaviour highlights a very poor dissipative capacity due to large slippage in the connections.

When changing the connecting system from the rivets to the screws, a better performance of the panel was observed, with an increase of both kinematic ductility and energy absorption capacity, as can be seen in Figure 69.

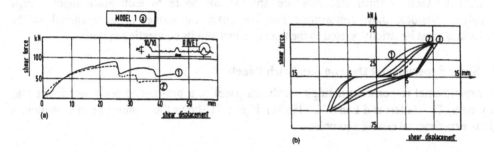

Figure 68. Monotonic (*a*) and cyclic (*b*) behaviour of riveted panels.

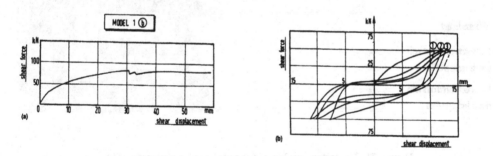

Figure 69. Monotonic (*a*) and cyclic (*b*) behaviour of screwed panels.

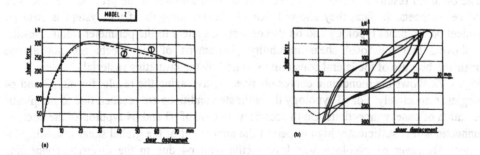

Figure 70. Monotonic (*a*) and cyclic (*b*) behaviour of welded panels.

Model 2 is obtained by using spot welds as connecting system and by inserting the complete panel into a perimeter frame. Figure 70 shows the results of monotonic and cyclic tests on this type of panel. It is clear an important reduction of the pinching effects in the hysteresis cycles, and consequently, an increase of both ductility and energy absorption capacity. The softening branch of the shear force versus shear displacement relationship arises from local buckling of the panel, which gives rise to the ultimate conditions.

These results illustrate that riveted and welded panels can be considered a lower bound and an upper bound, respectively, for the panel behaviour. In particular, the results concerning with the welded panels confirm that they are able to contribute to earthquake input energy dissipation. Besides, they demonstrate that the analytical model to be assumed in the calculations must be strictly related to the type of panel and its connecting system.

8.3 Some Experimental Data on Sandwich Panels

The experimental response of a single sandwich panel in a pin-jointed perimeter frame was analysed by De Matteis and Landolfo (1998a). Figure 71 shows the examined panel typologies and the experimental testing equipment.

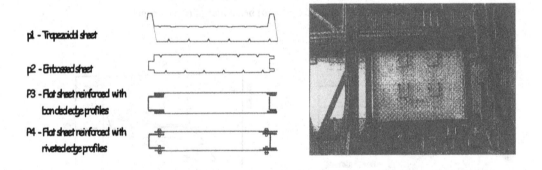

Figure 71. The tested typologies (a) and the testing apparatus (b).

The obtained results substantially agree with those illustrated in the previous section with reference to sheets. In fact, they showed that the lateral strength of the system is strongly dependent on the panel typology and on its connecting system to the perimeter frame. Besides they showed that the local shear instability phenomena of the sheets are not always determinant, because of the lateral constrain exercised by the insulating material.

Figure 72 shows the monotonic curves obtained by averaging the results for p2, p3 and p4 typologies, respectively. For p1 typology the ultimate conditions are reached due to the plastic deformation of panel connections. On the contrary, in case of p3 and p4 typologies the strength of connections was sufficiently high to permit the attainment of the global shear instability for the panel. However p3 typology was less ductile than p4 due to the different connecting system.

The cyclic behaviour of the sandwich panels is illustrated with reference to p4 typology in Figure 73. It was affected by strong pinching effects due to the plastic deformations developed around the holes of the connecting bolts and the consequent slip of the system under reversal of deformation.

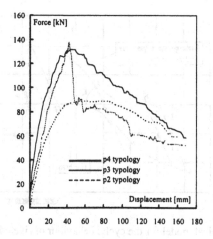

Figure 72. Average monotonic curves for p2, p3 and p4 typologies.

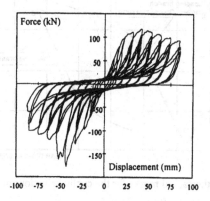

Figure 73. Cyclic behaviour of p4 typology.

8.4 Mathematical Modelling of Panel Cyclic Behaviour

In the previous Section it has been shown that the main feature of the cyclic behaviour of riveted and screwed sheets is the pinching effect due to large slippage in connections. The analytical model of the mechanical behaviour of this panel typology should consider this aspect. Therefore, an ad-hoc mathematical model was developed (Mazzolani and Piluso, 1990). The general features of this model are illustrated in Figure 74.

As the slip effects for the welded panels could be neglected, the corresponding model is illustrated by Figure 75.

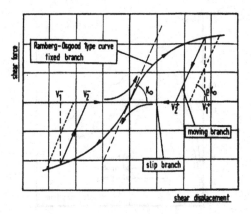

Figure 74. Mathematical model for the cyclic behaviour of riveted and screwed panels.

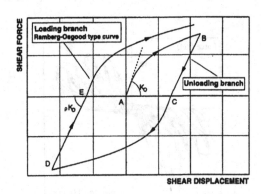

Figure 75. Mathematical model for the cyclic behaviour of welded panels.

Recently, more refined mathematical models were developed for both monotonic and cyclic loading conditions.

As a first step, a finite element model was developed in order to interpret panel monotonic behaviour (De Matteis et al., 1996). The numerical code ABAQUS was used and both material and geometrical non-linearity were considered, by adopting a simplified procedure which distinguishes two different behavioural phases, the pre-critical and the post-critical one. The comparison between the numerical simulation and the test data is shown in Figure 76.

The monotonic load-deformation curve obtained from experimental data or from the *FEM* model can be simplified and then approximated by means of mathematical formulas. These results has been used as the basis for the description of the cyclic behaviour of panels, by defining the loading and unloading curves as shown in Figure 77.

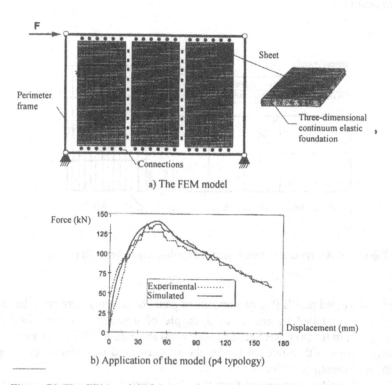

a) The FEM model

b) Application of the model (p4 typology)

Figure 76. The *FEM* model of the tested panels (*a*) and the obtained results (*b*).

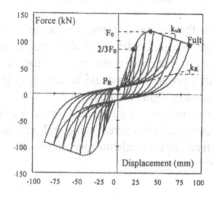

Figure 77. Mathematical model for the cyclic behaviour of sandwich panels.

8.5 Analytical Investigation on Seismic Response of Structures with Bracing Panels

In order to investigate the applicability of panels, in the form of trapezoidal sheets, as bracing system in relation with the seismicity level of the site, dynamic inelastic analyses were carried out, by using the structural scheme illustrated in Figure 78, where the collaborating panels are located in given positions.

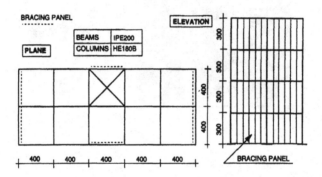

Figure 78. Analysed structure braced with light-gauge steel infill panels.

Both riveted and welded panels (the two bounds of panel behaviour) were considered in the analyses. They were modelled by means of a couple of diagonals, characterised by the hysteretic curves given in the previous Section (Mazzolani and Piluso, 1990). The experimental results previously shown with reference to the sheets were used to calibrate the empirical parameters of the mathematical models.

The seismic input has been generated from the elastic design response spectrum given in the proposal of the Italian seismic code CNR-GNDT (1984) for soil type S1, with peak ground accelerations of 0.15g, 0.25g and 0.35g for low-, medium- and high-seismicity zones, respectively.

Figures 79a and 79b illustrate the results in terms of maximum required interstorey drift and top sway displacement. The required interstorey drift has to be compared with the available ductility of panels, which is given by the experimental data. From this comparison, it can be concluded that riveted and screwed panels could be used only in low-seismicity zones, while welded panels can be also used in higher seismicity areas.

In order to demonstrate the convenience of pin-jointed structures braced by collaborating trapezoidal sheet panels, the previous considered structural scheme was redesigned by assuming a classical X-braced lateral resisting scheme (see Figure 80), with X-bracings located in the same positions of the collaborating panels.

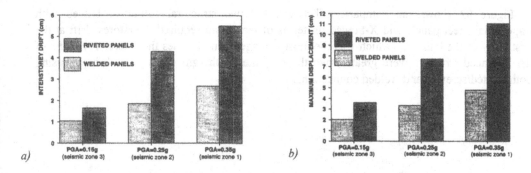

Figure 79. Maximum required interstorey drift (*a*) and top sway displacement (*b*).

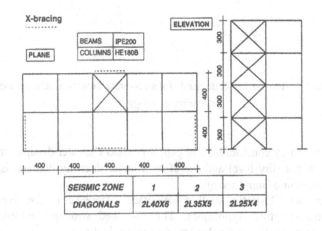

Figure 80. Cyclic model for the X-braced scheme.

The slenderness of the diagonal bracing members allowed a simplified modelling of their hysteretic behaviour neglecting the compressive strength, as illustrated in Figure 81.

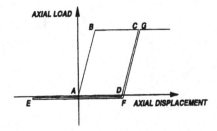

Figure 81. Cyclic model for braces.

Figure 82 shows the comparison between the two structural solutions: braced with trapezoidal sheet panels and X-braced, in terms of maximum required interstorey drift at the first story of the building (which was the most damaged). In all cases the interstorey drifts of the X-braced structure were greater than those of the diaphragmed pin-jointed frame, using both riveted/screwed and welded connections.

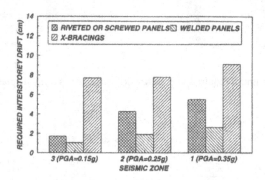

Figure 82. Comparison of interstorey drift demands for an X-braced classical scheme and a panel braced innovative typology.

This can be explained by considering that panels possess a greater dissipation capacity than X-bracings at a given ductility level and therefore a lesser ductility demand is necessary for dissipating a given amount of input energy.

Moreover, Figure 83 illustrates another important difference in the dynamic inelastic response of the two structural typologies. The X-braced structure exhibited a residual permanent displacement, while the diaphragmed structure had a symmetric behaviour. This is obviously a consequence of the hysteretic behaviour of the bracing system.

According to these results, greater ductility is required to a X-braced structure with respect to a panel braced pin-jointed frame.

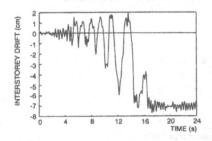 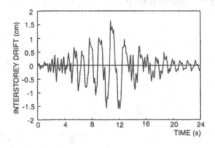

Figure 83. First storey drift time history for the X-braced structure (*a*) and the structure braced with panels (*b*).

A broader analytical investigation on the applicability of sandwich panels as collaborating lateral resisting structural component was carried out by De Matteis and Landolfo with reference to both industrial buildings (1998*b*) and multistorey frames (1998*c*). The following objectives were pursued:

1. Comparison between the panel braced and the classical X-braced frames, by using the experimental data collected on the cyclic behaviour of sandwich panels;

2. Characterisation of the dynamic response of panel braced systems by means of ductility spectra over the period range 0 – 3 sec, which cover the practical range of application;

3. Investigation on the economic convenience of the application of panels as lateral resisting structural component in case of both industrial buildings and multistorey frames.

The single mass system of Figure 84 was studied in the step 1 by comparing the inelastic dynamic response of panel braced and X-braced systems. In the case of panel braced structure a type p4 sandwich panel (see Section 8.3) was assumed and the correspondent experimental ultimate ductility was used to judge the survival to a give earthquake. In the case of X-braced structure a zero compressive strength was assumed for bracings, as shown in Figure 84 by the adopted hysteresis model. In the same Figure the equivalence criteria between the models adopted for panels and bracings is also shown.

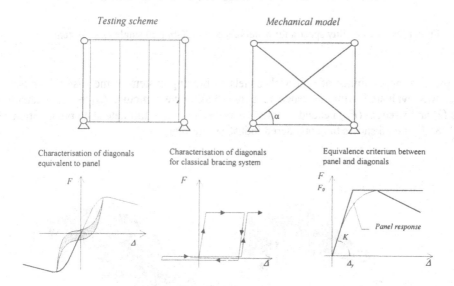

Figure 84. The analysed single mass system.

The system period of vibration was varied so that to cover the period range 0.2-3 s, and the design *q*-factor (q_d) was also varied considering the values 1, 1.5, 2, 2.5, 3, 4, 5. Then, dynamic inelastic analyses were carried out for evaluating the ductility demand for each period and each q_d value. Analyses were repeated considering several natural accelerograms recorded during *italian* earthquakes. Figure 85 shows the results obtained in the case of ten accelerograms recorded during the earthquake of Tolmezzo. In the same Figure is also plotted the ultimate value of ductility coming from the experimental results previously recalled. It can be observed that for short periods of vibration the ductility demand is often greater than capacity, and the value $q_d = 2$ can be assumed as an upper bound for the design behaviour factor.

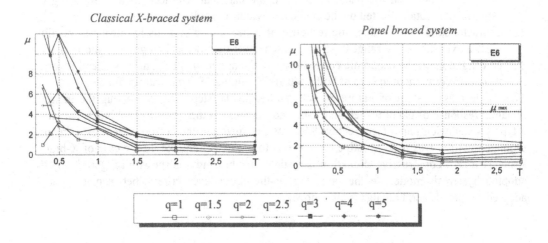

Figure 85. The ductility spectra for panel braced and X-braced single mass system.

The practical applicability of sandwich panels as bracing system in the case of industrial buildings was evaluated with reference to a non-dissipative structure ($q_d = 1$), evaluating stiffness (k) and strength (F_{el}) needed to satisfy serviceability and ultimate state prescription of Eurocode 8. Figure 86 shows the considered industrial building.

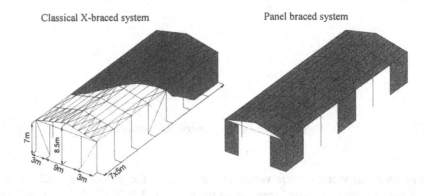

Figure 86. The analysed industrial building.

The obtained values of k and F_{el} were compatible with the characteristic values of panels usually adopted in such type of structure. Moreover, an 18% material weight saving was reached with respect to the classical X-bracing structural solution. The latter is not a good solution in case of moderate seismic action, because of the impossibility to take advantage from the higher value of q_d due to the limit values of bracing slenderness which condition their design.

Finally, the possibility of using panels as stiffening elements for the satisfaction of serviceability limitations in multistorey rigid frames was investigated. With reference to a civil building of five story (Figure 87 illustrate the building plan), a design methodology alternative to the current one has been proposed (De Matteis and Landolfo, 1998c).

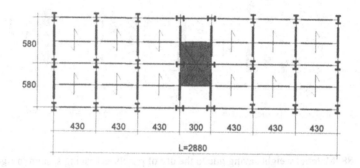

Figure 87. Examined building plan.

The main structure is designed at the ultimate limit state, without considering any contribution of panels. After this, the stiffness and strength requisites of panels, needed to satisfy deformability limitations under service earthquake, are evaluated. Frames were designed with reference to a 0.2g PGA value and by considering three values of the design q-factor (q_d = 4, 6 and 8). Second order effects were taken into account when evaluating internal forces, but no limits were assumed for the value of the coefficient θ_i defined as follows by Eurocode 8:

$$\theta_i = \frac{N_{tot,i} d_{r,i}}{V_{tot,i} h_i} \tag{46}$$

In this relationship $N_{tot,i}$ is the total weight over the i-th storey, $d_{r,i}$ is the relative displacement between the i-th and the (i-1)-th floor (i-th interstorey drift), $V_{tot,i}$ is the storey shear and h_i is the i-th storey height. As it is well known, a limit value of 0.3 is given by EC8 for θ_i in order to consider acceptable the frame sensitivity to second order effects. After the design of the main structure, panel stiffness and strength were evaluated in order to obtain an elastic behaviour with maximum interstorey drift limited to $0.004h_i$, $0.006h_i$ and $0.008h_i$. The computed values of strength and stiffness were always belonging to the ranges F_{el} = 28 - 170 kN and k = 2.1 - 11.3 kN/mm. The mean values of F_{el} and k obtained by the experimental campaign belong to these ranges, being 78 kN and 4.6 kN/mm, respectively. Besides, the difference between maximum required values of F_{el} and k and the experimental values is considered acceptable because in actual cases panels are connected also on the vertical sides to the perimeter frame, while in the experiments they were free on this sides. Therefore, the commonly adopted typologies of panels can be used as structural elements. Moreover, Figure 88 shows that the use of panels as drift limiting structural component could produce a 50 - 60 % material cost saving.

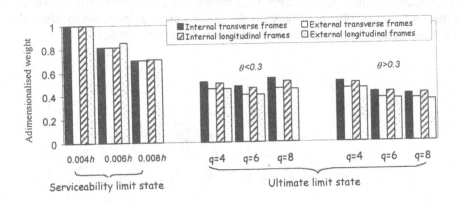

Figure 88. Material weight saving due to the use of panels as bracing system in rigid frames.

References

AIJ (1990). Standard for Limit State Design of Steel Structures (Draft). Architectural Institute of Japan (English version, October 1992).

AISC (1997). Seismic Provisions for Structural Steel Buildings. American Institute of Steel Construction.

Akiyama H. and Yamada S. (1997). Seismic input and damage of steel moment frames - General report. In *Proceedings of the 2ʰᵗ International Workshop on "Behaviour of Steel Structures in Seismic Areas", STESSA '97*, Kyoto, Mazzolani F. M., Akiyama H. editors, published by 10/17, Salerno, pp. 789-800.

Aribert J.M and Grecea D, (1999). Dynamic behaviour control of steel frames in seismic areas by equivalent static approaches. In *Proceedings of the Stability and Ductility of Steel Structures Conference (SDSS'99)*, Timisoara, September. D. Dubina and Ivanyi M. editors, Elsevier Science Ltd.

Astaneh A. (1995a). Seismic Design of Bolted Steel Moment-Resisting Frames. *Structural Steel Educational Council, Technical Information and Service.*

Astaneh A. (1995b). Seismic behaviour and design of steel semi-rigid structures. In Proceedings of *Behaviour of Steel Structures in Seismic Areas, STESSA '94*, Timisoara, Romania, F.M. Mazzolani, and V. Gioncu editors, published by E & FN Spon, pp. 547-556.

Astaneh A., Nader M. (1992). Shaking table tests of steel semi-rigid frames and seismic design procedures. In *Proceedings 1ˢᵗ State of the Art Workshop COST C1*, Strasbourg.

Azizinamini A. and Radziminski J. B. (1988). Prediction of moment-rotation behaviour of semi-rigid beam-to-column connections. In R. Bjorhovde, J. Brozzetti and A. Colson eds., *Connections in steel structures: behaviour, strength and design.* London and New York: Elsevier Applied Science.

Ballio G. (1985). ECCS approach for the design of steel structures against earthquakes. In *Symposium on steel in buildings,* Luxembourg, 1985; *IABSE-AIPC-IVBH Report,* Vol. 48, pp.373-380.

Ballio G., Calado L., De Martino A., Faella C., Mazzolani F.M. (1986). Steel beam-to-column joints under cyclic loads: experimental and numerical approach. In *Proceedings of the 8th European Conference on Earthquake Engineering.* Lisbon, September.

Ballio G., Calado L., De Martino A., Faella C., Mazzolani F.M. (1987). Cyclic behaviour of steel beam-to-column joints: experimental research. *Costruzioni Metalliche,* N. 2.

Ballio G. and Castiglioni C.A., (1993). An approach to the seismic design of steel structures based on cumulative damage criteria. In *Paper accepted for publication on Earthquake Engineering and Structural Dynamics.*

Ballio G. and Chen Y. (1993a). An experimental research on beam-to-column joints: interior connections. In *Proceedings of the XIV CTA Congress.* Viareggio, Italy.

Ballio G. and Chen Y. (1993b). An experimental research on beam-to-column joints: exterior connections. In *Proceedings of the XIV CTA Congress.* Viareggio, Italy.

Bertero V. V., Anderson J. C., Krawinkler H. (1994). Performance of steel building structures during the Northridge Earthquake. In *Earthquake Engineering Research Center, Report No. UBC/EERC-94/09,* University of California, Berkeley.

Bjorhovde R., Colson A. and Brozzetti J. (1990). Classification system for beam-to-column connections. *Journal of Structural Engineering,* ASCE, Vol. 116, N. 11, November.

Bruneau M., Uang C.M. and Whittaker A. (1998). *Ductile design of steel structures.* McGraw-Hill.

Calado L. and Azevedo J., (1989). A model for predicting the failure of structural steel elements. In *Journal of Constructional steel research,* Vol. 14, pp. 41-64, 1989.

Calderoni B., Ghersi A. and Mazzolani F.M. (1996). Critical analysis of EC8 approach to face the problem of structural regularity. In *Proceedings of the European Workshop on Seismic Behaviour of Asymmetric and Setback Structures.* Anacapri, Italy, October 4-5, Roberto Ramasco and Avigdor Rutemberg editors.

CNR-GNDT (1984). Norme tecniche per le costruzioni in zone sismiche. *Gruppo Nazionale per la Difesa dai Terremoti.* Dicembre.

Commission of the European Communities (1993). Eurocode 3: Design of Steel Structures, ENV.

Commission of the European Communities (1994). Eurocode 8: Structures in Seismic Regions, ENV.

Como M. and Lanni G. (1983). Aseismic toughness of structures. In *Meccanica.* N. 18, pp. 107-114.

Cosenza E., De Luca A., Faella C. and Piluso V., (1988). A rational formulation for the q-factor in steel structures. In *9th World Conference on Earthquake Engineering,* Tokyo, August.

Cosenza E., Faella C. and Piluso V. (1989). Effetto del degrado geometrico sul coefficiente di struttura. In *Proceedings of the IV Convegno Nazionale "L'ingegneria sismica in Italia",* Milano, Ottobre.

De Matteis G. and Landolfo R. (1998a). Structural behaviour of sandwich panel shear walls: an experimental analysis. *Materials and structures,* vol. 32, pp. 331-341.

De Matteis G. and Landolfo R. (1998b). Diaphragm effect for industrial buildings under earthquake loading. In *Proceedings of 2nd World Conference on Steel Construction,* San Sebastian, Spain.

De Matteis G. and Landolfo R. (1998c). Dynamic response of infilled multistorey steel frames. *Proceedings of 11th Conference on Earthquake Engineering,* Paris La Defense.

De Matteis G., Landolfo R., Mazzolani F.M. (1996). On the shear flexibility of corrugated shear panels. *Steel Structures, Journal of Singapore Structural Steel Society.*

De Matteis G., Landolfo R., Dubina D. and Stratan A., (1999). Influence of the structural typology on the seismic performance of steel framed buildings. In the *RECOS Copernicus project report,* Chapter 7.3.

ECCS (European Convention for Constructional Steelwork) (1977). European recommendations for the stressed skin design of steel structures.

ECCS (European Convention for Constructional Steelwork) (1986). Recommended testing procedure for assessing the behaviour of structural steel elements under cyclic loads.

ECCS (European Convention for Constructional Steelwork) (1988). European Recommendations for Steel Structures in Seismic Zones.

Faella C., Mazzarella O. and Piluso V. (1993). L'influenza della non-linearità geometrica sul danneggiamento strutturale. In *Proceedings of the VI Convegno Nazionale "L'ingegneria sismica in Italia"*, Perugia, 13-15 Ottobre 1993.

Faella C., Piluso V. and Rizzano G. (1994). Connection influence on the seismic behaviour of steel frames. In *Proceedings of International Workshop and Seminar on Behaviour of Steel Structures in Seismic Areas, STESSA 94*, Timisoara, Romania, F.M. Mazzolani, and V. Gioncu editors, published by E & FN Spon, 26 June-1 July.

Faggiano B. and Mazzolani F.M. (1999). Interpretazione del criterio di gerarchia nel dimensionamento dei telai in acciaio. In *Proceedings of the IX National Conference "L'Ingegneria sismica in Italia"*. Torino, Italy.

Ghersi A., Marino E. and Neri F., (1999). A simple procedure to design steel frames to fail in global mode. In *Proceedings of the Stability and Ductility of Steel Structures (SDSS) Conference 1999*, Timisoara, Romania, September.

Gioncu V. (1997). Ductility demands (General Report). In *Proceedings of the STESSA 97 Conference on "Behaviour of Steel Structures in Seismic Areas"*, Kyoto, F.'M. Mazzolani & H. Akiyama editors, published by 10/17 Salerno.

Gioncu V., Mazzolani F. M. (1995). Alternative methods for assessing local ductility In *Proceedings of the 1st International Workshop on "Behaviour of Steel Structures in Seismic Areas" STESSA '94*, Timisoara, Romania, F.M. Mazzolani, and V. Gioncu editors, published by E & FN Spon, an Imprint of Chapman & Hall, London.

Giuffrè A. and Giannini R., (1982). La duttilità nelle strutture in cemento armato. In *ANCE-AIDIS*, Roma.

Guerra C. A., Mazzolani F. M., Piluso V. (1990a). On the seismic behaviour of irregular steel frames. In *Proceedings 9th ECEE*, Moscow.

Guerra C. A., Mazzolani F. M. and Piluso V. (1990b). Overall stability effects in steel structures. In *Proceedings of the 9th European Conference on Earthquake Engineering*, Moscow, 11-16 September.

Horne M.R. and Morris L.J., (1973). Optimum design of multi-storey rigid frames. In *Chapter 14 of "Optimum Structural Design - Theory and Application"*, edited by R.H. Gallagher and O.C. Zienkiewicz, Wiley.

Horne M.R. and Morris L.J., (1981). Plastic design of low-rise frames. In *Constrado, Collins Professional and technical books*, London.

Kato B. (1995). Development and design of seismic resistant steel structures in Japan. In *Proceedings of the 1st International Workshop on "Behaviour of Steel Structures in Seismic Areas" STESSA '94*, Timisoara, Romania, F.M. Mazzolani, and V. Gioncu editors, published by E & FN Spon, an Imprint of Chapman & Hall, London.

Kato B. and Akiyama, (1982). Seismic design of steel building. In *Journal of Structural Division, ASCE*, August.

Kishi N., Chen W.F., Goto Y. and Hasan R. (1996). Behaviour of tall buildings with mixed use of rigid and semi-rigid connections. In *Computer and Science*, No 6, Elsevier Science Ltd, 1993-1206.

Krawinkler H. (1995). Systems behaviour of structural steel frames subjected to earthquake ground motion. In *Background Reports SAC 95-09*.

Krawinkler H. and Nassar A.A., (1992). Seismic design based on ductility and cumulative damage demands and capacities. In *Non linear analysis and design of reinforced concrete buildings*, Eds. P. Fajfar and H. Krawinkler, Elsevier, London.

Kuwamura, H., Kato B. (1989). Effect of randomness in structural members' yield strength on the structural systems' ductility. *Journal of Constructional Steel Research*. N. 13.

Landolfo R., Mazzolani F. M. (1990). The consequence of design criteria on the seismic behaviour of steel frames. In *Proceedings 9th ECEE*, Moscow.

Lee H-S. (1996). Revised Rules for Concept of Strong-Column Weak-Girder Design. In *Journal of Structural Engineering*. Vol.122, No 4, April, 359-364.

Matsui C. and Sakai J. (1992). Effect of collapse modes on ductility of steel frames. In *Proceedings of the Tenth World Conference on Earthquake Engineering*. Balkema, Rotterdam.

Mazzolani F. M. (1988). The ECCS activity in the field of recommendations for steel seismic resistant structures. In *Proceedings of the 9th World Conference on Earthquake Engineering, WCEE*. Tokyo, Kyoto.

Mazzolani F. M. (1995a). Design of seismic resistant steel structures. In *Proceedings of the 10th ECEE*. Vienna 1994, published by Balkema, Rotterdam.

Mazzolani F. M. (1995b). Eurocode 8 - chapter "Steel": background and remarks. In *Proceedings of 10th European Conference on Earthquake Engineering, ECEE*. Vienna, 1994, published by Balkema, Rotterdam.

Mazzolani F. M. (1998). Design of steel structures in seismic regions: the paramount influence of connections. In *Proceedings of the COST C1 International Conference on "Control of the semi-rigid behaviour of civil engineering structural connections"*. Liege, September 17-18.

Mazzolani F.M. (1999). Principles of design of seismic resistant steel structures. In *Proceedings of the National Conference on Metal Structures*. Ljiubljana, May 20.

Mazzolani F.M. (edr., 2000). Moment resisting connections of steel frames in seismic areas: design and reliability. Published by E & FN SPON, London, in press.

Mazzolani F. M., Mele E. and Piluso V. (1990a). Statistical features of mechanical properties of structural steels. *ECCS Document TC13.26.90*.

Mazzolani F. M., Mele E. and Piluso V. (1990b). Analisi statistica del comportamento inelastico di telai in acciaio con resistenza casuale. In *Proceedings of the V Convegno Nazionale "L'Ingegneria Sismica in Italia"*. Palermo 29 Settembre-2 Ottobre.

Mazzolani F. M. and Piluso V. (1990). Skin-effect in pin-jointed steel structures. *Ingegneria sismica*. N. 3.

Mazzolani F. M., Mele E. and Piluso V. (1991). On the effect of randomness of yield strength in steel structures under seismic loads. *ECCS Document*. TC13.01.91.

Mazzolani F. M. and Piluso V. (1991). Influence of panel connecting system on the dynamic response of structures composed by frames and collaborating claddings. In *Proceedings of the Second International Workshop "Connections in steel structures: behaviour, strength and design"*. Pittsburgh, Pennsylvania, April, 10-12.

Mazzolani F. M., Mele E. and Piluso V. (1992). The seismic behaviour of steel frames with random material variability. In *Proceedings of the X World Conference on Earthquake Engineering*. Madrid, July.

Mazzolani F. M., Piluso V. (1993a). Member behavioural classes of steel beams and beam-columns. In *Proceedings of the XIV CTA Congress*, Viareggio.

Mazzolani F. M., and Piluso V. (1993b). Failure mode and ductility control of seismic resistant MR frames. In *Proceedings of the XIV CTA Congress*, Viareggio.

Mazzolani F. M., Piluso V. (1993c). P-Δ effect in seismic resistant steel structures. In *Proceedings of SSRC Annual Technical Session & Meeting*, Milwaukee, April.

Mazzolani F. M., Piluso V. (1995a). Seismic design criteria for moment resisting steel frames. In *Proceedings of the 1st European Conference on Steel Structures, EUROSTEEL*. Athens, published in "Steel Structures", (editor Kounadis, Balkema).

Mazzolani F. M., Piluso V. (1995b). An attempt of codification of semi-rigidity for seismic resistant steel structures. In *Proceedings 3rd International Workshop on Connections in Steel Structures*, Trento, Pergamon 1996.

Mazzolani F.M. and Piluso V. (1995c). A new method to design steel frames failing in global mode including P-Δ effects. In *Proceedings of the 1st International Workshop on "Behaviour of Steel Structures in Seismic Areas" STESSA '94*, Timisoara, Romania, F.M. Mazzolani, and V. Gioncu editors, published by E & FN SPON, an Imprint of Chapman & Hall, London.

Mazzolani, F.M. and Piluso, V. (1996a). *Theory and Design of Seismic Resistant Steel Frames*. E & FN Spon, an Imprint of Chapman & Hall, London.

Mazzolani F. M., and Piluso V. (1996b). Stability issues in seismic design of rigid and semirigid frames. *SSRC Technical Meeting*. Chicago, April.

Mazzolani F. M., and Piluso V. (1996c). Behaviour and design of set-back steel frames. In *Proceedings of the European Workshop on the Seismic Behaviour of Asymmetric and Setback Structures*. Anacapri, Isle of Capri, Italy, 4-5 October.

Mazzolani F. M., Piluso V. (1997a). Plastic Design of Seismic Resistant Steel Frames. In *Earthquake Engineering and Structural Dynamics*, Vol. 26, 167-191.

Mazzolani, F.M. and Piluso, V. (1997b). The influence of design configuration in the seismic response of moment-resisting frames. In *Proceedings of the 2st International Workshop on "Behaviour of Steel Structures in Seismic Areas"*, Kyoto, Mazzolani F. M., Akiyama H. (editors), 10/17, Salerno.

Mazzolani F. M., Piluso V. (1997c). Review of code provisions for vertical irregularity. In *Proceedings of the STESSA 97 Conference on "Behaviour of Steel Structures in Seismic Areas"*, Kyoto. F. M. Mazzolani and H. Akiyama editors, published by 10/17, Salerno.

Mazzolani F. M., Piluso V., Rizzano G. (1998). Design of full-strength extended end-plate joints for random material variability. In *Proceedings of COST C1 Congress*, Liège.

Newmark N.M. and Hall J.W., (1973). Procedures and criteria for earthquake resistant design. In *Building practice for disaster mitigation, Building science series 45, National Bureau of Standards*, Washington, pp. 90-103, Feb.1973.

Mazzolani F. M. and Sylos Labini F. (1984). Skin-frame interaction in seismic resistant steel structures. *Costruzioni Metalliche*, N. 4.

Palazzo B. and Fraternali F., (1987a). L'uso degli spettri di collasso nell'analisi sismica: proposta per una diversa formulazione del coefficiente di struttura. In *3° Convegno nazionale "L'ingegneria sismica in Italia"*, Roma.

Palazzo B. and Fraternali F., (1987b). L'influenza dell'effetto P-Δ sulla risposta sismica dei sistemi a comportamento elasto-plastico: proposta di una diversa formulazione del coefficiente di struttura. In *Giornate italiane delle costruzioni in acciaio, C.T.A.*, Trieste.

SAC 96-03 (1997). Interim guidelines. FEMA 267/A, SAC Join Venture, California, USA.

Sanpaolesi L., Biolzi L. and Tacchi R. (1983). Indagine sperimentale sul contributo irrigidente di pannelli in lamiera grecata. In *Proceedings of the IX CTA Congress*, Perugia, Italy.

Sanpaolesi L. (1984). Indagine sperimentale sulla resistenza e duttilità di pannelli-parete in lamiera grecata. *Italsider, Quaderno Tecnico N. 7*.

Setti P. (1985). Un metodo per la determinazione del coefficiente di struttura per le costruzioni metalliche in zona sismica. In *Costruzioni metalliche*, No 3.

Sedlacek G. and Kuck J., (1993). Determination of q-factors for Eurocode 8. In *Aachen den* 31.8.1993.

Shen J. (1996). A new dual system for seismic design of steel buildings. In *Proceedings of Advanced in Steel Structures*, ICSASS'96, Hong Kong, Vol. 2, Edited by S.L. Chan and J.G. Teng, Pergamon, Elsevier Science Ltd, 1027-1033.

UBC (Uniform Building Code, 1997). Structural Engineering Design Provisions. Volume 2.

CHAPTER 5

DESIGN OF BRACED FRAMES

I. Vayas
National Technical University of Athens, Athens, Greece

1. INTRODUCTION

As well known, structures in seismic regions are designed so that:
- collapse is prevented during strong rare earthquakes and
- the extend of damage is limited during moderate, frequent earthquakes.

Seismic building Codes introduce two limit states beyond which the structure no longer satisfies the design performance requirements, an ultimate limit state associated with no collapse and a serviceability limit state associated with damage limitation.

The criteria set up in the above limit states lead to a design of the structure for stiffness and strength. The former ensures the limitation of deformations that may cause damage to non-structural elements and possibly high 2nd order effects. The latter ensures that the structure is able to resist the applied seismic action effects. The provision of stiffness and strength is sufficient for the design of structures that are supposed to remain elastic when subjected to seismic actions. However, during strong earthquakes specific structural parts are allowed to yield in order to dissipate part of the energy that is input in the structure. This results in the development of smaller applied action effects as compared to an elastic response, which may lead to a more economic design. Structural parts that are supposed to yield as well as the entire structure shall exhibit, when cyclic loaded, stable hysteretic behaviour without considerable reduction in stiffness and strength. The relevant criteria are therefore extended to a design for ductility under cyclic loading conditions, in addition to stiffness and strength.

During the conceptual design of building structures appropriate systems for the provision of lateral stability must be selected. For steel structures those systems are usually composed of moment resisting frames, concentrically braced frames, eccentrically braced frames, reinforced concrete or composite steel-concrete walls and cores, or combinations of the above. Such structural systems, appropriately positioned in layout and connected through the floor diaphragms ensure the overall stability of the entire building frame.

The purpose of this paper is to present appropriate design criteria for earthquake resistant braced frames. Due to the large number of bracing configurations, the presentation of the material is oriented towards the design criteria rather than the structural systems. This allows a better comparison between the characteristics of the various bracing systems in respect to each criterion. Accordingly, the paper includes chapters on the design for stiffness, strength and ductility. Reference is made to in-plane response. Three-dimensional effects are not covered in this paper.

In order to justify the proposed rules, an effort was made for the analytical derivation of the design expressions as far as possible. The proposed expressions don't necessarily coincide with corresponding rules of a specific design Code. However, the methodology adopted, follows to a large extend the Eurocodes, especially Eurocodes 3 and 8. The notation follows also the Eurocodes as far as possible. Analytical expressions are compared to corresponding provisions from Codes in order to detect any deviations. The material concentrates on the main issues and doesn't cover all aspects of a seismic design. Obviously, it is not intended to serve as a substitute of a design Code, but as a tool for a better understanding of the structural behaviour and the design targets.

2. NOTATION

A = area
A_{net} = net area at bolt holes
A_v = shear area
A_{vb} = shear area of the beam section
D = damage index
E = modulus of elasticity
$F_{b,Rd}$ = bearing resistance of bolts
$F_{v,Rd}$ = shear resistance of bolts
$F_{s,Rd}$ = slip resistance of bolts
G = shear modulus
H = overall height of frame
M = bending moment
M_{NRd} = moment resistance allowing for axial forces
M_{VRd} = moment resistance allowing for shear forces
N = axial force
N = number of cycles
N_{cRd} = compression resistance
N_{tRd1} = tension yield resistance
N_{tRd2} = tension tearing resistance at bolt holes
I = second moment of area
I_b = second moment of area of the beam
I_B = second moment of area at the base
I_d = second moment of area of the diagonal (brace)
S_v = shear stiffness
S_{vB} = shear stiffness at the base
V = shear force
V_{pl} = plastic shear resistance

α = angle between a diagonal and a column
α^* = atan(h_0/a)
α = aspect ratio of a plate panel
α_{CD} = overstrength parameter
γ = link rotation angle
δ = lateral displacement
δ_e = lateral displacement from elastic analysis
ε = e/h_0 relative eccentricity
$\varepsilon = \sqrt{235 / f_y}$, f_y in Mpa
ε_y = yield strain
ε_u = ultimate strain
$\xi = x / H$
λ = slenderness
$\overline{\lambda} = \sqrt{f_y / \sigma_{cr}}$ = relative slenderness
ν = Poisson ratio
σ_{cr} = critical buckling stress
φ = shear deformation
$\varphi = r\,(1-\varepsilon) + \varepsilon$ = geometric and stiffness parameter
χ = reduction factor for flexural buckling
χ_V = reduction factor for shear buckling
ψ = ratio of end-moments

a = storey height
d = length of the diagonal, depth of a beam
e = eccentricity, length of link
e_v = vertical eccentricity
f_y = yield strength, nominal value
f_{yr} = yield strength, actual value

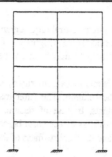

Figure 1: Moment resisting frames

Concentric braced frames are braced by linear bracing elements, connected without eccentricity to the adjacent members. The frames are characterized by the geometrical configuration of the bracing system as following (Figure 2):

- Diagonally braced frames (Figure 2a)
- X- braced frames (Figure 2b)
- V- or inverted V- braced frames (Figures 2c and d)
- K- braced frames (Figure 2e)

The braces may be placed within one bay and storey or extend over more bays or storeys of the structure (Figures 2b and f). Concentric braced frames respond to lateral loads primarily by axial forces in the braces and the beams and to overturning moments by axial forces in the columns. Their lateral stiffness is accordingly provided by the axial stiffness of the braces, the beams and the columns.

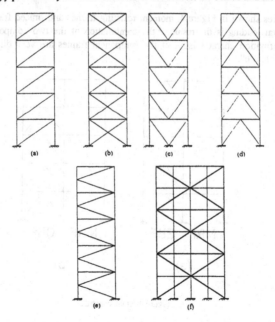

Figure 2: Concentric braced frames

Eccentrically braced frames are characterized by the eccentric connection of the braces to the beams and the columns (Figure 3). The short part of the beam between braces or braces and columns is called **shear link**, due to the fact that high shear is developed in it during loading. The position of the braces may vary in connection to the static and architectural requirements, the most usual configurations being shown in Figure 3. Shear links may be horizontal (Figures 3 a, b, c) or vertical (Figure 3d). For the most usual case of horizontal links, distinction may be made between inverted V-, D- and V- braced frames.

Eccentric braced frames respond to lateral loads primarily by shear and bending in the shear links and axial deformations in the braces and the columns. The latter may be subjected to bending, if they are rigidly connected to the adjacent elements. In case of seismic loading, inelastic action is essentially limited in the shear links. The lateral stiffness of eccentric braced frames is primarily provided by the bending and shear stiffness of the links and the axial stiffness of the braces, the beams and the columns.

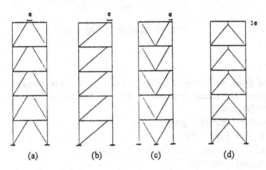

Figure 3: Eccentrically inverted V-, D- and V- braced frames (a to c) and d) eccentrically braced frames with vertical shear links

In **dual structures** shown in Figure 4, moment resisting frames and braced frames act jointly. Their response to lateral loading is the result of the combination of the two component frames. As well known, the deformation characteristics of the component frames are very different and justify the notation "dual structure".

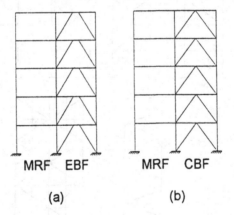

Figure 4: Dual structures

Infilled moment resisting frames shown in Figure 5 are characterized as **mixed structures.** The infill material may be of concrete, of steel panels, of composite panels from steel and concrete or of masonry walls. As before, the response to lateral loading is the result of the combination of the two structural components, the moment resisting frame and the infill material.

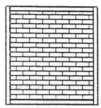 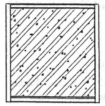

Figure 5: Mixed structures

Concluding, five typologies of frames in connection with their response to lateral loading may be distinguished:

+ Moment resisting frames
+ Concentric braced frames
+ Eccentric braced frames
+ Dual structures (or dual frames)
+ Mixed structures (or mixed frames)

It is possible to laterally stabilize an unstable frame by a separate frame as illustrated in Figure 6. In this Figure the unstable frame is designated as frame A, the stable frame as frame B. The stable frame B may be of any of the above referred frame types, a concrete shear wall, a concrete core etc.

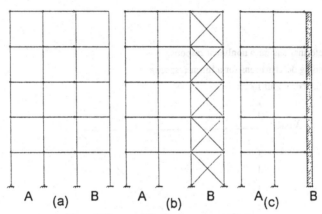

Figure 6: Frame stabilization by a separate frame

Sometimes there is the need to enhance the lateral stiffness of a stable frame. Figure 7 illustrates some examples in which the softer frame is designated as frame A and the stiffer as frame B.

3.2 Classification of frames according to Eurocode 3

The classification of frames as proposed by Eurocode 3 (1992) differs from the one given before. This will be discussed in the following in order to avoid misunderstandings, as this Code is used as a basic document in this chapter. The definition for braced frames according to Eurocode 3 is as following:

"a frame may be classified as braced if its sway resistance is supplied by a bracing system with a response to in-plane horizontal loads which is sufficiently stiff for it to be acceptable accurate to assume that all horizontal loads are resisted by the bracing system".

4. DESIGN FOR STIFFNESS

4.1 Introduction

Frames subjected to seismic loading are designed for stiffness in order to limit their lateral deformations. The limitation refers to both the serviceability and the ultimate limit states, it applies therefore for moderate and strong earthquakes. In the serviceability limit state lateral deformations have to be controlled primarily in order to limit the damage of non-structural elements. In the ultimate limit state deformations must be limited in order to avoid instabilities due to second order effects.

The structural response to lateral loads may be represented by the base shear – top displacement curve as shown in Figure 9. The initial slope of the curve expresses the elastic stiffness of the structure. For the serviceability limit state, the lateral deformations calculated on the basis of the elastic stiffness correspond to the actual deformations of the structure, as the structural response is expected to be nearly elastic in the event of moderate earthquakes. However, the actual deformations at the ultimate limit state are larger than those calculated on the assumption of an elastic structural response. This is due to the nonlinear behaviour of the structure, parts of which yield during strong earthquakes. However, in order to avoid nonlinear analysis, most seismic Codes, as Eurocode 8 (1994), propose a simple relationship between the nonlinear elastic-plastic and the elastic deformations. The two types of deformations are related through the value of the behaviour factor assumed in the analysis as following:

$$\delta = q \cdot \delta_e \tag{1}$$

where:

δ = lateral deformations for a nonlinear response

δ_e = corresponding deformations for a linear response

q = value of the behaviour factor

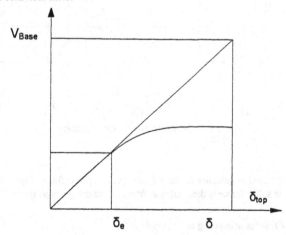

Figure 9: Base shear – top displacement curve of a frame

Although the above methodology constitutes an approximation of the real behaviour, it is evident that the knowledge of the elastic lateral stiffness of a frame is important in design. This stiffness may be determined by substitution of the frame by an equivalent beam having the same deformation properties. The equivalent beam is subjected to deformations due to bending moments and shear forces. It has correspondingly a bending stiffness, EI, and a shear stiffness, S_v. In the following, the two stiffness values of the equivalent beam representing various braced frame configurations will be determined.

4.2 Diagonally and X- braced frames

A diagonally braced frame, shown in Figure 10, behaves basically as a truss. Bending moments are resisted by axial deformations of the columns acting as flanges. The bending stiffness of the frame may be accordingly determined from:

$$EI = E \cdot A_c \cdot h_0^2 / 2 \tag{2}$$

where:

A_c = area of the cross section of one column

h_0 = width of the frame (= distance between gravity centers of the columns)

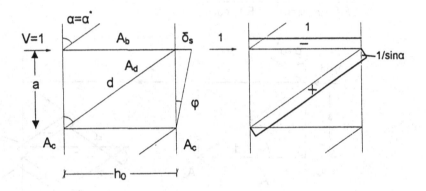

Figure 10: Notation and forces in diagonally braced frames

For the evaluation of the shear stiffness, a unit transverse force V=1 is applied and the forces in the members are determined, as shown in Figure 10. The resulting lateral deformation is determined by means of eq. (3):

$$\delta_s = \Sigma \int \frac{N^2}{EA} dx \tag{3}$$

where:

N = axial forces in the truss members due to the unit force

A = areas of the corresponding cross sections

The application of eq. (3) to the specific frame of Figure 10 gives:

$$\delta_s = \frac{d}{(\sin \alpha)^2 \, EA_d} + \frac{h_0}{EA_b} \tag{4}$$

The shear deformation, angle of sway, of the truss is given by:

$$\varphi = \delta_s / a \tag{5}$$

The shear stiffness is determined from eq. (6):

$$S_v = V / \varphi = 1/\varphi \tag{6}$$

Substituting (4) and (5) in (6) and eliminating the lengths of the members leads to following expression for the shear stiffness:

$$S_v = \cfrac{1}{\cfrac{1}{EA_d \sin^2 \alpha \cos \alpha} + \cfrac{1}{EA_b \cot \alpha}} \tag{7}$$

For X-braced frames where two diagonal participate in the force transfer (Fig. 11), the relevant expression may be written as:

$$S_v = \cfrac{1}{\cfrac{1}{2EA_d \sin^2 \alpha \cos \alpha} + \cfrac{1}{EA_b \cot \alpha}} \tag{8}$$

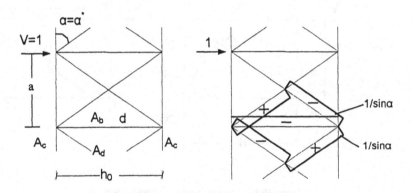

Figure 11: Notation and forces in X-braced frames

However, if in X-braced frames the contribution of the compression diagonal due to buckling is neglected, eq. (7) instead of (8) shall be used.

The variation of the non-dimensional shear stiffness for X-braced frames, divided by the axial stiffness EA_d of one diagonal is presented as a function of the angle between the diagonal and the column in Figure 12. The ratio ρ between the cross sections of the diagonal and the beam serves as a parameter. It may be seen that the optimum stiffness is reached for angles between 40^0 and 50^0. As expected, the shear stiffness is more influenced by the axial stiffness of the diagonals than the stiffness of the columns. In effect, if the area of the beams is doubled, the shear stiffness is increasing by around 20%, while if the area of the diagonals is doubled the increase accounts for 80%. The diagrams of Figures 12 may be also used for diagonally braced frames. In such a case half of the area of the diagonal shall be used in the determination of the parameter ρ.

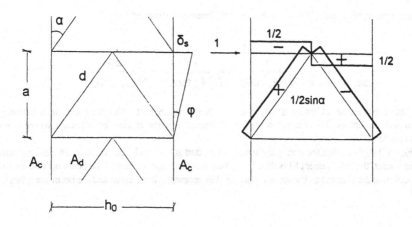

Figure 13: Notation and forces in inverted V- braced frames

4.4 Eccentrically inverted V-braced frames

The bending stiffness is given by eq. (2) as before. For the determination of the shear stiffness, it is assumed that the braces are simply connected to the beam. The distribution of forces and moments in the members of the frame due to the application of a unit force V = 1 is shown in Fig. 14.

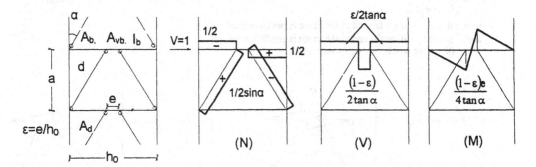

Figure 14: Notation and forces in inverted V eccentrically braced frames

The lateral deformations may be determined by application of eq. (11):

$$\delta_s = \Sigma \int \frac{N^2}{EA} dx + \Sigma \int \frac{V^2}{EA_v} dx + \Sigma \int \frac{M^2}{EI} dx \qquad (11)$$

The terms in eq. (11) represent the contribution of axial forces, shear forces and bending moments respectively. The deformation is then calculated as:

$$\delta_s = \frac{h_0 - e}{2^2 EA_b} + \frac{2d}{(2\sin\alpha)^2 EA_d} + \frac{(h_0 - e)e^2}{(2h_0 \tan\alpha)^2 GA_{vb}} + \frac{e(1 - e/h_0)^2}{(2\tan\alpha)^2 GA_{vb}} + \frac{2(1 - e/h_0)^3 e^2}{3 \, 2 \, (4\tan\alpha)^2 EI_b} + \frac{(1 - e/h_0)^2 e^3}{3(4\tan\alpha)^2 EI_b}$$

$$(12)$$

By application of (5) and (6) in (12), the shear stiffness is determined from:

$$S_v = \cfrac{1}{\cfrac{1}{2EA_d \sin^2 \alpha \cos \alpha} + \cfrac{1}{2EA_b \cot \alpha} + \cfrac{\varepsilon}{2GA_{bv} \tan \alpha} + \cfrac{(1-\varepsilon)e^2}{24EI_b \tan \alpha}} \qquad (13)$$

The above expression differs from (10) in the last two terms that introduce the eccentricity between the beam and the braces. These terms obviously disappear in case of zero eccentricity, where $\varepsilon = e/h_0$ becomes zero.

Eq. (13) applies when the braces are simply connected at their ends. In case they are rigidly connected to the beams, the link moment M is distributed between them and the part of the beam outside the link in proportion to the stiffness as shown in Figure 15. The moments in the brace and the beam are given in eq. (14).

$$M_d = M \frac{k_d}{k_d + k_b} \qquad (14a)$$

$$M_b = M \frac{k_b}{k_d + k_b} \qquad (14b)$$

The stiffness of the braces and the beam may be written as:

$$k_d = \mu \frac{I_d}{d} \qquad (15a)$$

$$k_b = \frac{I_b}{(h_0 - e)/2} \qquad (15b)$$

where $\mu = 1$ if the far end of the brace is pin connected
and $\mu = 4/3$ if the far end of the brace is rigidly connected.
     ($\mu = 0$ for doubly pinned connections)

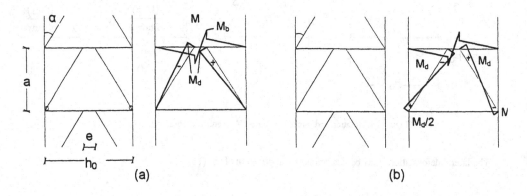

Figure 15: Distribution of link moment for rigid connection of the braces

Due to the new moment distribution, eq. (12) has to be modified in order to include terms due to bending of the brace that reduce the contribution of bending of the beam. Considering modifications only due to moment redistribution and not shear, the new shear stiffness may be written as:

$$S_v = \cfrac{1}{\cfrac{1}{2EA_d \sin^2 \alpha \cos \alpha} + \cfrac{1}{2EA_b \cot \alpha} + \cfrac{\varepsilon}{2GA_{bv} \tan \alpha} + \phi \cfrac{(1-\varepsilon)e^2}{24EI_b \tan \alpha}} \tag{16}$$

φ is a reduction factor determined from:

$$\varphi = r \cdot (1 - \varepsilon) + \varepsilon \tag{17a}$$

$$r = \cfrac{1}{1 + \mu \cdot i \cfrac{(h_0 - e)/2}{d}} = \cfrac{1}{1 + \mu \cdot i \cdot \sin \alpha} \tag{17b}$$

$\mu = 0$ if the braces are pin connected at their ends
$\mu = 1$ if the braces are rigidly connected to the beam and pin connected at the far end
$\mu = 4/3$ if the braces are rigidly connected at their ends
$i = I_d / I_b$

Eq. (16) is more general than (13). For pin connected braces, the parameters r and φ are equal to unity and eq. (16) is identical to (13).

Assuming that the height of the beam is equal to h_b and that the beam consists of a double-flange cross section so that:

$$I_b = A_b \cdot h_b^2 / 4 \tag{18a}$$

new non-dimensional parameters may be introduced as following :

$$A_{bv} / A_b = 0,8 \tag{18b}$$

$$\lambda = h_0 / h_b \tag{18c}$$

$$\rho = A_d / A_b \tag{18d}$$

When the connection of the beam to the columns is rigid, the frame behaves as a dual frame. Its elastic stiffness is determined under consideration of the rigidities of the moment resisting frame and the eccentrically braced frame with pin connected beams. The moment frame action will develop at higher deformation stages due to the fact that the braced frame is usually stiffer and will initially resist most of the applied actions.

Figure 16 shows the variation of this stiffness in dependence on various parameters. For a direct comparison between eccentrically and X- braced frames, the angle α^* of the diagonal is introduced. Low angles correspond to slender frames, high angles to wide frames. For $\alpha^* = 45^0$ the height and the width of the frame are equal. Figure 16a shows that the optimum angle α^* is larger than for X- braced frames, varying between 65^0 and 70^0. This corresponds to a brace angle α between 45^0 and 55^0, similar as for X-braced frames.

The influence of the eccentricity is illustrated in Figure 16b. The Figure shows that the stiffness reduces almost linear with increasing eccentricity. However, the reduction is a function of the angle α^*. It takes its largest values for angles α^* between 60^0 and 70^0, i.e in the range of the optimum values. At higher angles, i.e. as the frame becomes stockier, the eccentricity has almost no effect on the stiffness. This is due to the fact that the transverse force that causes bending in the beam becomes smaller.

Figures 16 c and d show the variation of the stiffness as a function of the parameters μ and λ. The invariability of the shear stiffness in respect to μ (Figure 17c) indicates that the type of connection of the braces does not influence the elastic global frame behaviour. A little more affected is the stiffness by the parameter $\lambda = h_0 / h_b$, which, under the assumption made above of a beam with double flange cross section, is expressing the stiffness of the beam. As expected (Figure 16d), the shear stiffness reduces with a reduction in stiffness of the beam, i.e. for higher values of λ.

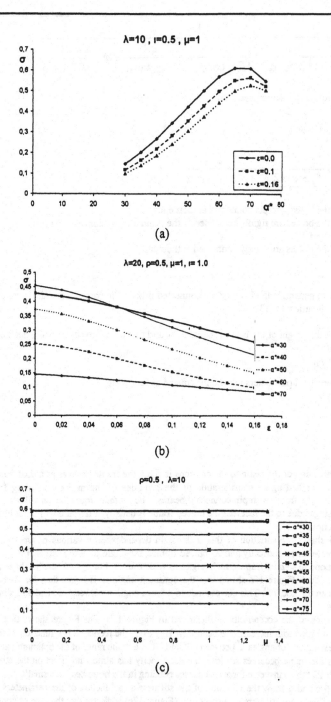

Figure 16: Shear stiffness $\sigma = S_v / (E \cdot A_d)$ of eccentrically inverted V-braced frames

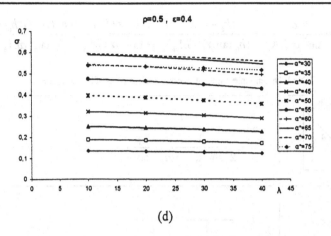

(d)

Figure 16 (continued): Shear stiffness $\sigma = S_v /(E \cdot A_d)$ of eccentrically inverted V-braced frames

4.5 Eccentrically D-braced frames

The bending stiffness is given by eq. (2) as for the other types of frames. For the determination of the shear stiffness, it is assumed in a first step that the beam and the braces are simply connected at their ends. The distribution of forces and moments in the members due to a unit force $V = 1$ is shown in Figure 17.

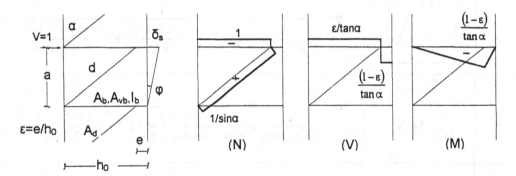

Figure 17: Forces and moments in an eccentrically D-braced frame

The horizontal displacement is determined by application of eq. (10) from:

$$\delta = \frac{h_0 - e}{EA_b} + \frac{d}{\sin^2 \alpha\, EA_d} + \frac{(h_0 - e)e^2}{(h_0 \tan \alpha)^2 GA_{vb}} + \frac{e(1 - e/h_0)^2}{\tan^2 \alpha\, GA_{vb}} + \frac{e^2(1 - \varepsilon)^2 h_0}{3\ \tan^2 \alpha EI_b} \qquad (19)$$

However, the beam to column connection at the near end must be rigid to allow for a link action. This moment is transferred to the column, which is subjected to bending in addition to axial forces. Assuming that the moments at the link ends are, as usually the case, equal and neglecting for the sake of simplicity the influence of the column stiffness, following expression for the lateral deformation is derived:

$$\delta_s = \frac{h_0 - e}{EA_b} + \frac{d}{\sin^2 \alpha \, EA_d} + \frac{(h_0 - e)e^2}{(h_0 \tan \alpha)^2 GA_{vb}} + \frac{e(1 - e/h_0)^2}{\psi \tan^2 \alpha \, GA_{vb}} + \frac{e^2(1-\varepsilon)^2 h_0 \psi}{3 \, \tan^2 \alpha EI_b} \tag{20}$$

Table 1: Dimensionless shear stiffness of various types of braced frames

1	A_b, h_0, a, d, $\alpha = \alpha^*$	$\sigma = \dfrac{1}{\dfrac{1}{2 \cdot \sin^2 \alpha \cdot \cos\alpha} + \dfrac{\rho}{\cot\alpha}}$
2	A_d, $\alpha = \alpha^*$	$\sigma = \dfrac{1}{\dfrac{1}{\sin^2 \alpha \cdot \cos\alpha} + \dfrac{\rho}{\cot\alpha}}$
3	A_d, α	$\sigma = \dfrac{1}{\dfrac{1}{2 \cdot \sin^2 \alpha \cdot \cos\alpha} + \dfrac{\rho}{2\cot\alpha}}$
4	e, α	$\sigma = \dfrac{1}{\dfrac{1}{2 \cdot \sin^2 \alpha \cdot \cos\alpha} + \dfrac{\rho}{2 \cdot \cot\alpha} + \dfrac{2.6 \cdot \varepsilon \cdot \rho}{2 \cdot 0.8 \cdot \tan\alpha} + \varphi \dfrac{(1-\varepsilon) \cdot \varepsilon^2 \cdot 4 \cdot \lambda^2 \cdot \rho}{24 \cdot \tan\alpha}}$
5	A_b, I_b, e, A_d, I_d, α	$\sigma = \dfrac{1}{\dfrac{1}{\sin^2 \alpha \cdot \cos\alpha} + \dfrac{\rho}{\cot\alpha} + \dfrac{2.6 \cdot \varepsilon \cdot \rho}{0.8 \cdot \tan\alpha \cdot \psi} + \varphi \cdot \psi \dfrac{(1-\varepsilon) \cdot \varepsilon^2 \cdot 4 \cdot \lambda^2 \cdot \rho}{3 \cdot \tan\alpha}}$
6		$\sigma = \dfrac{3 \cdot \tan\alpha}{\rho \cdot \lambda^2}$

structures 1,2,3,6: $\alpha = \alpha^*$

structure 4: $\alpha = \arctan\left(\dfrac{1}{2}\tan\alpha^*\right)$

structure 5: $\alpha = \arctan\left((1-\varepsilon) \cdot \tan\alpha^*\right)$

$\sigma = \dfrac{S_V}{E \cdot A_d}$, $\varepsilon = \dfrac{e}{h_0}$, $\rho = \dfrac{A_d}{A_b}$

$\lambda = \dfrac{h_0}{h_b}$, $i = \dfrac{I_d}{I_b}$, $\mu = 0$ or 1 or 1.33

$\psi = 1 - \varepsilon$

where:

$$\psi = 1 - \varepsilon \tag{21}$$

In case that the brace is rigidly connected at its ends, the link moment is distributed between the part of the beam outside the link and the diagonal. This contribution may be analogously accounted for as for V-braced frames. After some numerical manipulation, the shear stiffness may be calculated as following:

$$S_v = \cfrac{1}{\cfrac{1}{EA_d \sin^2 \alpha \cos \alpha} + \cfrac{1}{EA_b \cot \alpha} + \cfrac{\varepsilon}{GA_{bv} \tan \alpha \psi} + \phi \psi \cfrac{(1-\varepsilon)e^2}{3EI_b \tan \alpha}} \tag{22}$$

The parameter φ is determined from eq. (17) but r from eq. (23). The parameter μ is taken as before equal to 0, 1 or 4/3 in dependence on the supporting conditions of the diagonal.

$$r = \cfrac{1}{1 + \mu \cdot i \cfrac{h_0 - e}{d}} = \cfrac{1}{1 + \mu \cdot i \cdot \sin \alpha} \tag{23}$$

The results are illustrated again as a function of the angle α^* of the diagonal in Figure 18. The optimum angle is between 55^0 and 65^0, i.e. a little higher than for X-braced frames (Figure 18a). The negative influence of the eccentricity is more pronounced than in inverted V-braced frames, especially for low angles α^* where the transverse force to the beam is large (Figure 18b). The influence of μ and λ, not presented here, is like for inverted V-braced frames of lower importance.

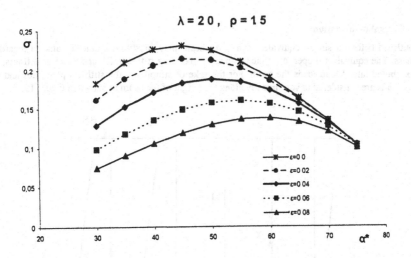

Figure 18: Shear stiffness $\sigma = S_v /(E \cdot A_d)$ of eccentrically D-braced frames

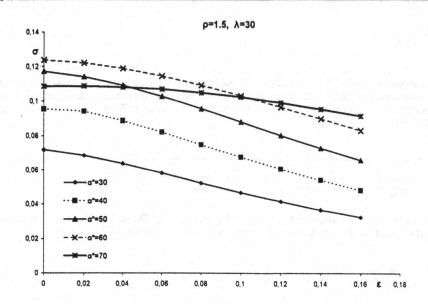

Figure 18(continued): Shear stiffness $\sigma = S_v /(E \cdot A_d)$ of eccentrically D-braced frames

Table 1 summarizes the expressions for the non-dimensional shear stiffness for all types of braced frames.

4.6 Lateral deformations

As outlined before, a simple equivalent cantilever beam may substitute a braced frame in regard to stiffness. The equivalent properties of the beam are its bending stiffness, EI, and its shear stiffness, S_v. A distributed lateral load loads the beam. For the sake of simplicity, the stiffness properties and the lateral loads are considered to vary linearly along the height of the beam as shown in Figure 19.

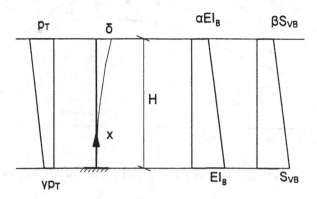

Figure 19: Equivalent cantilever beam

The relevant expressions for the stiffness properties and the lateral loading may be written as:

$$p(\xi) = p_T (\gamma + \xi - \gamma \cdot \xi) \tag{24a}$$

$$EI(\xi) = EI_B (1 - \xi + \alpha \cdot \xi) \tag{24b}$$

$$S_{vB}(\xi) = S_{vB}(1 - \xi + \beta \cdot \xi) \tag{24c}$$

where $\xi = x / H$

The lateral deformations may be determined by application of the force method. The shear forces and bending moments due to the external loads and respectively a unit horizontal force at the position x_0 where the displacement is to be determined are equal to:

- Due to external loads:

Shear force: $$V(\xi) = \frac{1}{2} p_T \cdot (1 + \gamma + \xi - \gamma \cdot \xi) \cdot (1 - \xi) \cdot H \tag{25}$$

Bending moment: $$M(\xi) = \frac{1}{6} p_T \cdot (2 + \gamma + \xi - \gamma \cdot \xi) \cdot (1 - \xi)^2 \cdot H^2 \tag{26}$$

- Due to a unit force at the position $\xi_0 = x_0 / H$:

Shear force: $$\overline{V}(\xi) = 0 \quad \text{if } \xi > \xi_0 \tag{27a}$$
$$\overline{V}(\xi) = 1 \quad \text{if } \xi \le \xi_0 \tag{27b}$$

Bending moment: $$\overline{M}(\xi) = 0 \quad \text{if } \xi > \xi_0 \tag{28a}$$
$$\overline{M}(\xi) = 1 \quad \text{if } \xi \le \xi_0 \tag{28b}$$

Using the force method, the lateral deformations at the position ξ_0 may be determined by appropriate integration by means of eq. (29).

$$\delta = \delta_b + \delta_s = \int_0^{x_0} \frac{M\overline{M}}{EI} dx + \int_0^{x_0} \frac{V\overline{V}}{S_v} dx \tag{29}$$

δ_b expresses the deformations due to bending moments, δ_s those due to shear forces. Substituting (25) to (28) in (29) and executing the integration. following expressions are found for the deformation at any position x:

$$\delta_b = \frac{H^4 \cdot p_T}{6 \cdot EI_B} D_B \tag{30a}$$

$$\delta_s = \frac{H^2 \cdot p_T}{2 \cdot S_{vB}} D_S \tag{30b}$$

where:

$$D_B = \frac{1}{12 \cdot (\alpha - 1)^5} \left\{ \begin{array}{l} (1 - a) \cdot \xi \cdot \{\alpha^2[-36 - 18(3 + 2\gamma) \cdot \xi + (2 + 16\gamma) \cdot \xi^2 + 3 \cdot (1 - \gamma) \cdot \xi^3] + \\ + \xi \cdot [-12 + 2\xi + \xi^2 - \gamma \cdot (6 - 4\xi + \xi^2)] + \\ + \alpha \cdot \xi \cdot [48 - 4\xi - 3\xi^2 + \gamma \cdot (24 - 14\xi + 3\xi^2)] + \\ + \alpha^3 \cdot [24 + 18\xi - \xi^3 + \gamma \cdot (12 + 18\xi - 6\xi^2 + \xi^3)]\} + \\ + 12\alpha^2 \cdot [-3 + \alpha \cdot (2 + \gamma)] \cdot [1 + (\alpha - 1) \cdot \xi] \cdot \log[1 + (\alpha - 1) \cdot \xi] \end{array} \right\} \tag{31a}$$

$$D_S = \frac{1}{2 \cdot (\beta - 1)^3} \left\{ \begin{array}{l} (\beta - 1) \cdot \xi \cdot [2 + \xi - \beta\xi + \gamma \cdot (2 - 4\beta - \xi + \beta\xi)] + \\ + 2\beta \cdot (-2 + \beta + \beta\gamma) \cdot \log[1 + (\beta - 1) \cdot \xi] \end{array} \right\} \tag{31b}$$

The total deformation may be written as:

$$\delta = \frac{p_T \cdot H^2}{2S_{vB}} \left(\rho_M \frac{D_B}{3} + D_S \right) \tag{32}$$

where:

$$\rho_M = \frac{S_{vB} \cdot H^2}{EI_B} \tag{33}$$

The base shear is equal to:

$$V_B = \frac{1}{2} p_T (1 + \gamma) \cdot H \tag{34}$$

Inserting eq. (34) in (32) the total lateral deformation may be written in a dimensionless form as:

$$\frac{\delta}{H} \cdot \frac{S_{vB}}{V_B} = \frac{1}{1 + \gamma} \cdot \left(\rho_M \frac{D_B}{3} + D_S \right) \tag{35}$$

Eqs. (33) and (35) indicate that for low values of ρ_M the shear deformations are prevailing, while for high values of ρ_M the bending deformations prevail.

The variation of the deformation along the height of the building for the case of triangular loading ($\gamma = 0$) is presented in Figure 20 for various values of the parameters ρ_M, α and β. It may be reminded that low values of the parameters α and β correspond to a reduction in stiffness, and therefore normally in strength too, along the height of the building. As expected, the lateral deformations are most influenced by the variation of the shear stiffness (parameter β) for buildings with prevailing shear deformations (Figure 20b) and by the variation of the bending stiffness (parameter α) in the opposite occurs (Figure 20c).

As well known, seismic Codes provide limitations for inter-storey drifts in order to limit the damage in non-structural elements, as well as 2nd order effects. Accordingly, an important design target is a uniform drift distribution over the height of the building, suggesting a linear variation of lateral deformations, as far as possible. Fig. 20a shows that for low values of ρ_M, where shear deformations prevail, the design target of linear variation of the lateral deformations is best achieved by a reduction in stiffness over the height of the structure. However, for increasing values of ρ_M better results are achieved for higher α and β values.

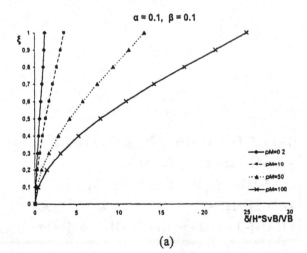

(a)

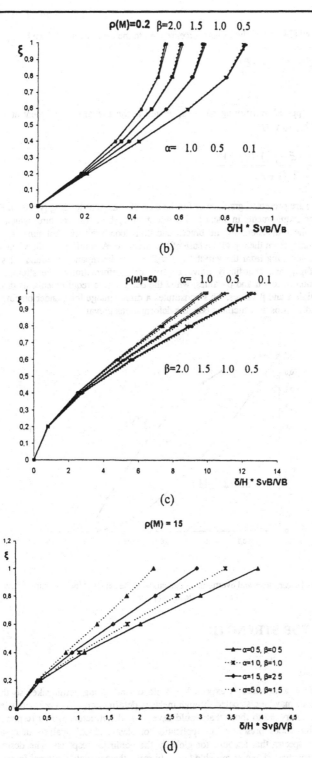

Figure 20: Lateral deformations over the height of the structure for triangular lateral loading (γ = 0)

Using eqs. (25) and (34) the applied shear force, related to the base shear, is given by eq. (36):

$$\frac{V}{V_B} = \frac{(1+\gamma+\xi-\gamma\xi)\cdot(1-\xi)}{1+\gamma} \tag{36}$$

Analogously, the applied overturning moment, related to the overturning moment at the base of the structure is given by eq. (37):

$$\frac{3M}{2V_B H} = \frac{(2+\gamma+\xi-\gamma\xi)\cdot(1-\xi)^2}{2\cdot(1+\gamma)} \tag{37}$$

Eqs. (36) and (37) are presented graphically for the case of triangular loading ($\gamma = 0$) in Figure 21. The curves express the requirements in shear and bending strength of the structure, which correspond to requirements for the dimensions of the braces and correspondingly the columns. As expected, the requirements decrease from the top to the base of the structure. Assuming that the variation of strength is not very much deviating from the variation in stiffness, a discrepancy is found for structures with high value of of ρ_M, and acordingly prevailing bending deformations. The strength requirements suggest the adoption of low α and β-values, while the deformation requirements, as discussed before, the adoption of high α and β-values. This constitutes a disadvantage for slender bracing systems with high height-to-width ratios, in which the bending deformations prevail.

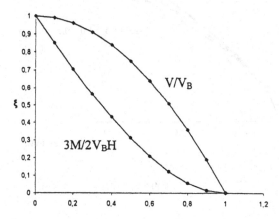

Figure 21: Applied shear forces and overturning moments over the height of the structure for triangular lateral loading ($\gamma = 0$)

5.   DESIGN FOR STRENGTH

5.1   Introduction

As discussed before, a structure is expected to yield during strong earthquakes so that part of the seismic input energy may be dissipated through inelastic deformations. This behaviour leads to smaller seismic forces as compared to those that would develop if the structure would respond in the elastic range. Most design codes prescribe the application of elastic global analysis in combination with inelastic response spectra that account for globally the nonlinear response. The design for strength ensures that the structural elements are able to safely resist the calculated internal forces and moments as determined from global analysis. The relevant design format may be written as:

$$S_d \leq R_d = R_k / \gamma_M \qquad\qquad (38)$$

where:

S_d = design action effects resulting from global analysis

R_d = corresponding design values for resistance

R_k = characteristic values for resistance

γ_M = partial safety factors for resistance

The condition (38) serves as a design criterion only for dissipative elements and modes of failure of the structure. Such elements are for the frames under consideration (Figure 22):

- the braces in concentrically braced frames and
- the shear links in eccentrically braced frames

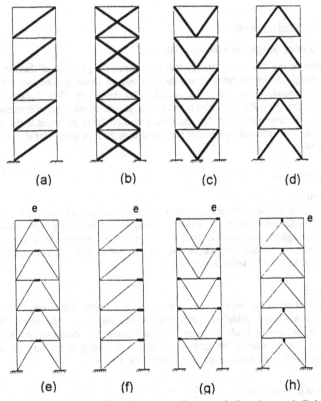

Figure 22: Dissipative (dark) and non-dissipative elements in concentrically and eccentrically braced frames

However, S_d represents the **minimum required** strength for dissipative elements. The **actual** strength R_u of these elements is much higher, mainly due to the following reasons:

- Due to standardization, other criteria, or because the most critical combination of actions is other than the one resulting from the seismic combination, the nominal value of the design strength R_d may be larger than S_d
- The nominal value of the characteristic strength R_k is larger than R_d as it is free from safety factors
- The actual value of the characteristic strength R_{kr} is larger than R_k, because the actual material strength is higher than its nominal value. This is always the case in steel structures, because the

actual yield strength of steel is always larger than the specified yield strength which is a minimum guaranteed value

- The ultimate maximal strength R_u is larger than R_k because of strain hardening and/or because the design formula for the calculation of the resistance is on the safe side

For all the above reasons, dissipative structural elements or modes of failure exhibit considerable overstrength in excess of their nominal design values. This overstrength will actually develop in a deformation-controlled situation, such as during a seismic event. However, not all elements shall exhibit inelastic behaviour. As shown in Figure 22, inelastic action shall be limited in certain elements and should be prevented in other elements, as it would possibly lead to non-ductile local or global structural response. Accordingly, the requirements and the relevant design criteria for non-dissipative elements are different from those of dissipative elements. These elements shall resist action effects not as determined by global analysis but as they arise when the adjacent dissipative elements develop their actual strength (their carrying capacity). This design methodology is called capacity design. The relevant criteria may be written as:

$$S_{cd} \leq R_d \tag{39}$$

where:

S_{cd} = capacity design action effects

R_d = corresponding design values for resistance

The capacity design serves as a strategy that shall ensure the provision of ductility for the overall system. Accordingly dissipative elements and ductile modes of failure are designed for strength, while non-dissipative elements or non-ductile modes of failure for overstrength. The relevant design criteria for strength and overstrength will be presented in this section. It must be noted that overstrength requirements refer actually to the provision of ductility and should be accordingly presented in the relevant section. However, they are both included for practical reasons here, as the right hand sides of inequalities (38) and (39) are identical.

5.2   X- braced frames

5.2.1   General

In X braced frames, horizontal shear forces are resisted by the diagonals and overturning moments by the columns. The dissipative elements are the diagonals, while columns and beams shall remain elastic. Therefore, the diagonals are verified for strength against the action effects as calculated by the structural analysis and detailed for ductility. The columns and the beams are designed for overstrength in order to resist the capacity design action effects.

5.2.2   Diagonals

As the direction of seismic action changes, the diagonals in X-braced frames are alternatively subjected to tension and compression. Additional secondary moments due to end restraint may be neglected, as the connections are usually made by means of gusset plates. For diagonals in tension, two failure modes are possible, yielding of the gross cross section or tearing of the net section. The relevant resistance formulae may be written according to Eurocode 3 as:

- Yielding of gross section: $N_{tRd1} = A f_y / \gamma_{M1}$ (40)
- Tearing of net section: $N_{tRd2} = 0,9 A_{net} f_u / \gamma_{M2}$ (41)

The relevant partial safety factors for resistance equal to $\gamma_{M1} = 1,10$ and $\gamma_{M2} = 1,25$.

As will be shown later, yielding shall precede tearing. Therefore, the tension resistance N_{tRd} shall be determined from eq. (40).

For diagonals in compression, two failure modes are possible, flexural buckling or distorsional buckling. For this reason, closed sections are preferable than open sections, as less prone to distorsional buckling, than open ones. For the same reason I-sections are more preferable than double angle sections. However, for the usual cross sections and slenderness values, flexural buckling is the most probable failure mode. The compression design resistance is then given by:

$$N_{cRd} = \chi A f_y / \gamma_{M1}$$
(42)

where:

χ = reduction factor for flexural buckling determined from the European buckling curves according to:

$$\chi = \frac{1}{\Phi + \sqrt{\Phi^2 - \overline{\lambda}^2}} \leq 1$$
(43a)

$$\Phi = 0,5[1 + \alpha(\overline{\lambda} - 0,2) + \overline{\lambda}^2]$$
(43b)

The imperfection parameter α corresponding to the appropriate European buckling curve is obtained from Table 2.

Table 2: Imperfection factor α

Buckling curve	a_0	a	b	c
α	0,13	0,21	0,34	0,49

The appropriate buckling curve may be selected as a function of the type of the cross section and the axis of buckling from the relevant Table of Eurocode 3. The relative slenderness $\overline{\lambda}_y = \dfrac{l_{0y}/i_y}{\lambda_1}$ and

$\overline{\lambda}_z = \dfrac{l_{0z}/i_z}{\lambda_1}$ (where $\lambda_1 = 93,9 \ \varepsilon$) as well as the reduction factors have to be determined for both strong

and weak axes buckling. The final value of χ is the lesser between χ_y and χ_z.

If the diagonals are not connected at their intersection, the buckling length for out-of- plane buckling may be conservatively taken equal to the system length. For in-plane buckling some restraint is provided by the gusset plates and the connections, so that the buckling length may be taken as equal to 0,9 of the system length.

If the diagonals are appropriately spliced at their intersection, they may be considered as continuous in that region. The buckling length for in-plane buckling may then be taken equal to the system length between the end support and the intersection, i.e. ½ of the total length. For out-of-plane buckling, it has to be considered that one diagonal is in tension and the other in compression. The tension diagonal supports the compression diagonal reducing its buckling length. Following the notation of Figure 23, the out-of-plane buckling length l_0 of the latter may be determined from:

$$l_0 = l \sqrt{\frac{1 - 0,75 \dfrac{N_{tSd} l}{N_{cSd} l_1}}{1 + \dfrac{I_1 l^3}{I l_1^3}}} \geq 0,5 \cdot l$$
(44)

If both diagonals have the same length ($l = l_1$), the same cross section ($I = I_1$) and are subjected to equal, and opposite, forces, the application of eq. (44) leads to $l_0 = 0,35 \cdot l < 0,5 \cdot l$. Therefore the out-of-plane buckling length shall be taken equal to ½ of the total length.

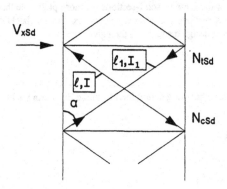

Figure 23: Buckling length for diagonals

As shown in Figure 24, the design storey shear resistance is equal to:

- $V_{xRd} = (N_{cRd} + N_{tRd}) \cdot \sin \alpha = (1 + \chi) \cdot A_d \cdot f_y \cdot \sin \alpha / \gamma_{M1}$ (45a)

if the design resistance of both diagonals is considered,

- $V_{xRd} = N_{tRd} \cdot \sin \alpha = A_d \cdot f_y \cdot \sin \alpha / \gamma_{M1}$ (45b)

if only the tension diagonal is considered as active,

- $V_{xRd} = 2 \cdot N_{cRd} \cdot \sin \alpha = 2 \cdot \chi \cdot A_d \cdot f_y \cdot \sin \alpha / \gamma_{M1}$ (45c)

if it is assumed that the external force is equally distributed to both diagonals.

In eq. (24), A_d is the area of the cross section of the diagonal and $\gamma_{M1} = 1.10$.

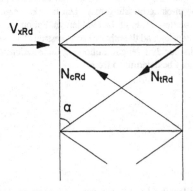

Figure 24: Forces in an X-braced frame

For a consistent design, the adoption of one or the other formula for resistance must be accompanied by a corresponding assumption in respect to the stiffness. Accordingly, if the contribution of the compression diagonal to the resistance is neglected, eq. (45b), the frame shall be analysed as a diagonally braced frame and its shear stiffness is calculated according to eq. (7). If the forces are considered as equally distributed to the two diagonals, eq. (45c), the shear stiffness may be determined according to eq. (8). In case that both diagonals are taken into account with their ultimate resistance, eq. (38a), the relation for the shear stiffness must be modified to take into account that the two braces are not stressed by the same axial force. Accordingly, the factor $2 \cdot A_d$ in the denominator of eq. (8) must be replaced by $(1+\chi)$ A_d. It may be noted that Eurocode 8 recommends to neglecting the contribution of the compression diagonals. This is however not in line with the Code provision for limitation of the slenderness as will be shown later.

The overstrength of the frame at a certain storey i, is the ratio between the acting design storey shear V_{xSd} to the actual ultimate storey shear resistance V_{xu}. It may be determined from:

$$\alpha_{CD_i} = (\frac{V_{xu}}{V_{xSd}})_i = \frac{V_{xRd}}{V_{xSd}} h \cdot r_y \cdot \gamma_{M1}$$

(46)

where:

V_{xSd} = acting design storey shear

V_{xRd} = design storey shear resistance from eq. (45)

$r_y = f_{yr} / f_y$ = ratio between the actual and the specified nominal yield strength of steel   (47)

h accounts for strain hardening and may be taken equal to 1,20

$\gamma_{M1} = 1,10$

5.2.3 Beams and Columns

As outlined before, beams and columns are designed to remain essentially elastic during the seismic event and must therefore proportioned to resist the capacity design forces. The latter are those as determined from the analysis, multiplied by the magnification factors α_{CD}. An appropriate selection of these factors is required, as they vary over the height of the building. The adoption of the calculated value of this factor at each storey would be too conservative due to the resulting accumulation of axial forces in the lower storeys. However, it is unlikely that overstrength develops simultaneously at all diagonals, especially when the number of floors is high. The application of an SRSS procedure may then be considered. The provision of Eurocode 8 that adopts the **smaller** value of the overstrength may be a little unconservative for the higher storeys, but is more appropriate for the lower, and more critical, storeys. Taking account of the axial forces due to gravity loading, the capacity design axial forces of the beams and the columns are finally determined from:

$$N_{Scd} = N_{Sd,G} + \alpha_{CD} \cdot N_{Sd,E}$$

(48)

where:

$N_{Sd,G}$ = axial forces due to gravity loading

$N_{Sd,E}$ = axial forces due to seismic actions

$\alpha_{CD} = \min \alpha_{CD_i}$ or $\sqrt{\sum (\alpha_{CD_i})^2} \leq q$

Obviously the beams and the columns shall be verified for strength and stability for the combined actions due to the above capacity design axial forces, as well as the bending moments resulting from the analysis.

It shall be noted that according to Eurocode 8, the capacity design axial forces are determined from eq. (49). It may be observed that Eurocode 8 incorporates the influence of strain hardening to both gravity and seismic effects.

$$N_{Scd} = 1.20 \cdot (N_{Sd,G} + \alpha_{CD} \cdot N_{Sd,E})$$

(49)

where:

$\alpha_{CD} = \min(N_{tRd}/N_{Sd})$ for all bracing members $\leq q$

N_{tRd} is to be determined for the actual yield strength of the material.

5.2.4 Connections

The connections of the diagonals to their adjacent members are usually formed by means of gusset plates and bolting. They are designed for the capacity design forces in order to preclude failure before yielding of the connected member. Their design resistance may be therefore exceed $\alpha_{CD} \cdot N_{Sd,E}$ or more conservatively $h \cdot r_y \cdot \gamma_{M1} \cdot N_{tRd}$, where $N_{Sd,E}$ and N_{tRd} denote the acting design axial force and design resistance of the tension diagonal.

5.3   V- and inverted V-braced frames

5.3.1 General

In V- and inverted V-braced frames, the horizontal shear force is resisted by the braces and the overturning moments by the columns. As in X-braced frames, inelastic behaviour is restricted to the braces, columns and beams be designed for overstrength in order to remain essentially elastic. The braces are accordingly verified against the forces as determined by global analysis and detailed for ductility. The columns and the beams have to resist the capacity design action effects that are higher than those determined from global analysis.

5.3.2 Braces

During seismic loading, the braces in V-type braced frames are alternatively subjected to tension and compression. Depending on the brace type, these forces are superimposed to axial compression (inverted V-braced frame) or tension (V-braced frame) due to vertical loading on the beams the are connected to, as shown in Figure 25. However, the axial forces due to vertical loading are neglected because, as explained later, the beams are designed to resist vertical loading without consideration of the support provided by the braces.

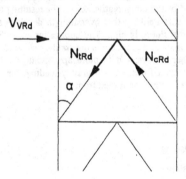

Figure 25: Brace forces in V-type braced frames

The design storey shear resistance may be determined from the buckling resistance of the compression braces according to:

$$V_{VRd} = 2 \cdot N_{cRd} \cdot \sin\alpha = 2 \cdot \chi \cdot A_d \cdot f_y \cdot \sin\alpha / \gamma_{M1} \tag{50}$$

This shows that the full carrying capacity of the tension braces is not exploited. This is due to the fact that as explained later an unbalanced force develops in this case. As this force is not considered in the dimensioning of the adjacent beam, it is not appropriate to take into account the full tension capacity of the braces.

The buckling length of the braces may be conservatively taken as the system length. However, some restraint is supplied by the gusset plates at the ends, which may lead to a reduction of the buckling length up to 30%. The amount of restraint is certainly affected by the detailing of the connection. If for instance cross-plates are welded to the gusset plates, the out-of-plane buckling length may be reduced. As a conservative rule, the buckling length of braces connected by means of simple gusset plates without any additional reinforcement at their ends may be taken equal to:

- 0,9 times the system length for in plane buckling and
- 1,0 time the system length for out of plane buckling.

5.3.3 Beams and Columns

As previously mentioned, the braces of V-type braced frames are supporting the intersecting beam when the frame is subjected to vertical loading. For horizontal loading, the braces are subjected to tension and compression. These forces are equal as long as the compression brace does not buckle. However, the situation changes at higher strain levels as shown in Figure 26. Due to the different response of the tension and compression braces, an **unbalanced force** develops at a pair of braces. The value of this force depends on the slenderness of the brace and the level of applied strain.

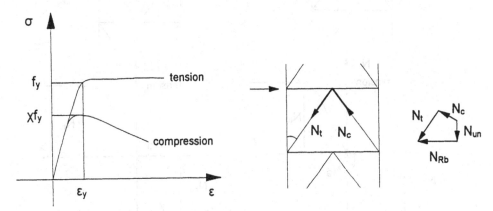

Figure 26: Response of tension and compression members and effects on the beam

The unbalanced force creates a transverse force, and therefore bending moments, to the beam. In order to accommodate with this effect, V-type frames are designed such that:

- The beams run continuous over their entire length and
- The beams resist alone all directly acting vertical loads, neglecting the support that may be supplied by the braces.

The unbalanced forces are generally different between the two bracing configurations. Indeed, due to the presence of gravity loading, some "pre-load forces" exist in the braces. These forces are in V bracing compressive, while in inverted V bracing tensile (Fig. 27). Accordingly buckling starts later in inverted V bracing and the corresponding unbalanced forces at the same level of strain are smaller than in V bracing. Therefore the requirements on the beams could be alleviated for this type of bracing. However, due to several uncertainties and the usually small level of axial forces due to vertical loading in comparison to the axial forces due to seismic loading, such provisions are not foreseen in the seismic Codes.

In respect to column design, the same rules apply as for X braced frames. Accordingly, columns shall resist the moments and shear forces as determined by the analysis and the capacity design axial forces calculated from eq. (48), where α_{CD} is to be determined analogously from eq. (46)

5.3.4 Connections

The connections of the diagonals to the beams shall be designed as for X braced frames for the capacity design forces. However for the connection design the probability of the development of the tension resistance may not be neglected. Therefore the design forces shall be determined according to the procedure outlined in 5.2.4.

5.4 K-braced frames

In K-braced frames, the braces are alternatively subjected to compression and tension. As outlined before, at higher loads the compression braces buckle and unbalanced forces develop. However, unlike before, the unbalanced force applies now at the midheight of the columns. This results a dramatic reduction of ductility due to the development of 2nd order effects and instability in the columns. Consequently, K-braced frames are not considered as ductile and should resist the elastic seismic forces without any reduction due to ductility. Their design is therefore governed exclusively from strength considerations and will be not presented in the following.

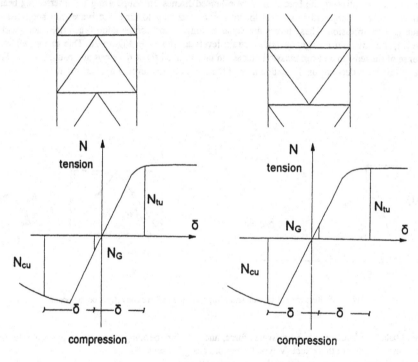

Figure 27: Unbalanced forces for preloaded braces

5.5 Eccentrically braced frames

5.5.1 General

The design of eccentrically braced frames (Kasai and Popov 1986) in respect to strength and ductility conceptually differs from the design of concentrically braced frames, due to the fact that inelastic behaviour is expected in the beam links rather than in the braces. Accordingly, the beam links are designed for the action effects as determined from global analysis, while the braces, the parts of the beam outside the links and the columns are supposed to remain elastic and shall resist the capacity design action effects.

5.5.2 Beam links

Beam links are supposed to resist the design forces and moments N_{Sd}, V_{Sd} and M_{Sd} as determined from global analysis (Figure 28). These action effects coexist simultaneously, so that the links are to be verified for the combined action. Figure 28 shows that due to equilibrium conditions following relation between acting moments and shear forces exists:

$$e = \frac{2M_{Sd}}{V_{Sd}} \tag{51}$$

For a certain link length, the cross section yields simultaneously in bending and shear. This length, called **critical length**, is determined from:

$$e_{crit} = \frac{2M_{pl}}{V_{pl}} = \frac{2M_{pl,Rd}}{V_{pl,Rd}} \tag{52}$$

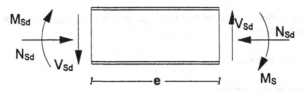

Figure 28: Design forces and moments at a beam link

As shown in Figure 29, three regions of the interaction M-V diagram of the link cross section may be distinguished.

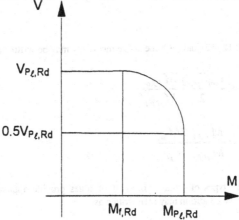

Figure 29: Interaction diagram of moments and shear forces of the link cross section

a)   Region of low shear forces

This is characterized by the condition:

$$V_{Sd} \leq 0,5 \cdot V_{pl,Rd} \tag{53}$$

In such a case the bending resistance of the cross section is not affected by the presence of shear forces. Accordingly, the acting moment may reach the plastic resistance of the cross section. It is therefore in the limit:

$$M_{Sd} = M_{pl,Rd} \tag{54}$$

Considering expressions (51), (52) and (54), condition (53) may be written as:

$$\frac{2M_{Sd}}{e} = \frac{2M_{pl,Rd}}{e} \le 0,5 \frac{2M_{pl,Rd}}{e_{crit}}$$

or

$$e \ge 2e_{crit} = \frac{4M_{pl}}{V_{pl}} \tag{55}$$

The above condition defines a range for the link length for which bending action is prevailing. Such a beam link is called **bending link.**

b)   Region of high shear forces

This is characterized by the condition:

$$V_{Sd} = V_{pl,Rd} \tag{56}$$

In such a case the acting moment is resisted only from the flanges of the cross section, as the web is exclusively devoted to resist shear. The relevant design moment is denoted as $M_{f,Rd}$. It is therefore:

$$M_{Sd} \le M_{f,Rd} \tag{57}$$

Introducing eqs. (51), (52) and (56), the above inequality may be written as:

$$\frac{eV_{Sd}}{2} = \frac{eV_{pl,Rd}}{2} \le \frac{e_{crit}V_{pl,Rd}}{2} \frac{M_{f,Rd}}{M_{pl,Rd}}$$

or

$$e \le \frac{M_{f,Rd}}{M_{pl,Rd}} e_{crit} = \frac{M_{f,Rd}}{M_{pl,Rd}} \frac{2M_{pl}}{V_{pl}} \tag{58}$$

For the most usual types of cross sections, the flanges provide approximately 80% of the bending resistance. The above condition may then written as:

$$e \le 1,6 \frac{M_{pl}}{V_{pl}} \tag{59}$$

which is the most usual expression introduced in Codes (AISC 1997). Inequality (57) defines a range for the link length for which shear action is prevailing. Such a beam link is called **shear link.**

c)   Region of intermediate shear forces

Considering the above expressions, this is characterized by the condition:

$$\frac{M_{f,Rd}}{M_{pl,Rd}} e_{crit} \le e \le 2e_{crit} \tag{60}$$

This constitutes an **intermediate link** length, where bending moments and shear forces are in interaction. The moment capacity is reduced to $M_{V,Rd}$ due to the presence of shear:

$$M_{Sd} \le M_{V,Rd} = (W_{pl} - \frac{\rho A_v^2}{4t_w}) f_y / \gamma_{M1}$$

(61)

where:

$$\rho = (2V_{Sd} / V_{pl,Rd} - 1)^2$$

A_v = shear area of the cross section

t_w = thickness of the web

W_{pl} = plastic moment of resistance of the cross section

It may be noted, that in all the above expressions plastic design resistances were used. This results from the fact that due to ductility requirements the link cross sections shall be compact, so that local buckling is excluded.

The possible influence of acting axial forces N_{Sd} in the link may be taken into account by appropriate reduction of the plastic design moment to a moment $M_{N,Rd}$. For rolled I-sections and strong axis bending, this moment may be approximately determined from:

$$M_{N,Rd} = 1{,}11 \cdot M_{pl,Rd} \cdot (1 - n) \le M_{pl,Rd}$$

(62)

where $n = N_{Sd} / N_{pl,Rd}$

The nominal shear resistance of the link is determined from:

♦ $V_{Rd} = \dfrac{2M_{pl,Rd}}{e}$ for a moment link   (63a)

♦

♦ $V_{Rd} = V_{pl,Rd}$ for a shear link and   (63b)

♦

♦ $V_{Rd} = \dfrac{2M_{V,Rd}}{e}$ for an intermediate link   (63c)

♦

The actual ultimate shear resistance of the link, accounting for the real yield strength of the material, as well as strain hardening effects, is equal to:

$$V_{ul} = V_{Rd} \cdot h \cdot r_y \cdot \gamma_{M1}$$

(64)

where the parameters h, r_y and γ_{M1} have the same meaning as in eq. (46). However, for the strain hardening parameter h a derivation will be given in section 5.7.

Considering the free body diagram (Figure 30a), the ultimate and the design storey shear resistance for a K-braced frame may be determined from the above calculated link resistances according to eq. (65):

$$V_{ecu} = \frac{h_0}{a} V_{ul}$$

(65a)

$$V_{ecRd} = \frac{h_0}{a} V_{Rd}$$

(65b)

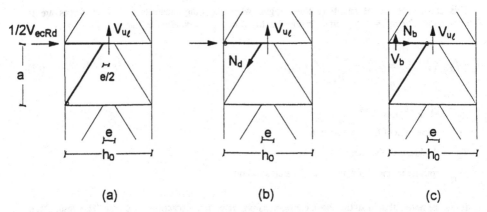

Figure 30: Member axial and shear forces

The overstrength of the frame at a certain storey i, is evaluated as the ratio between the acting design storey shear V_{ecSd} in the seismic situation to the actual ultimate storey shear resistance V_{ecu}, and may, analogously to eq. (46), be determined from:

$$\alpha_{CDi} = (\frac{V_{ul}}{V_{Sd}})_i = \frac{V_{Rd}}{V_{Sd}} \cdot h \cdot r_y \cdot \gamma_{M1}$$ (66)

The parameter α_{CDi} generally varies along the height of the frame. However, a uniform distribution of α_{CDi} is beneficial to the overall behaviour since it is more likely that more links will exhibit inelastic deformations and participate in the energy dissipation (Kasai and Han 1997). For a uniform overstrength distribution, the link (and beam) sections must be reduced over the height of the frame.

5.5.3 Beams outside links and braces

As referred to before, the parts of the beams outside the links and the braces are supposed to remain elastic during the seismic action. These members shall therefore resist the capacity design action effects resulting from the developing overstrength of the links. The axial and shear forces in the beams and the braces due to seismic loading may be determined from the free body diagram (Figure 30 b,c), or from Figure 14, and are given expressed by eq. (67):

$$N_{Scd,E,b} = \frac{h_0}{2 \cdot a} V_{ul} = \alpha_{CD} \cdot N_{Sd,E,b}$$ (67a)

$$N_{Scd,E,d} = \frac{h_0}{h_0 - e \cos\alpha} V_{ul} = \alpha_{CD} \cdot N_{Sd,E,d}$$ (67b)

$$V_{Scd,E,b} = \frac{e}{h_0 - e} V_{ul} = \alpha_{CD} \cdot V_{Sd,E,b}$$ (67c)

As outlined before, the factor α_{CD} may be taken equal to:

$$\alpha_{CD} = \min(\alpha_{CDi}) \text{ or } = \sqrt{\sum(\alpha_{CDi})^2} \leq q$$

The final capacity design axial and shear forces result from the addition of the forces due to gravity and seismic loads, according to:

$$N_{Scd} = N_{Sd,G} + N_{Scd,E}$$ (68)

$$V_{Scd,b} = V_{Sd,G,b} + V_{Scd,E,b} \tag{69}$$

At the end of the link, a moment M_{link} develops that is equal to:

$$M_{link} = V_{ul} \cdot (e/2) = h \cdot r_y \cdot \gamma_{M1} \cdot V_{Rd} \cdot (e/2) \tag{70}$$

where:

V_{Rd} = nominal shear resistance of the link from eq. (63)

and the parameters h, r_y and γ_{M1} as in eq. (46).

The moment M_{link} is transferred from the link to the beam outside the link and the diagonal. This moment may be reduced, if the beam to brace connection is eccentric as shown in Figure 31. The relevant expression may be written as:

$$M_{trans} = M_{link} - N_d \cdot \sin\alpha \cdot \cdot e_v \tag{71}$$

where:

N_d = axial force of the diagonal and

e_v = vertical eccentricity between diagonal and beam.

Eq. (71) shows that a negative eccentricity results in an increase in the transfer moment and should be avoided. Therefore the intersection of the brace and beam centerlines shall be either at the end of the link or in the link.

The moment M_{trans} is fully transferred to the beam section outside the link, if the brace to beam connection is designed as a pinned connection. Otherwise, this moment is distributed to both beam and brace in proportion to their stiffness as suggested by eq. (14). If the beam outside the link is allowed to yield, the brace to beam connection has to be a moment connection. In this case, the brace will have to resist the moment difference determined from:

$$M_d = M_{trans} - M_{b,Rd} \tag{72}$$

where $M_{b,Rd}$ is the moment capacity of the beam section adjacent to the link.

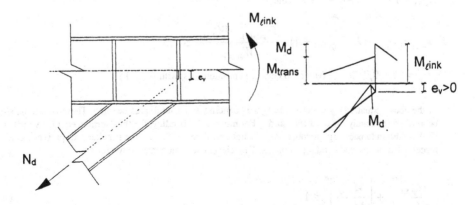

Figure 31: Moment transfer from the link to the adjoining members

The beams outside the link shall be verified for the combination of the following action effects:

- N_{Scd} from eqs. (66a) and (67)

- $V_{Scd,b}$ from eq. (69)

- M_{Scd} from eq. (14b) in combination with eqs. (70) and (71)

The braces shall be verified for buckling due to compression when pinned connected or compression and in plane bending if rigidly connected at their ends. The axial force is determined from (66b) and (67) and the bending moment from (14a) in combination with eqs. (70) to (72).

The buckling length coefficient may be taken for pinned connections equal to 0,9 for in plane and 1,0 for out-of-plane buckling. For rigid connections it may be taken equal to 0,5 to 0,7 for both buckling modes.

5.5.4   Columns

Columns shall be verified for buckling generally due to compression and biaxial bending. Bending moments in columns are generally low to moderate. However, axial forces must be increased to account for the link overstrength. The capacity design forces may be approximately determined from:

$$N_{Scd} = N_{Sd,G} + \alpha_{CD} \cdot N_{Sd,E} \tag{73}$$

where:

$N_{Sd,G}$ = axial forces due to gravity loading

$N_{Sd,E}$ = axial forces due to seismic actions

$$\alpha_{CD} = \min\ (\alpha_{CD_i})\ \text{ or }\ = \sqrt{\sum (\alpha_{CD_i})^2} \leq q$$

and α_{CD_i} from eq. (65)

The buckling length coefficient for columns may be conservatively taken equal to 1,0, i.e. equal to the storey height.

Buckling verifications may be performed according to eq. (74):

$$\frac{N_{Sd}}{\chi_{min} N_{pl,Rd}} + \frac{k_y M_{y,Sd}}{M_{ply,Rd}} + \frac{k_z M_{z,Sd}}{M_{plz,Rd}} \leq 1 \tag{74}$$

where:

$$k_{y(z)} = 1 - \frac{\mu_{y(z)} \cdot N_{Sd}}{\chi_{y(z)} \cdot N_{pl}} \leq 1,5 \quad \text{and}$$

$$\mu_{y(z)} = \overline{\lambda}_{y(z)} \cdot (2 \cdot \beta_{My(Mz)} - 4) - \frac{W_{pl,y(z)} - W_{el,y(z)}}{W_{el,y(z)}} \leq 0,9$$

$$\beta_M = 1,8 - 0,7 \cdot \psi \quad (\psi = \text{ratio between end-moments, with } |\psi| \leq 1)$$

In the above equations, y-denotes strong axis bending and z weak axis bending. These equations shall be applied separately for the axes y and z. For mono-axial bending, the third term in eq. (74) vanishes.

In addition to buckling, member cross sections shall be verified for the simultaneous presence of axial forces, shear forces and bending moments. The relevant condition may be written as:

$$\left[\frac{M_{y,Sd}}{M_{y,Rd}}\right]^\alpha + \left[\frac{M_{z,Sd}}{M_{z,Rd}}\right]^\beta \leq 1 \tag{75}$$

where the moments M_{Rd} are determined from (61) and (62) to account for the presence of axial and shear forces and the parameters α and β depend on the shape of the cross section.

5.5.5 Connections

Beam-to-column connections away from links may be designed as pinned by means of welded or bolted plates or angles connected to the web of the beam (Figure 32a). They have to transfer the capacity design forces of the beam. However, in D- or V-braced frames where the link is adjacent to the column, they must be rigid to transfer the full link moment. The link web should be preferably welded to the connecting plate, otherwise slippage could occur between the web and the plate, imposing a shear force transfer through the flange welds (Figure 32b). In case of moment links, the connection must be also full strength, in order to allow for the development of the link moment capacity. Accordingly it is formed as a moment resisting frame rigid connection, either welded or bolted (Figure 32c). Appropriate stiffeners must reinforce the column web opposite to the beam flanges. A haunch near the column may also reinforce the beam section. In such a case the link is the not reinforced part of the beam (Figure 32d).

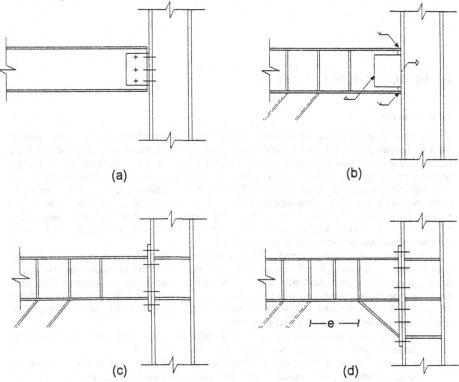

Figure 32: Link beam-to-column connection

Pinned brace-to-link connections are formed by means of gusset plates and must be designed for the axial strength of the brace (Figure 33 a). The dimensions of the gusset plates, especially their b/t-values, must be such that plate buckling is avoided. Otherwise they may be stiffened by cross plates. Rigid brace-to-link connections may resist part of the transfer moment as discussed before. Rigid connections may be formed, by full welding between the brace section and the beam (Figure 33 b). The brace is then spliced outside the connection.

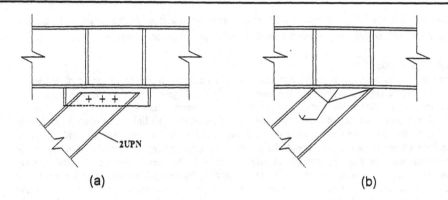

Figure 33: Pinned and rigid brace-to-link connections

6. DESIGN FOR DUCTILITY

6.1 Introduction

As referred to before, structures are allowed to exhibit inelastic deformations during a seismic event in order to dissipate part of the input seismic energy. This result in a reduction of earthquake induced forces compared to those that would develop under a linear response. Structures are accordingly designed by means of inelastic response spectra by application of a force reduction factor, the behaviour factor q according to the notation of Eurocode 8. The behaviour factor expresses the ability of the structure to sustain cyclic inelastic deformations without considerable reduction in stiffness and strength. This ability is described by the generic term ductility. Accordingly, structures subjected to seismic actions have to be designed for ductility, in addition to stiffness and strength, which assures that the assumed seismic forces, will be indeed realized.

The design for ductility is governed by the observation that although steel is ductile as a material, steel members are not necessarily. This is due to the fact that certain types of failure, mainly caused by instability, are non-ductile. Additionally, it may be noted that under certain circumstances (high strain rates, low temperature, triaxiality etc.) the material itself does not behave in a ductile manner. The design for ductility is therefore aimed to create the conditions under which the overall structural response is ductile. This is achieved by means of a set of rules that:

- Allow the structure to yield only in certain dissipative zones or elements by providing the remaining parts with adequate overstrength in order to remain elastic
- Promote in the dissipative zones or elements only ductile modes of failure by appropriate detailing measures and the provision of overstrength for the other failure modes.

This chapter presents the rules that provide ductility and govern the design of dissipative elements. For the reasons explained before, the provisions for overstrength of non-dissipative elements were included in the previous section. Non-dissipative elements designed to resist the capacity design action effects, are not necessarily required to comply with the requirements for ductility.

In regions of low seismicity, or if other actions, e.g. wind action, prevail, structures may be designed as non-dissipative. Non-dissipative structures are supposed to resist earthquakes through elastic behaviour so that they must not necessarily comply with the requirements for ductility and overstrength. However, the complete absence of ductility provisions, especially for structures in regions of high seismicity, should be avoided. This is due to the uncertainties involved in the field of earthquake engineering, including the accurate evaluation of seismic actions. Such uncertainties include peak responses during real earthquakes that are not completely covered by the spectral values, definitions of soil types, values of the peak ground acceleration assumed in Codes that may be exceeded, simplifications in the assumptions during analysis etc. It is therefore recommended to comply, at least partly, with some ductility rules as a second line of defense.

6.2 Connections in dissipative zones

As previously demonstrated, connections in dissipative zones have to be designed for overstrength to allow for yielding of the connected dissipative elements rather than connection failure. The relevant capacity design forces are given in the previous section.

Welded connections may be formed by means of full or partial penetration butt welds, or fillet welds. The use of partial penetration welds should be avoided as far as possible and be restricted only to cases where full penetration is for reasons of execution not possible. Full penetration welds, whose execution is controlled according to the relevant standards, are supposed to possess the required overstrength without any additional calculation. Fillet welds and partial penetration butt welds shall be verified for the capacity design forces.

Bolted connections shall be designed for overstrength to resist the capacity design forces. As an extra safety, bearing failure should precede shear failure, as it is more ductile. This rule should apply as an extra safety to all bolted connections, independently on the element type, dissipative or not. Using the notation of Eurocode 3, the relevant expressions may be written as:

$$F_{Sd} \text{ (or } F_{Scd}) \leq F_{b,Rd} \tag{76a}$$

$$F_{b,Rd} \leq F_{v,Rd} \tag{76b}$$

where:

$F_{b,Rd}$ = bearing resistance of bolts

$F_{v,Rd}$ = shear resistance of bolts

It is further recommended to pre-load the bolts in order to restrict slip deformations, at least up to the level of the analysis forces and by application of the partial safety factors in the serviceability limit state. The relevant expressions, again with the notation of Eurocode 3, may be written as:

$$F_{Scd} \leq F_{s,Rd,ser} \tag{76c}$$

where:

$F_{s,Rd,ser}$ = slip resistance of bolts for the serviceability limit state

6.3 Members subjected to tension

Tension members may fail due to yielding of the gross cross section or tearing of the net section. As well known, yielding of the gross section is a ductile mode of failure. However, if the member has holes due to its connections, yielding starts at lower stresses due to the stress concentration near the hole and extends progressively over the cross section. The redistribution of stress within the cross section with the hole happens at the expense of the ductility of the member when it fail due to tearing of the net section. In order to assure a ductile mode failure, yielding of the gross cross section must therefore precede tearing of the net section.. Accordingly, a ductility condition, as proposed by Eurocode 3, must be satisfied:

$$N_{tRd1} \leq N_{tRd2} \quad \text{or}$$

$$A_{net} / A \geq 1.11 \frac{f_y \, \gamma_{M2}}{f_u \, \gamma_{M1}} \tag{77}$$

where:

N_{tRd1} = design resistance for yielding of the gross cross section

N_{tRd2} = design resistance for tearing of the net cross section

$\gamma_{M1} = 1,10$

$\gamma_{M1} = 1.25$

Considering that for normal steels the ratio between tensile strength and yield strength is usually equal to 1,5, eq. (77) implies that the net area has to be limited to 85% of the gross area. Evidently, due to strain hardening, the ductility criterion does not absolutely guarantee that tearing will not actually take place. However it is expected that if the axial deformations are limited, sufficient yielding and energy dissipation will take place prior to cracking or collapse of the member.

6.4 Members subjected to compression

Members subjected to compression may fail due to local buckling, global buckling or their combination. Local buckling refers to plate buckling of the walls of the cross sections, global buckling to flexural or lateral torsional buckling. Lateral torsional byckling seldom occurs for the usual types of cross sections and member dimensions. The governing mode of failure depends on the plate slenderness of the walls of the cross section and the overall slenderness. These are determined from:

$$\bar{\lambda}_p = \sqrt{\frac{f_y}{\sigma_{cr}}} = \frac{b/t}{28,4\varepsilon\sqrt{k_\sigma}} \tag{78}$$

$$\bar{\lambda} = \frac{l_0/i}{93,9\varepsilon} \tag{79}$$

where:

b, t = width and thickness of the compressed plated parts

σ_{cr}, k_σ = plate buckling stress and coefficient for the compressed plated part

l_0 = buckling length of the member

i = radius of gyration of the cross section

ε = $\sqrt{235/f_y}$ (f_y = yield strength in MPa)

Monotonic load-shortening-curves for compression members with various values of the overall column slenderness $\bar{\lambda}$ and the local plate slenderness are illustrated in Figure 34 (Vayas 1996). The member exhibits ductile behaviour as long as its local and overall slenderness is low. It may be seen that the response is mainly influenced by the local slenderness for small values of $\bar{\lambda}$, while for larger values of $\bar{\lambda}$ this influence is almost non-existent. This is due to the fact that when the overall slenderness increases the maximal acting compression stress becomes lower, which leads to a reduction in the nonlinear behaviour of individual plated elements. Ductility is accordingly a member rather than a cross section property. It is therefore of little benefit to increase the compactness of the cross sections by reducing the b/t-ratio of its walls, as long as the overall slenderness is high. Eurocode 8 proposes a general condition for the overall slenderness, expressed by $\bar{\lambda} \leq 1,5$, which is probable not very conservative as Figure 34 shows.

The cyclic response of a compression member is illustrated in Figure 35 (Black et. al 1980). It may be seen that the member undergoes progressively larger, slip type displacements before it is able to resist tension. This is due to the fact that during compression large out-of-plane displacements occur which have to be "flattened out" before the member is stressed again. The buckling resistance progressively deteriorates due to the fact that any out-of-plane displacements act as geometrical imperfections in subsequent cycles and lead to a reduction in strength (eq. (43)). These phenomena become more pronounced with increasing slenderness of the member (ECCS 1986). The member response improves for fixed end-supports, as the number of developing plastic hinges becomes three, one at midspan and two at the ends, compared with one for pin supported ends.

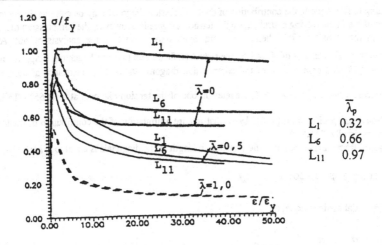

Figure 34: Load-shortening curves of compression members

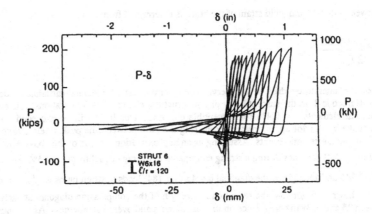

Figure 35: Cyclic response of compression members

6.5   X- braced frames

As shown in Figure 35, the response of a member to compression and tension substantially differs. Placing diagonals in only one direction leads therefore to an accumulation of lateral deformations in one direction for the overall frame. In order to achieve a symmetric response, diagonals with similar cross sectional properties should be placed in both directions. The relevant provision of Eurocode 8 is:

$$\frac{\left|A^+ - A^-\right|}{A^+ + A^-} \leq 0,05 \tag{80}$$

where:
A^+ and A^- = areas of the horizontal projections of the cross sections of the tension diagonals in the two directions in the plane of the frame.

According to Eurocode 8, the contribution of the compression diagonal may be completely neglected, the strength of the frame relying entirely upon the tension diagonals. However, the compression diagonals participate in the force transfer, whether they are taken into account in the analysis or not. At the maximally allowed slenderness of $\overline{\lambda} = 1,5$, the reduction factor due to overall buckling is approximately equal to $\chi = 0,35$. The contribution of the compression diagonal to the total resistance at the peak is therefore at least equal to $\dfrac{0,35}{1,35} \approx 25\%$, the relevant one of the tension diagonal at most 65%. This is in line with American specifications (AISC 1997) that require a limitation of the contribution of the tension diagonal to maximally 70%.

For a lateral frame displacement δ, the variation in length of the diagonals is equal to (Figure 36):

$$\Delta d = \delta \cdot \cos\alpha = \varphi \cdot \cdot a \cdot \cos\alpha \tag{81}$$

The resulting axial strain is equal to:

$$\varepsilon = \frac{\Delta d}{d} = \frac{\varphi \cdot a \cdot \cos\alpha}{a / \sin\alpha} = \frac{1}{2}\varphi \cdot \sin 2\alpha \tag{82}$$

Considering the value of the yield strain,

$$\varepsilon_y = f_y / E \tag{83}$$

the ratio between ultimate and yield strain of the brace is determined from:

$$\frac{\varepsilon_u}{\varepsilon_y} = \frac{E \cdot \varphi_u \cdot \sin 2a}{2 f_y} \tag{84}$$

The limit of the interstorey drift at the serviceability limit state (moderate earthquake) depends according to Eurocode 8 on the susceptibility of non-structural elements to deformations. It is equal to 0,6% for normal elements and 0,4% for susceptible elements. This limit will be exceeded during a strong earthquake by a factor approximately expressing the ratio between the peak accelerations of the strong to the moderate seismic events. Assuming as an approximation a value of the above ratio equal to 3, the maximal drift becomes during a strong earthquake in average equal to $\varphi_u = 1,5\%$. For a usual steel quality S 235 and an angle of the diagonal $\alpha = 45^{\circ}$, eq. (84) leads to strain ratios $\varepsilon_u / \varepsilon_y \approx 6,5$ for the brace. As shown in Figure 34, the reduction in strength of the compression diagonal at such strain varies between 25 and 50% in dependence on its compactness and overall slenderness. As the increased capacity of the tension diagonal due to strain hardening cannot balance this reduction, the carrying capacity of the frame is reduced. The design storey shear resistance should therefore be preferably determined from eq. (45c).

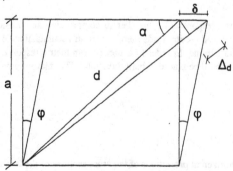

Figure 36: Variation in length of the diagonals

6.6 V- and inverted V- braced frames

For an inverted V- braced frame subjected to a horizontal displacement δ, the variation in length of the braces may be determined from eq. (81). However, at higher drifts the beam is pulled down due to the development of the transverse unbalanced force discussed before. This results in a shortening of both diagonals (Figure 37) by the amount:

$$\Delta d = w \cdot \cos \alpha \qquad (85)$$

Combining eqs. (81) and (85), the variation in length of the two diagonals is determined from:

Tension diagonal: $\Delta d_t = \delta \cdot \sin \alpha - w \cdot \cos \alpha$ (86a)

Compression diagonal: $\Delta d_c = -\delta \cdot \sin \alpha - w \cdot \cos \alpha$ (86b)

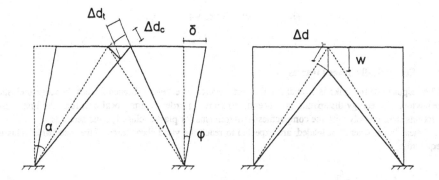

Figure 37: Variation in length of the diagonals

The tension diagonal can accordingly not be stretched to a point where it can develop its full capacity, unless the beam resists the unbalanced force. Obviously, w is decreasing with increasing stiffness of the beam. On the other side the value of the unbalanced force resisted by the beam is a function of its moment capacity $M_{p,beam}$. For beams acting composite with the concrete slab, this moment is higher.

Concluding it may be stated that the overall frame response is a function of the slenderness of the braces and the stiffness and strength of the beam. Reduction of the slenderness of the braces and provision of stiff and strong beams lead to higher ductility for the frame.

The braces do not generally yield simultaneously during a seismic excitation. Energy dissipation may therefore be limited only to a low number of braces. Oversizing the braces at lower floors does not provide a satisfactory solution, as higher excitation modes may be critical for particular earthquakes. A new configuration was therefore proposed, in which all brace-to-beam intersection points are tied together by a vertical strut (Figure 38). This ensures a redistribution of forces among braces, so that energy dissipation and damage will be more uniformly distributed among them. The ties shall resist the unbalanced forces as described above and shall be designed to remain elastic during the seismic excitation. The capacity design force is equal to the vertical projection of the difference between the tension and compression capacities of the braces.

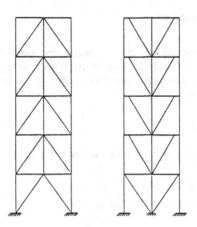

Figure 38: Vertically tied V-braced frames

6.7 Eccentrically braced frames

The response of the links is crucial for the performance of eccentric braced frames, because inelastic behaviour and energy dissipation is concentrated there. In order to limit local buckling, the link cross sections shall comply with the compactness requirements for plastic, class 1, cross sections.

Shear links, when first loaded, are expected to reach the yield shear force of the web panel. This is equal to:

$$V_y = d \cdot t_w \cdot f_{yw} / \sqrt{3} \tag{87}$$

However, with increasing applied shear deformations the shortened diagonal buckles and a tension field develops along the elongated diagonal (Figure 39). The carrying mechanism changes to a truss having the transverse stiffeners as posts and the tension field as diagonal. The tension field stress is determined as the reserve strength of the web panel beyond the shear buckling stress τ_u, in accordance to:

$$\sigma_{tf} = \sqrt{f_{yw}^2 - 3\tau_u^2 + \sigma_0^2} - \sigma_0 \tag{88}$$

where:

σ_{tf} = tension field stress

$\sigma_0 = 1{,}5 \cdot \tau_u \cdot \sin 2\varphi$

φ = inclination of the tension field ($\tan \varphi \approx \dfrac{d}{e} = \dfrac{1}{\alpha}$)

$\alpha = e/d$ = aspect ratio of the panel

For a compact web panel considered here, the shear strength is equal to the shear yield stress ($\tau_u = f_y/\sqrt{3}$). Inserting this value in eq. (88), the resulting tension field stress is equal to zero ($\sigma_{tf} = 0$). However, for large strains the tension field stress may be set due to strain hardening equal to the difference between the tensile strength and the yield strength of the material ($\sigma_{tf} = f_{uw} - f_{yw}$).

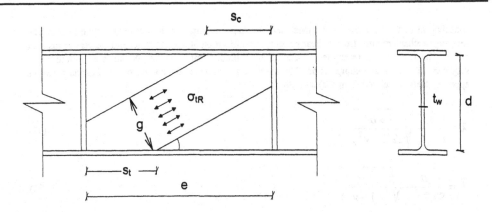

Figure 39: Web panel deformations and tension field development

The additional shear force that may be resisted by truss action of the link is equal to:

$$V_{tf} = g \cdot t_w \cdot \sigma_{tf} \cdot \sin \varphi \tag{89}$$

From geometric considerations and setting approximately the anchorage lengths s_c and s_d equal to $e/2$, the width of the tension field may be determined from:

$$g = d \cdot \cos\varphi - (e - s_c - s_t) \cdot \sin \varphi \approx d \cdot \cos\varphi \tag{90}$$

The strength ratio of the link in excess of the yield strength is determined from:

$$\frac{V_{tf}}{V_y} = \frac{\sqrt{3}}{2} \sin 2\phi (\frac{f_u}{f_y} - 1) \tag{91}$$

The above relation implies that the link excess strength improves by a reduction of the aspect ratio of the panel by application of more transverse stiffeners. For panels with aspect ratios $\alpha = 3$ and $\alpha = 1$, the resulting overstrength ratios are equal to 25% and 43% correspondingly.

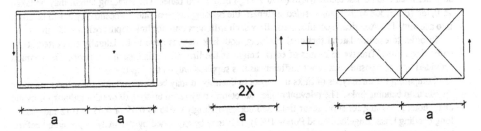

Figure 40: Carrying mechanisms of a web panel before and after the development of the tension field

For cyclic loading the above analysis has to be modified to take into account that for every new loading cycle the panel has buckled during the previous cycles. These buckles act during any subsequent loading as initial imperfections, so that the ultimate shear strength τ_u is progressively reduced. The contribution of the truss action becomes therefore more important at higher deformation cycles. The carrying mechanism turns from a shear panel to a combination of a shear panel and a truss (Vayas et al 1994). The truss action resembles to that of an X-braced frame whose diagonals are alternatively subjected to tension and compression. The absolute and relative slenderness of the diagonals may be approximately calculated from (92) and (93):

$$\lambda = \frac{l}{i} = \frac{d}{t_w}\sqrt{\frac{1+\alpha^2}{12(1-v^2)}} \tag{92}$$

$$\overline{\lambda} = \frac{d}{93,9\varepsilon\,t_w}\sqrt{\frac{1+\alpha^2}{12(1-v^2)}} \tag{93}$$

The application of (93) for a panel with d/t$_w$ = 69 ε (limit for no shear buckling), leads to $\overline{\lambda} = 0,71$ when α = 3 and $\overline{\lambda} = 0,31$ when α = 1. These figures must be further increased due to the fact that the tension field strain increases beyond the yield strain. At larger inelastic deformations tearing of the web panel due to low-cycle-fatigue occurs, similarly as that observed in knee welded joints with slender web panels (Vayas et al 1995). The above analysis suggests that, as for monotonic loading, the addition of transverse stiffeners limits the slenderness of the diagonals of the X frame and leads to a more ductile behaviour. The American specifications (LRFD 1995) require the provision of equally spaced intermediate stiffeners at distances (30 to 52 $t_w - d/5$) for link rotation angles between 0,09 and 0,03 radians. For the truss action to develop and an overall ductile behaviour to be achieved, the web thickness should not be very high compared to the flange thickness, as the flange would not be capable to providing the necessary anchorage. Therefore the reinforcement of the web by means of doubler plates is not allowed. Obviously no web openings are permitted in the area of the link.

Bending links differ from shear links in that inelastic deformations are not due to shear yielding and tension field formation in the web, but due to plastic hinges at the link ends. Their response is therefore similar to beams in moment frames and no addition of transverse stiffeners is required.

The response of **intermediate links** is between the above extreme cases. Plastic hinge rotation and the need for some tension field anchorage, lead to out-of-plane deformations of the flanges. Transverse stiffeners provide a restraint against these deformations. Taking into account that the length of the flange buckles is approximately equal to $3 \cdot b_f$, the transverse stiffeners shall be placed at distance $1,5 \cdot b_f$ from the link end, as provided by the US Specifications.

Transverse stiffeners shall resist compression as posts of a truss in the post-buckling state and must therefore be verified for buckling (Figure 41). The cross section of the stiffeners consists from the stiffener itself and an associated width (= 30 $\varepsilon \cdot t_w$) of the web panel. The buckling length may be taken ≥ 0,75d (Eurocode 3). For single-sided stiffeners, the bending moment due to eccentricity has to be taken into account. The American specifications allow such stiffeners only for link depths below 635 mm.

In order to exclude lateral torsional buckling, both link flanges have to be laterally supported at the ends of the links. Transverse beams of equal height as the link can realize such a support. The restraint provided by a composite deck is not sufficient as it is supplied only to the top flange.

Comparing the three types of links in respect to ductility, it may be observed that shear links are more ductile than bending links. The relevant angles of rotation may reach up to 0,10 or 0,20 radians for cyclic or monotonic loading for short shear links and correspondingly 0,015 to 0,09 or 0,03 to 0,12 radians for long bending links (Engelhardt and Popov 1989). This may be explained by the analysis presented before. In short links, a stable X-braced frame mechanism develops in addition to the shear mechanism at higher deformation cycles. In bending links ductility is attained through high axial strains in the flanges of the link cross section. However, beyond a certain strain the flanges are subjected to local buckling even if they comply with the requirements for plastic, class 1, cross sections. This is due to the fact that at higher strains buckling is strain-controlled, so that high equivalent plate slenderness are attained that lead to local buckling and consequently to limitation of ductility.

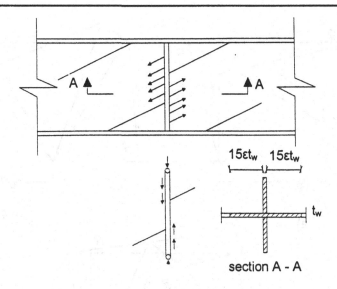

Figure 41: Anchorage of the tension field and cross section of transverse stiffeners

Once the limit link rotation angles have been established, the relevant values of the storey drifts may be derived. Neglecting elastic deformations, the relationship between link rotation and storey drifts may be derived from geometric relationships as illustrated in Figure 42 (Bruneau 1998).

- D-braced frame: $\quad\quad\quad \varphi \cdot (h_0 - e) = \varphi' \cdot e$

$$\varphi + \varphi' = \gamma \quad \rightarrow \quad \varphi = \gamma \cdot \frac{e}{h_0} \quad\quad\quad\quad\quad (94a)$$

- V-braced frame: $\quad\quad\quad \varphi \cdot (\frac{h_0}{2} - e) = \varphi' \cdot e$

$(=1/2 \text{ D-frame}) \quad \varphi + \varphi' = \gamma \quad \rightarrow \quad \varphi = \gamma \cdot \frac{2 \cdot e}{h_0} \quad\quad\quad (94b)$

- Inverted V-braced frame: $\quad \varphi \cdot \left(\frac{h_0}{2} - \frac{e}{2}\right) = \varphi' \cdot \frac{e}{2}$

$$\varphi + \varphi' = \gamma \quad \rightarrow \quad \varphi = \gamma \cdot \frac{e}{h_0} \quad\quad\quad\quad\quad (94c)$$

Eq. (94) shows a hyperbolic relationship between the ratios γ/φ and e/h_0 (Figure 43) and may be applied for the determination of the relative link length if the storey drifts and the available link rotation are known. For example, if the storey drift φ_u is 1,5 % and the available link rotation is 10% radians, as calculated before, the relative length shall be equal to 7,5% for a V- braced frame and 15% for a D- or an inverted V- braced frame. However, it must be reminded that with increasing link length the available link rotation γ decreases, as the link may become a bending link, and the storey drift φ increases with increasing storey stiffness. Short, shear links are therefore preferable than bending links.

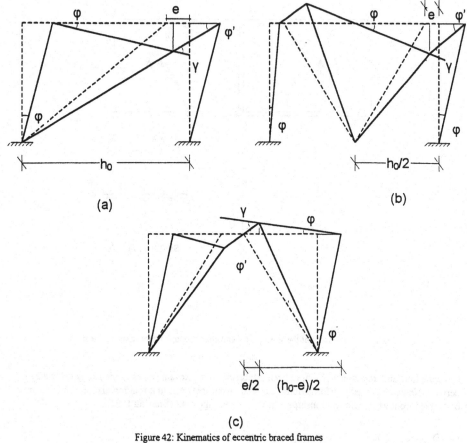

(a)   (b)

(c)

Figure 42: Kinematics of eccentric braced frames

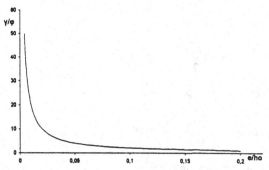

Figure 43: Link rotation to storey drift as a function of the relative link length

A uniform link deformation γ over the height of the building would optimize the design. Accordingly, the strength of the links should increase from top to the bottom of the structure. However, this does not necessarily lead to uniform distribution of plastic deformations, as higher modes can be dominating for specific earthquakes. It is therefore recommended (Lu et al 1997) to tie, as for V-braced frames, the link ends together by means of vertical struts (Figure 44). The links at different floors are then forced to undergo the same shear deformation. Studies of the inelastic response for such frames indicate lower

References

AISC. (1995). Load and Resistance Factor Design (LRFD), AISC, Chicago

AISC. (1997). Seismic Provisions of Structural Steel Buildings, AISC, Chicago.

Akiyama, H. (1985). Earthquake-Resistant Limit-State Design for Buildings, University of Tokyo Press.

Black, R.G., Wenger, W.A. and Popov, E.P. (1980). Inelastic Buckling of Steel Struts Under Cyclic Load Reversal. *Report No. UCB/EERC-80/40*. Berkeley: Earth. Eng. Research Center. Univ. of California.

Bruneau, M., Uang, C-M and Whittaker, A. (1998). Ductile Design of Steel Structures, McGraw-Hill.

ECCS. (1986). Study on Design of Steel Building in Earthquake Zones, *ECSC Technical Research Agreement* 7210-ZZ 437.

Eurocode 3. (1992). Design of steel structures, Part 1.1: General rules and rules for Buildings. *European Committee for Standardization (CEN)*, ENV 1993-1.1.

Eurocode 8. (1994). Design provisions for earthquake resistance of structures- Part 1.3: General rules – Specific rules for various materials and elements. *European Committee for Standardization (CEN)*, ENV 1998-1-3.

Engelhardt, M.D. and Popov, E.P. (1989). Behaviour of Long Links in Eccentrically Braced Frames, *Report No. UCB/EERC-89/01*. Berkeley: Earth. Eng. Research Center. Univ. of California.

Kasai, K. and Han, X. (1997). New EBF design method and application: Redesign and analysis of US-Japan EBF. In Mazzolani, F., and Akiyama, H., eds., *Behaviour of Steel Structures in Seismic Areas, Stessa '97*, Edizioni 10/17, 242-249.

Kasai, K. and Higgins, C. (1997). Real-time and full-scale tests of a viscoelastically damped steel frame under large seismic and gravity loads. In Mazzolani, F., and Akiyama, H., eds., *Behaviour of Steel Structures in Seismic Areas, Stessa '97*, Edizioni 10/17, 708-715.

Kasai, K. and Popov, E.P. (1986). A Study of Seismically Resistant Eccentrically Braced Frames. *Report No. UCB/EERC-86/01*. Berkeley: Earth. Eng. Research Center. Univ. of California.

Lu, L.-Wu, Ricles, J.M. and Kasai, K. (1997). Global performance, General report . In Mazzolani, F., and Akiyama, H., eds., *Behaviour of Steel Structures in Seismic Areas, Stessa '97*, Edizioni 10/17, 361-381.

Mazzolani, F.M. and Serino, G. (1997). Viscous energy dissipation devices for steel structures: Modelling, analysis and application. In Mazzolani, F., and Akiyama, H., eds., *Behaviour of Steel Structures in Seismic Areas, Stessa '97*, Edizioni 10/17, 724-733.

Nakashima,M., Mitani, T. and Tsuji, B. (1997). Control of maximum and cumulative deflections in steel building structures combined with hysteretic dampers. In Mazzolani, F., and Akiyama, H., eds., *Behaviour of Steel Structures in Seismic Areas, Stessa '97*, Edizioni 10/17, 744-751.

Ricles, J.M and Bolin, S.M. (1991). Seismic performance of eccentricity braced frames, *Report 91-09 Str. Sys. Research Proj.*, University of California, San Diego.

Serino, G. (1994). Design methodologies for energy dissipation devices to improve seismic performance of steel buildings. In Mazzolani, F., and Gioncu, V., eds., *Behaviour of Steel Structures in Seismic Areas, Stessa '94*,E & FN SPON, 703-713

Takayama, M., Tsujii, T., Ogura, K.,Izumi, M. and Tsujita,O. (1997). Seismic design for framing structure equipped with energy absorbing systems, In Mazzolani, F., and Akiyama, H., eds., *Behaviour of Steel Structures in Seismic Areas, Stessa '97*, Edizioni 10/17, 770-777.

Vayas, I., Pasternak, H., Schween, T. (1995). Cyclic Behavior of beam-to-column steel joints with slender web panels, *ASCE, J. of Struct. Engn*. Vol 121, No 2, 240-248.

Vayas, I., Pasternak, H. Schween, T. (1994). Beanspruchbarkeit und Verformung von Rahmenecken mit schlanken Stegen, *Bauingenieur* 69, 311-317.

Vayas, I. (1996). Stregnth and Ductility of Axially Loaded Members with Outstand Plated Elements, In Rondal, J. et al., eds. *Proc. Coupled Instabilities in Metal Structures*, CIMS '96, Liege, Imperial College Press,189-198.

Vayas, I. (1997). Stability and Ductility of Steel Elements, *J. of Constructional Steel Research* Vol. 44, 1-2, 23-50.

Wada, A, Huang,Y.H., Yamada, T., Ono, Y., Sugiyama, S., Baba, M., Miyabara, T. (1997). Actual size and real time speed tests for hysteretic steel dampers, In Mazzolani, F., and Akiyama, H., eds., *Behaviour of Steel Structures in Seismic Areas, Stessa '97*, Edizioni 10/17, 778-785.

Yamada, M. (1980). Bauen in erdbebengefährdeten Gebieten, *Deutsche Bauzeitung 11/80*

CHAPTER 6

SEISMIC RESISTANT COMPOSITE STRUCTURES

A. Plumier
University of Liège, Liège, Belgium

1   Context, scope and concepts of the proposed design guidance.

1.1   Context

In the beginning of the years 1990's very little design guidance did exist for the seismic design of composite steel concrete structure.   The available test data were rather scarce, like were the numerical analysis.

A proposal was made at the time to introduce guidance in Eurocode 8 concerning composite structures, but the document proposed was judged so weak that it was kept only as an informative annex. It was the only chapter on material in the case.

To modify this situation, a strong research push was launched by DGIII and DG XII of the European Union, which involved major experimental programs, funding of researchers, numerical modeling of structures, under the ECOEST, TMR and COPERNICUS funding.

As a result of that research push, design guidance started to be developed within the ICONS project, which under its Topic 4, dealt with composite steel concrete structures. A group of researchers took advantage of the mobility founds of the project to create under the coordination of Prof. PLUMIER an extended working group and to produce the draft of a chapter on composite steel concrete structures for Eurocode 8. The text presented in this chapter reflects this draft, which is now in a very mature stage. It is presented in the form of main text and Comments. All the contributions to this draft are thanked for this work. They are:

A.   Elnashai (Imperial College London)
A.   Elghazouli (Imperial College London)
J.   Bouwkamp (Technical University of Darmstadt)
B.   Broderick (Trinity College Dublin)
J.M. Aribert (Institut National des Sciences Appliquées de Rennes)
A.   Lachal (Institut National des Sciences Appliquées de Rennes)
R.   Zandonini (University of Trento)
O.   Bursi (University of Trento)
C.   Cosenza (University Federico II of Napoli)
G.   Manfredi (University Federico II of Napoli)
C.   Doneux (University of Liege)

1.2. Design concepts

(1) Earthquake resistant composite buildings shall be designed according to one of the following concepts:

Concept a   Dissipative structural behaviour with composite dissipative zones.
Concept b   Dissipative structural behaviour with steel dissipative zones.
Concept c   Non-dissipative structural behaviour.

(2) In concepts a and b, the capability of parts of the structure (dissipative zones) to resist earthquake actions out of their elastic range is taken into account. When using the design spectrum defined in clause 4.2.4 of Part 1-1, the behaviour factor q is taken greater than 1,5 (see 3.2.).

(3) In concept b, structures are not meant to take any advantage of composite behaviour in dissipative zones; the application of concept b is conditioned by a strict compliance to positive measures that prevent involvement of the concrete in the resistance of dissipative zones; then the composite structure is designed to Eurocode 4 under non seismic loads and to section 3 of this chapter to resist earthquake action; the positive measures preventing involvement of the concrete are defined at paragraph 6.4., with possible additional details for each structural typology at paragraphs 7. to 11.

(4) In non dissipative structures (concept c), the action effects are calculated on the basis of an elastic analysis without taking into account non-linear material behaviour but considering the reduction in moment of inertia brought by the cracking of concrete in part of beam spans, according to general structural analysis data defined in 4. and specific ones related to each structural type at paragraphs 7. to 11. When using the design spectrum defined in clause 4.2.4 of Part 1-1, the behaviour factor q is taken equal to 1,5. The resistance of the members and of the connections should be evaluated in accordance with Part 1-1 of Eurocode 4, eventually supplemented by detailing rules enhancing the available ductility.

(5) The design rules for dissipative composite structures (concept a), aim at the development in the structure of reliable local plastic mechanisms (dissipative zones) and of a reliable global plastic mechanism dissipating as much energy as possible under the design earthquake action.
For each structural element or each structural type considered in this Section 6, rules allowing this general objective to be reached are given at paragraphs 5. to 11. with reference to what is called the specific criteria. These criteria aim at the development of a design objective that is a global mechanical behaviour for which design indications can be given at present.

C. Implicitly, this last sentence means that there may be other reasonable design objectives, but that the present state of the art is not such that a real design guidance can be provided.

2. MATERIALS

2.1. Concrete

In dissipative zones, the use of concrete class lower than C20/25 or higher than C40/50 is not allowed.

2.2. Reinforcing steel

(1) Reinforcing steel considered in the plastic resistance of dissipative zones should satisfy the requirement of table 2.1 of Section 2. Ductility class M or H applies to dissipative structures and to regions with high stresses of non dissipative structures. This applies to bars and to welded meshes.

(2) Except for closed stirrups or cross ties, only ribbed bars are allowed as reinforcing steel in regions with high stresses.

(3) Welded meshes not complying with the ductility requirements of (1) may be used in dissipative zones. In regions with high stresses, the sections of reinforcement in the mesh must then be duplicated by ductile reinforcements.

C. The problem behind the above statements is that in moment frames a reliable negative plastic moment in the connection zone can only be based on the presence of reinforcements guaranteed ductile, while the beam plastic moment considered in the capacity design of column must include all possible contributions of the reinforcement, non ductile welded mesh included for instance.

2.3. Steel sections

(1) Steel sections, welding and bolts shall conform to the requirements specified in clause 3 of Part 1-1 of Eurocode 3, unless specified otherwise hereafter.

(2) In non dissipative bolted connections fully prestressed high strength bolts in category 8.8 or 10.9 should be used in order to comply with the needs of capacity design (see 1.2.(5)). Bolts of category 12.9 are only allowed in shear connections.

(3) In dissipative zones the following additional rules apply :
    a) Structural steel should conform to EN 10025.
    b) The value of the yield strength, which cannot be exceeded by the material used in the fabrication of the structure and the tensile strength of the steel used, should be specified.

3. STRUCTURAL TYPES AND BEHAVIOUR FACTORS

3.1 Structural types

The design rules apply to structures that belong to one of the following structural types:
- moment resisting frames with limited sway
- concentrically braced frames
- eccentrically braced frames
- composite steel beams with reinforced concrete walls
- composite steel plates – concrete shear walls.

These structural types should comply with the criteria defined in 7. to 11.

C. The moment resisting frames with limited sway correspond to the ones classified "braced" or "non sway" in Eurocode 4; they are such that second order (P – D) effects need not be considered, because the following condition is satisfied at every storey.

$$\theta = \frac{P_{tot}.d_r}{V_{tot}.h} \le 0,10$$

This condition is defined in 4.2.2. of Part 2 of this Eurocode 8; it is similar to the condition in 4.9.4.2. of Eurocode 4. When SLS (Serviceability Limit State) requirements on drift are strong, this criteria is practically always fulfilled

3.2 Behaviour factors

(1) The behaviour factor q, introduced in clause 4.2.4 of Part 1-1 to account for energy dissipation capacity, takes as maximum the values given in figure 3 for various structural types, provided that the regularity requirements laid down in Part 1-2 and the detailing rules given in this Section 6 are met.

(2) The following parameters are used in Figure 3.1.
α_1 multiplier of the horizontal design seismic action, while keeping constant all other design actions, which corresponds to the point where the most strained cross-section (taking into account locations of stiffeners) reaches its plastic moment resistance.
α_u multiplier of the horizontal design seismic action, while keeping constant all other design actions, which corresponds to the point where a number of sections sufficient for the development of overall structural instability reach their plastic moment resistance.

(3) When calculation are not performed in order to evaluate the multiplier α_u, the approximate values of the ratio   presented in Figure 3.1. may be used.

If the building is non-regular in elevation (see clause 2.2.3 of Part 1-2), the q-values listed in Figure 3 .1.should be reduced by 20 % (but need not be taken less than q = 1,0).

C. These numbers are similar to those of the chapter on steel and reinforced concrete structures. It has been shown [9] that the values of q does not differ significantly from one material type to another, as long as a correct evaluation of the "first yield" in the structure is made. It should correspond to the loading for which the first plastic hinge starts rotation. However, some factors could influence these values of q:
- *— - the concentration of plastic rotations in the first hinges formed, because of the difference between $M_{Rd}{}^+$ and $M_{Rd}{}^-$ of composite sections*
- *— - the effect of reduced low cycle fatigue resistance of T sections, in comparison to symmetrical sections*
- *— - the effect of increased damping, in comparison to steel structure, due to cracking and to friction at steel concrete interface.*

At Figure 3.1, the values of q are given. Most values are similar to those of steel structures. However, for structures type 5,6,7 these values of steel structures have been revised in the way indicated at the 3rd page of Figure 3.1

1. Frame structures $\frac{\alpha_u}{\alpha_1}\sim 1.20$ $\frac{\alpha_u}{\alpha_1}\sim 1.10$	high regularity	medium regularity
 diss. zones diss. zones = bending zones in the beams	$q = 5\,\frac{\alpha_u}{\alpha_1}$ [*]	$q = 4\,\frac{\alpha_u}{\alpha_1}$ [*]
2. Concentric truss bracings Diagonal bracings 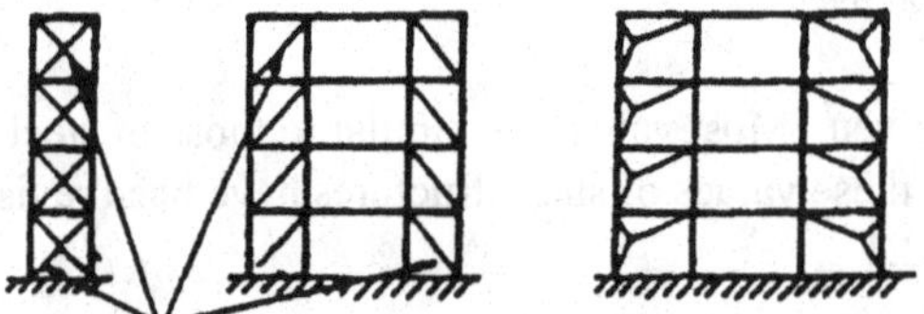 diss. zones = tension diagonals only	$q = 4$	$q = 3$
V - bracings 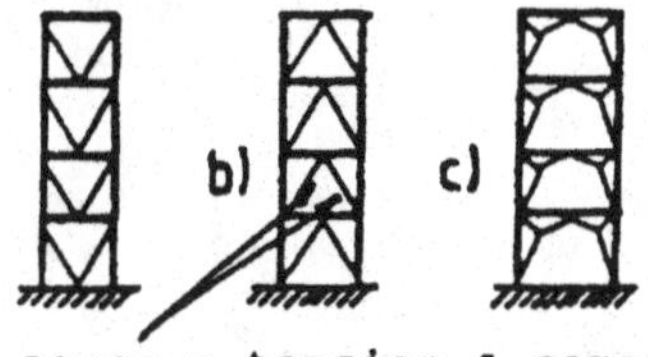 a) b) c) diss. zones = tension & compression diagonals	$q = 2$	$q = 1.5$
K -bracing 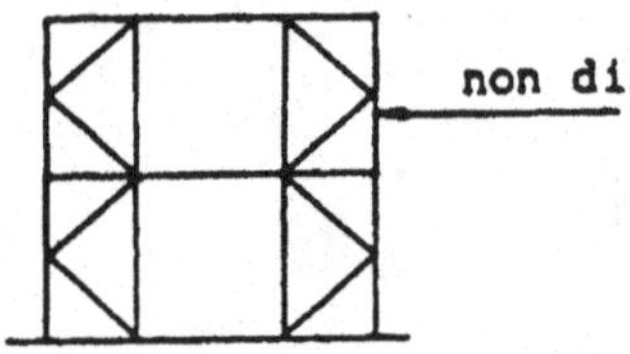 non dissipative	$q = 1$	$q = 1$

[*] higher values of $\alpha u/\alpha 1$ can be used, if justified (1.60 is the maximum)

Figure 3.1. : Structural types and behaviour factors.

3. Eccentric truss bracings	high regularity	medium regularity
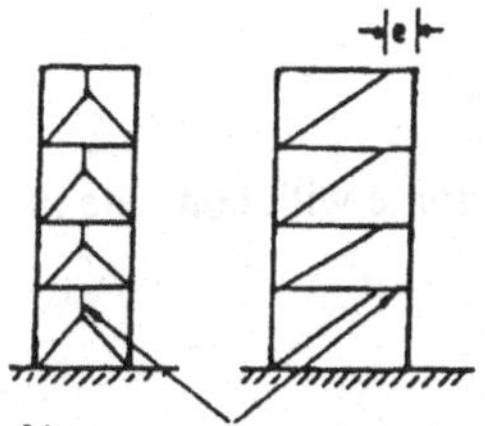 $\frac{a_u}{a_1} \sim 1.10$ diss. zones = bending or shear zones	$q = 5\,\frac{a_u}{a_1}$ *	$q = 4\,\frac{a_u}{a_1}$ *
4. Cantilever structures 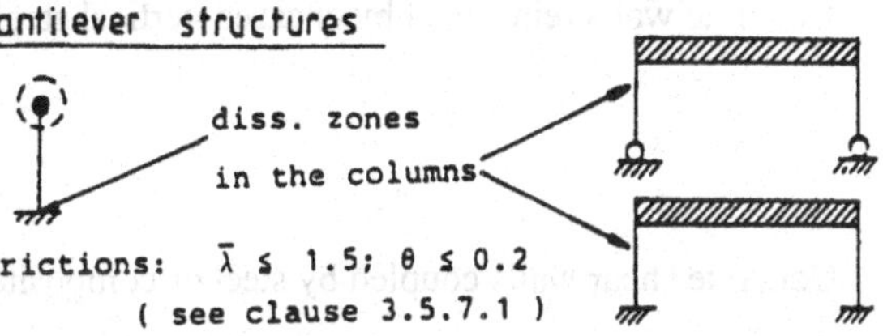 diss. zones in the columns Restrictions: $\bar{\lambda} \leq 1.5$; $\theta \leq 0.2$ (see clause 3.5.7.1)	$q = 2$	$q = 1.5$
5. Cores or walls in reinforced concrete 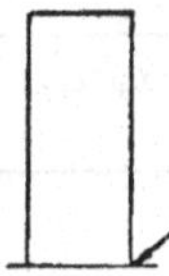 diss. zones	chapter 7	chapter 7
6. Dual structures 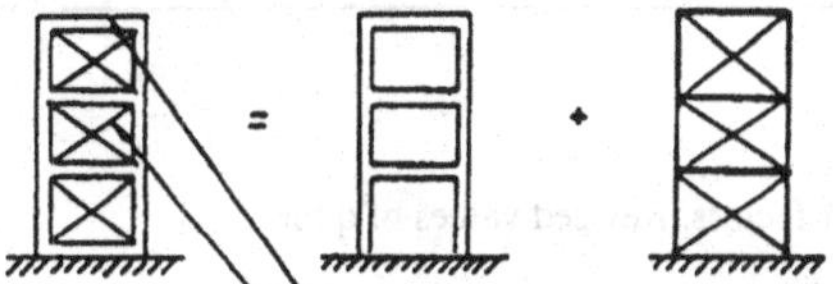 frames with diss. bending zones bracings with diss. tension zones	$q = 5\,\frac{a_u}{a_1}$ *	$q = 4\,\frac{a_u}{a_1}$ *
7. Mixed structures made from steel frames with reinforced concrete infills 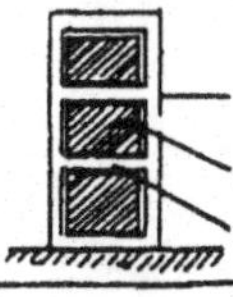 frames with diss. bending zones diss. shear walls diss. joints	$q = 2$	$q = 1.5$

* higher values of $\alpha u/\alpha 1$ can be used, if justified (1.60 is the maximum)
**Structures of types 5, 6 and 7 have revised versions of q given on next page

Figure 3.1. (continued) : Structural types and behaviour factors.

5. Reinforced concrete shear wall elements

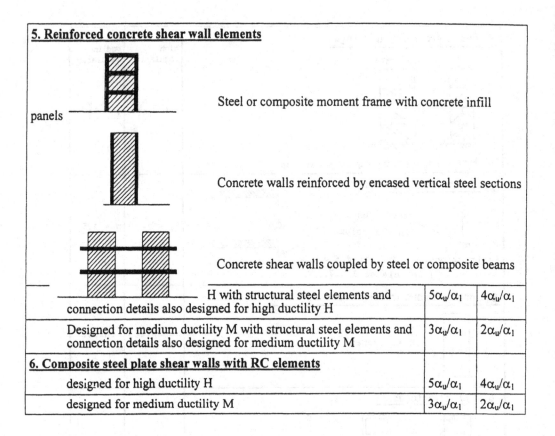

panels — Steel or composite moment frame with concrete infill		
Concrete walls reinforced by encased vertical steel sections		
Concrete shear walls coupled by steel or composite beams		
H with structural steel elements and connection details also designed for high ductility H	$5\alpha_u/\alpha_1$	$4\alpha_u/\alpha_1$
Designed for medium ductility M with structural steel elements and connection details also designed for medium ductility M	$3\alpha_u/\alpha_1$	$2\alpha_u/\alpha_1$
6. Composite steel plate shear walls with RC elements		
designed for high ductility H	$5\alpha_u/\alpha_1$	$4\alpha_u/\alpha_1$
designed for medium ductility M	$3\alpha_u/\alpha_1$	$2\alpha_u/\alpha_1$

Figure 3.1. (continued) : Structural types and behaviour factors. Revised values of q for structures with shear walls

4. STRUCTURAL ANALYSIS

4.1 Scope

The following rules apply for a static equivalent dynamic elastic analysis of the structure under earthquake action.

4.2 Stiffness of sections

1. The stiffness of composite sections in which the concrete is in compression is computed considering a modular ratio n
$$n = E_a / E_c = 7$$

2 For composite beams with slab in compression; the moment of inertia of the section is computed considering the effective width of slab defined in 6.3. and it is referred to by the symbol I1.

C. In EC4 wording, I1 is the second moment of area of the effective equivalent steel section calculated assuming that concrete in tension is uncracked. It is usually used for the second moment of area under sagging bending.

3. The stiffness of composite sections in which the concrete is in tension is computed considering that the concrete is cracked and that only the steel parts of the section are active.

4. For composite beams with slab in tension, the moment of inertia of the section is computed considering the effective width of slab defined in 6.3. and it is referred to by the symbol I2.

C. In EC4 wording, I2 is the second moment of area of the effective equivalent steel section calculated neglecting concrete in tension but including reinforcement. It is usually used for the second moment of area under hogging bending.

5. The structure is analysed considering the presence of concrete in compression in some zones and concrete in tension in other zones; the distribution of the zones is given in paragraphs 7. to 11. for the various structural types.

4.3. Diaphragms

The analysis of the structure can be made considering floor diaphragms rigid without further verification, if they consist of reinforced concrete according to section 2 of this Part, if their openings do not significantly affect the overall in-plane rigidity of the floor and if they are adequately linked to the steel elements in order to transmit the seismic forces of floors to the vertical earthquake resistant structure.

5. DESIGN CRITERIA AND DETAILING RULES FOR DISSIPATIVE STRUCTURAL BEHAVIOUR COMMON TO ALL STRUCTURAL TYPES.

5.1. General.

1. The design criteria given in 5.2. apply for earthquake-resistant parts of structures, designed according to the concept of dissipative structural behaviour.

2. The design criteria given in 5.2 are deemed to be satisfied, if the detailing rules given in 5.3 - 5.5 and in paragraphs 6. to 11. are observed.

5.2 Design criteria for dissipative structures

1. Structures with dissipative zones shall be designed such that these zones develop in those parts of the structure where yielding or local buckling or other phenomena due to hysteretic behaviour do not affect the overall stability of the structure.

2. Structural parts of dissipative zones shall have adequate ductility and resistance. The resistance shall be verified according to Eurocode 3 (concept b) and to Eurocode 4 (concept a).

3. Non-dissipative parts of dissipative structures and the elements connecting dissipative to non dissipative parts of the structure shall have sufficient overstrength to allow the development of cyclic yielding of the dissipative parts.

5.3. Plastic resistance of dissipative zones

1. Two plastic resistances of dissipative zones are considered in the design of composite steel concrete structures: a lower bound plastic resistance (index pℓ, Rd) and an upper bound plastic resistance (index u, Rd).

2. The lower bound plastic resistance of dissipative zones is the one considered in design checks concerning sections of dissipative elements ;
Example: $M_{Sd} < M_{pl,Rd.}$
The lower bound plastic resistance of dissipative zones is computed considering the concrete component of the section and only the steel components of the section, which are certified ductile.

3. The upper bound plastic resistance of dissipative zones is the one considered in the capacity design of elements neighbour of the dissipative zone.
Example: at the intersection of beams and columns of constant section

$$1.2(M^{+}_{u,RD,beam} + M^{-}_{u,Rd,beam}) < 2\,M_{pl,Rd,column}$$

The upper bound plastic resistance is established considering the concrete component of the section and all the steel components present in the section, including those that are not certified ductile.

4. Action effects, which are directly related to the resistance of dissipative zones, must be determined on the basis of the resistance of the full composite dissipative sections.
Example: the design shear force at the end of a dissipative composite beam or seismic link must be determined on the basis of the full moment of resistance of the full composite section.

5.4 Detailing rules for connections in dissipative zones

5.4.1. Composite connections

1. Composite connections in dissipative zones designed according to Concept a as defined in 1.2.(1) should exhibit sufficient overstrength to allow for yielding of the connected parts.

2. The integrity of the concrete shall be maintained during the seismic event. Yielding of the reinforcing bars should be allowed only if beams are designed to comply with clause 6.2.(7)

3. Connections of the steel parts made by means of butt welds or full penetration welds are deemed to satisfy the overstrength criterion.

4. For fillet weld connections or bolted connections, the following overstrength requirement should be met :

$R_d \geq 1,20\ R_{fy}$

where R_d resistance of the steel part of the connection according to clause 6 of Part 1-1 of Eurocode 3,
R_{fy} plastic resistance of the connected steel part.

5. For overstrength verifications an appropriate estimation of the actual strength value of the connected parts should be made.

6. For ultimate limit state verifications the partial safety factors for material properties γ_M as specified in clause 2.3.3.2 of Eurocode 4 for fundamental load combination apply.
   C. This clause should not be applied to composite connections only.

7. The overstrength condition for connections need not apply if the connections are designed in a manner enabling them to contribute significantly to the energy dissipation capability inherent to the chosen q-factor and if the contribution of such connections to the global flexibility can be properly assessed.

8. The stiffness, strength and rotation capacity of composite connections under cyclic loading should be established by tests. Under static loading, they may be calculated using the subsection 8.3. and 8.4. of prEN 1994-1-1, in agreement with the sections 6. and clause 5.4.1.(6). The rotation capacity should not be less than 35 mrad for negative bending and 20 mrad for positive bending.

   C. Annex J of Eurocode 4 means the Cost report on "Composite steel-concrete joints in frames for buildings: design provisions". Currently, we are not able to quantify the effect of hysteresis on stiffness degradation. None the less, we feel that the clause 5.4.1.(6) is adequate to account for strength degradation. The requirements on rotation capacity derive from [10]. This type of requirement need to be homogenised with the Section on steel structures.

9. For bolted shear connections, bearing failure should precede shear failure. The bolts should be tightened as prescribed in clause 6.5.3 of Part 1-1 of Eurocode 3 for connections of Category B or C to prevent loosening of the nuts.

5.4.2. Connections of the steel sections

1. Connections of the steel sections in dissipative zones designed according to Concept b as defined in 1.2.(1) should exhibit sufficient overstrength to allow for yielding of the connected parts. Clauses 5.4.1.3, 4 and 5 apply too.

2. For ultimate limit state verifications the partial safety factors for material properties γ_M as specified in Part 1-1 of Eurocode 3 apply.
   C. *This clause may be necessary to avoid confusion with the clause 5.4.1.6.*

3. The overstrength condition for connections need not apply if the connections are designed in a manner enabling them to contribute significantly to the energy dissipation capability inherent to the chosen q-factor and if the contribution of such connections to the global flexibility can be properly assessed.

4. The stiffness, strength and rotation capacity of steel connections under cyclic loading should be established by tests. The stiffness and strength of steel connections may be computed by means of the Annex J of Eurocode 3 in agreement with the clause 6.5.4.2.(2). The rotation capacity should not be less than 35 mrad.
   C. *We feel that both the stiffness and the strength of steel joints are not so sensitive to hysteretic effects. The requirements on rotation capacity derive from [11].*

5. For bolted shear connections, bearing failure should precede shear failure. The bolts should be tightened as prescribed in clause 6.5.3 of Part 1-1 of Eurocode 3 for connections of Category B or C to prevent loosening of the nuts.

5.4.3. Connections between concrete and steel structural elements in composite structural systems.

1. Connections between concrete and steel structural elements in composite systems will be designed such that yielding takes place in the steel elements.

2. Local design of the connection zone of the concrete structural elements will be justified by struts and ties mechanisms.

3. Two classes of ductility of the structural details in such connection zones can be defined. Ductility class M details should allow a 35mrad cyclic rotation capacity in the steel element to be developed, without significant degradation of the concrete. Ductility class H details should allow an 80 mrad cyclic rotation capacity in the steel element.

4. For each type of composite system, there is a relation between the ductility class of the connection details and the maximum behaviour factor that can be considered in the analysis of the structure. See Table 6.1.

5.5. Detailing rules for foundations.

1. The design values of the action effects E_{Fd} on the foundations should be derived as follows:

$$E_{Fd} = 1{,}20 \, (E_{F,G} + \alpha \, E_{F,E})$$

where $E_{F,G}$, action effect due-to the non-seismic actions included in the combination of actions for the   seismic design situation (see clause 4.4 of Part 1-1),

$E_{F,E}$ action effect due to the design seismic action multiplied by the importance factor,
α value of (R_{di}/S_{di}) of the dissipative zone or element i of the structure which has the highest influence on the effect EF under consideration,

where: - R_{di} design resistance of the zone or element i

- S_{di} design value of the action effect on the zone or element i in the seismic design situation.

6. DETAILING RULES FOR MEMBERS COMMON TO ALL STRUCTURAL TYPES

6.1. General and classes of steel sections

1   Composite members, which are part of the earthquake resistant structures, must comply with the rules of Part 1-1 of Eurocode 4 and with additional rules defined in this Section of Eurocode 8.
    C. Encased beams are at present out of the scope of Eurocode 4; partially encased beams are considered. The same option is chosen here.

2. The earthquake resistant structure is designed with reference to a global plastic mechanism involving local dissipative zones; there are members part of the earthquake resistant structure in which dissipative zone are located and other members where no dissipative zone should be present.

3. The members in which dissipative zones are located must be made of sections belonging to a class of cross sections defined in Eurocode 4. The relation between the maximum behaviour factor q considered in the analysis of the structure and the class of section is indicated in Table 6.1.

Table 6.1. Relation between classes of section and behaviour factor q.

Class of sections	q	Ductility class of the structure
1	$5\dfrac{\alpha_u}{\alpha_1}$	H
2	4	M

C. This Table makes explicit a fact which was implicit in Part 1.3 of previous versions of Eurocode 8: for steel structures, ductility classes were already defined through the relationship between local behaviour (members) and global behaviour (behaviour factors q). The third column in the above Table allocates a symbol to the ductility classes of structures.
C. For moment frames, the Seismic Provisions for Steel and Composite Buildings drafted by AISC [1] distinguish similarly between ordinary, intermediate and special moment frames, with increasing ability to dissipate energy in plastic mechanisms.

4. The members part of the earthquake resisting structure must belong to the class 1 or 2 of
    cross sections defined in Eurocode 4.
    *C. The background for rules 3 and 4 is really weak; this explains the safe side character
    of the proposal.*
    *C. The requirements on section classes implicitly means that class 1 sections have cyclic
    rotation capacities over 35 mrad (for instance), without great strength and stiffness
    degradation. Section of class 2 would correspond to rotation capacities over 20 mrad.*

5. Sections 6.2. to 6.8. apply to members where dissipative zones are located.

6.2. Beams made of a steel section composite with a slab.

For beams in which dissipative zones are located and for which it is intended to take advantage
of the composite character of the section for energy dissipation, the rules hereafter apply. They
correspond to the following design objective: the integrity of the concrete is maintained during
the seismic event; yielding takes place in the bottom part of the steel section and/or in the
rebars of the slab.
If it is not intended to take the advantage of the composite character of the beam section for
energy dissipation, paragraph 6.6.4. has to be considered.

1. Beams conceived to behave as composite elements in dissipative zones of the earthquake
   resistant structure shall be designed for full or partial shear connection according to 6.2.1.of
   Eurocode 4.

2. If connectors are ductile as defined in 6.1.2. of Eurocode 4, partial shear connection may
    be adopted with a minimum connection degree of 0,8. Full shear connection is required
    when non ductile connectors are used.
    C. The indication 0,8 comes out from research works [2] and [3].

3. The design resistance of connectors in dissipative zones is obtained from the design
    resistance provided in 6.3. of Eurocode 4 applying a reduction factor equal to 0.75.
    C Recent experimental evidences indicate such a value [4].

4. In the absence of more accurate design rules, the design resistance of shear connectors
    should be determined from push-type tests in accordance with the procedures established
    in 10.2 of Eurocode 4. In detail, the testing equipment should provide symmetric boundary
    conditions when testing specimens are subjected to cyclic loading. The cyclic slip capacity
    δu can be obtained from the aforementioned tests.
    For headed stud shear connectors, the cyclic slip capacity can be derived from the one
    relevant to a monotonic push-type test applying a reduction factor equal to 0.5.

5. To achieve ductility in plastic hinges, the distance from the concrete compression fibre to the plastic neutral axis should not be more than 15 % of the overall depth of the composite cross section.

C. This is the Eurocode 4 rule aiming at the development of extensive yielding in the bottom flange of the steel section when the beam is submitted to positive moments. Other formulations are possible.
In the AISC draft, a complete formula is given:
The distance from the maximum concrete compression fibre to the plastic neutral axis shall not exceed:

$$\frac{Y_{con} + d_b}{1 + \left(\dfrac{1{,}700\,F_y}{E_s}\right)}$$

where

Y_{con}	=	*the distance from the top of the steel beam to the top of concrete, in.*
d_b	=	*the depth of the steel beam, in.*
F_y	=	*the specified minimum yield strength of the steel beam, ksi.*
E_s	=	*the elastic modulus of the steel beam, ksi.*

It corresponds to a steel deformation and a concrete deformation
$\varepsilon_{cu} = 3\ ‰$ ε_s

A proposal by Plumier [5] was a formulation also directly linked to the physics of the problem:

$$\varepsilon_s = \mu\,\varepsilon_{sy} = 6\frac{f_y}{E} \qquad when \qquad \varepsilon_c = 2.10^{-3}$$

The formula is $\qquad \dfrac{x}{d} < \dfrac{2.10^{-3}}{2.10^{-3} + 6\dfrac{f_y}{E}}$

For a Fe 510 steel $(f_y = 355\ MPa)\ \dfrac{x}{d} < 0{,}16$

Which is similar to $\dfrac{x}{d} < 0,15$ *(Eurocode 4).*

6   For beams with slab in which dissipative zones are located, specific reinforcements of the slab called "seismic rebars" may have to be present in the zone of connection of the beam to the column. The design conditions of these re-bars are defined in Annex J to this document.

C. For beams with slab in which dissipative zones are located, specific reinforcements of the slab called "seismic rebars" must be present in the zones of connections of the beams to the columns. Their sections As and their lay out must be designed to reach ductility. It has been established that there cannot be one simple single rule which would guarantee that this goal is achieved. In fact, the design is different for exterior and interior columns; it is also different when beams transverse to those primarily bent by the earthquake action are, or not, present, that is when beams are present, or not, on two orthogonal axis directions at their connections to the columns. This has led to the definition of an "Annex j" on slab design in the connection zones.

Figure 6.6.1. : Layout of "Seismic Rebars"

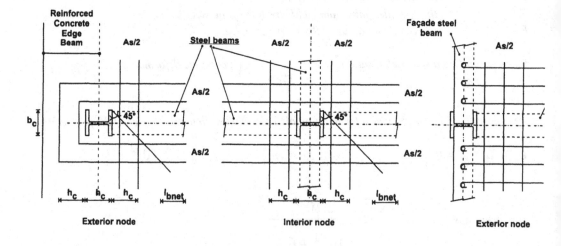

Exterior node   Interior node   Exterior node

7   When profiled steel sheeting is used, ribs should preferably be parallel to the beam. In case of sheeting with ribs transverse to the supporting beams the reduction factor k_t of the design shear resistance of connectors given by the relation 6.16 of Eurocode 4 should be reduced by applying the following rib shape efficiency factor k_r.

Figure 6.2. Value of the rib shape efficiency factor.

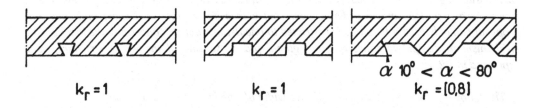

$k_r = 1$ $k_r = 1$ α $10° < \alpha < 80°$
                                                                $k_r = [0,8]$

C. Recent experimental evidences have shown that steeldeck with trapezoidal shapes having positive slopes α induce uplift forces in the slab which can generate a concrete cone failure at the connector; the problem is not yet calibrated and the above k_r are only suggested.
C. Furthermore, the additional reduction factor k_r might also be influenced by other parameters as the number of shear connectors per rib and the procedure of welding (either through the steel sheeting or directly on the steel flange for sheeting with holes).

C. Preferably, the re-bars of the slab should be positioned under the level of the head of the connectors. This has 2 positive effects :
-it equalises the required displacement of the connectors, contributing to a better correspondance between the design hypothesis of equal values of resistance provided by connectors
-it contributes to prevent uplift of the slab by improving the resistance of the slab under the level of connector 's heads.

Effective width of slab.

C. This paragraph has to tackle several problems, because its content should provide design data bringing fair estimate of:
the periods of the structure, which are a matter of elastic stiffnesses, having implications on the design forces (action effects) in the standard design method using design spectra
the plastic resistances of dissipative zones and of the structure as a whole, for which reliable lower bound estimates are needed

the capacity design of non dissipative elements, which should be based on upper bounds resistances of the dissipative zones.

The matter as such is vast, especially if one considers the many various possible design of "composite steel concrete" structures and the possible variety of design concepts for dissipative zones. As a first step, and in order to come out with a design tool in a short time, the code proposal can only exist if its scope is restricted to one definite option concerning the local plastic mechanism in dissipative zones and the global plastic scheme of the whole structure, as it expressed in 1.2. (5).

C. The following text only deals with effective width of slab in composite T beams.

(1)   The total effective width b_{eff} of concrete flange associated with each steel web should be taken as the sum of effective widths b_e of the portion of the flange on each side of the centreline of the steel web (Figure 6.3.). The effective width of each portion should be taken as b_e given in Table 2 but not greater than b defined hereunder in (2).

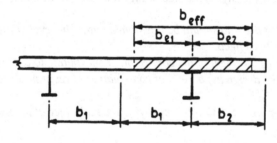

Figure 6.3.

(2) The actual width b of each portion should be taken as half the distance from the web to the adjacent web, measured at mid-depth of the concrete flange, except that at a free edge the actual width is the distance from the web to the free edge.

(3) The portion b_e of effective width of slab to be used in the determination of the elastic and plastic properties of the composite T sections made of a steel section connected to a slab are defined in Table 6.2. and Figure 6.4.

These values are valid when the seismic re-bars defined in the Annex J to the present main text are present.

Figure 6.4. : Definition of Elements in Moment Frame Structures.

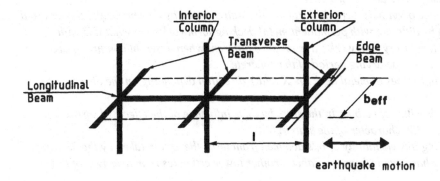

TABLE 6.2.

DEFINITION OF EFFECTIVE WIDTH OF SLAB

b_e	Transverse beam	b_e for M_{Rd} (PLASTIC)	b_e for l (ELASTIC)
At interior column	Present, fixed to the column, with connectors for full shear	For M^- : 0,1 ℓ, For M^+ : 0,075 ℓ	0,05 ℓ 0,0375 ℓ
At interior column	Not present, or present and not fixed to the column, or not having connectors for full shear	no proposal	no proposal
At exterior column	Present as an edge beam fixed to the column in the plane of the columns, with connectors for full shear and specific detailing for anchorage of re-bars exterior to the column plane, with re-bars of the hair pin type	For M^- : 0,1 ℓ, For M^+ : 0,075 ℓ	0,05 ℓ 0,0375 ℓ
At exterior column	Not present or no re-bars anchored	For M^- 0 For M^+ : $b_c/2$ or $h_c/2$	0 $b_c/4$ or $h_c/4$

C1. At present, the values given here in Table 2 have been established on the basis of the Ispra 3D test, the Darmstadt tests and the Saclay test.

The analysis of the test results indicated quite repetitive values for what concerns the evaluation of the plastic moment M_{Rd} in all the test performed, in spite of the differences existing in dimensions, type of formwork, layout of re-bars...

On the contrary, the effective widths deduced from the elastic part of the tests were quite variable from test to test or from X to Y direction in the Ispra-3D test. This may be the result of the influence of several factors, which is more important on the elastic stiffness than on the plastic resistance. These factors are, for instance:

- *the existence of an initial cracked state of the slab induced by the shrinkage; this cracked state can be different with permanent metal decking formwork from what it is with classical temporary formwork ; it is for sure different when different conditions are realised during the drying period of the concrete*
- *different influences of metal deck in X and Y direction depending on the direction of waves*
- *different densities of re-bars in the slab, by their influence of the effects of shrinkage*
- *the 2D or 3D character of the slab in tests.*
- *Considering this variability of b_e's deduced from tests, the option taken by the drafting committee has been to keep for Table 1 rather lower estimates than averages of test results.*

C2. In Eurocode 4, the effective width b_e of the portion of the flange on each side of the centreline of the steel web is linked with the bending moment diagram and should be taken as $b_e = l_0/8$ ($< b$).

The length l_0 is the approximate distance between points of zero bending moment. For simply-supported beams it is equal to the span. For typical continuous beams, l_0 may be assumed to be as shown in Figure C1, in which values at supports are shown above the beam, and midspan values below the beam.

Figure C1   Equivalent spans, for effective width of concrete flange.

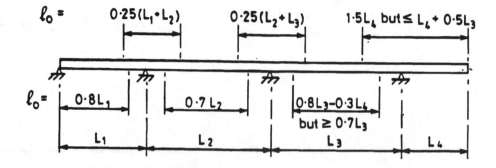

The same definition can be used for earthquake loading while remaining consistent with the experimental results, if appropriate lengths l_0 are defined for horizontal loading situation. Figure C2 presents the equivalent of Figure 3 for earthquake loading:

Figure C2 : Equivalent spans for effective width of concrete flange – horizontal loading case

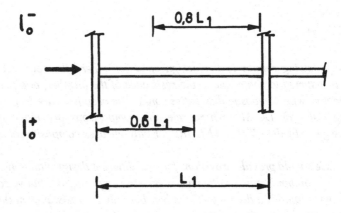

The partial effective widths deduced from Figure C2 are consistent with the experimental values:

$$b_e^+ = \frac{\ell_o^+}{8} = 0{,}075\,L$$

$$b_e^- = \frac{\ell_o^-}{8} = 0{,}1\,L$$

6.4   Conditions for disregarding the composite character of sections in dissipative zone.

In elements intended to act as pure steel sections in dissipative zones, as required for a design according to concept b, there must be a complete mechanical disconnection between the concrete, part of the structural resistance of the composite beam, and the neighbour structural elements; this disconnection is defined for each structural typology in paragraphs 6.7. to 6.11.

6.5.   Fully Encased Composite Columns

6.5.1. General

C1. The following relates to the design condition where the reinforced concrete part of the section is assumed to contribute to the resistance of the member, and the member is designed to be dissipative with composite dissipative zones. The proposals are largely based on the EC8 RC detailing rules for DC'M'. Where relevant, comparisons are made with the corresponding guidance given by the AISC and NEHRP for fully-encased composite columns in the US.

C2. The code should provide provisions for four different design situations:
  - *a) the column is a dissipative member with composite dissipative zones*
  - *b) the column is a dissipative member, but only the resistance of the steel part is considered in the dissipative zones*
  - *c) the column is a non-dissipative composite member*
  - *d) the column is a non-dissipative steel member*

Situation (a) requires most design guidance and the following proposals relate mainly to this case. However, the majority of column members are non-dissipative and the elastic cyclic response needs to be assured. Non-dissipative columns (the majority) may be designed as composite if it is assured that their cyclic elastic response is adequate. Only dissipative columns (the minority) need to be detailed to ensure an adequate inelastic cyclic response before they can be designed as composite members. Situation (d) may often provide a simple workable solution requiring only a few simple earthquake-resistant rules.
Three cross-section types are considered: fully-encased, partially-encased and filled. Many of the design rules are the same or very similar for two or all of these. Many of the comments given after individual rules fully-encased columns apply to the other types as well. In addition, the design rules for partially-encased beams are similar to those for partially-encased columns.

1   The design of fully-encased composite columns may consider the resistance of the steel section alone or the composite resistance of the steel section and concrete encasement.

2. The design of fully uncased composite columns in which only the steel section is assumed
    to provide the member   resistance may be carried out according to the provisions of
    Chapter 3 of this part and the capacity design provisions of 5.3.
    C. Clause 5.3 ensures that the design shear force is determined on the basis of the moment
    of resistance of the full composite section.

3. In moment-resisting frames, the following inequality must apply for all composite
    columns:

$$N_{sd} / N_{pl,Rd} < 0,30$$

*C. This is a conservative rule which could be relaxed with more test data. However, it
should not be too restrictive as moment capacity will decrease significantly above this
level of axial load. A cross-reference to this clause should be given in the rules for
moment-resisting frames (7.1). In braced frames, concrete infill is used to increase the
axial resistance of the member to prevent buckling. Some bending moment may also be
present and this is taken into account using interaction curves. No further limit on axial
load should be imposed in these cases. In moment-resisting frames, column response is
primarily flexural, and may be dissipative. To ensure a satisfactory cyclic response, it
may be necessary limit column axial force below a certain value. This is especially
important for dissipative zones.*

4. In dissipative columns, the shear resistance shall be determined on the basis of the
    structural steel section alone.
    *C. This rule is a conservative proposal in the absence of better research. Eurocode 4
    allows the shear to be divided between the steel and concrete, and both the AISC and
    NEHRP codes determine shear strength on the basis of the steel section and the tie
    reinforcement. In the test series described above, however, the contribution of the
    concrete was seen to reduce drastically at even moderate displacements, while the
    contribution of the tie-reinforcement was not fully mobilised. As a result, the authors
    recommend the proposal 4 above.*

5. In non-dissipative members, the shear resistance of the column shall be determined
    according to the rules given in Eurocode 4.

6.5.2. Rules for composite behaviour.

1. When the concrete encasement is assumed to contribute to the resistance of the member,
    the design rules hereunder apply. These rules ensure full shear interaction between the
    concrete and steel parts and protect the dissipative zones against premature inelastic
    failure.

C1. The following relates to the design condition where the reinforced concrete part of the section is assumed to contribute to the resistance of the member, and the member is designed to be dissipative with composite dissipative zones.
The proposals are largely based on the EC8 RC detailing rules for DC'M'.
Where relevant, comparisons are made with the corresponding guidance given by the AISC and NEHRP for fully-encased composite columns in the US.

C2. Dissipative zones normally only occur at the base of columns. However, the EC8 detailing rules for RC columns consider critical zones at the upper and lower ends of all column lengths. It is proposed that the same principle be applied to fully- and partially-encased columns in which the concrete encasement will need protection.

2.   Critical regions at both ends of a column extend for a distance, l_{cr}, equal to:

$$l_{cr} = \max (1.5d_c, l_{cl}/6, 450mm)$$

   where:
         d_c is the largest cross-sectional dimension of the column,
         l_{cl} is the clear length of the column.

C. This is the existing EC8 rule for RC columns of DC'M'. The rule for DC'H' is: $max(1.5d_c, l_{cl}/5, 600mm)$. The rule for DC'L' is: $(d_c, l_{cl}/6, 450 mm)$.
In comparison, the AISC and NEHRP rules for fully-encased composite columns both stipulate that l_{cl} be taken as the maximum of $(d_c, l_c/6, 457 mm)$, i.e. a similar distance to that proposed above. This applies to columns in the higher Seismic Performance Categories (NEHRP) and to Intermediate and Special Moment Frames (AISC).

3.   To satisfy plastic rotation demands and to compensate for loss of resistance due to spalling of cover concrete, the following shall be satisfied within the critical regions defined above:

$$\alpha \, \omega_{wd} \geq k_o \, \mu_{1/r} \, v_d \, \varepsilon_{sy,d} \, (0{,}35A_c/A_o + 0{,}15) - 10\varepsilon_{cu}$$

   in which the variables are as defined in 2.8.1.3(5) of Section 2 of this Chapter 1.3. with $k_o = 60$ and $\mu_{1/r} >= 9$, but with the normalised design axial force defined as

$$v_d = N_{sd}/(A_c \, f_{cd} + A_a \, f_y)$$

C. The above is the EC8 rule which defines the minimum mechanical volumetric ratio of confining hoops within the critical regions of RC columns of DC'M'.
(For DC'L', $k_o = 55$ and $\mu_{1/r} >= 13$, while for DC'L', $k_o = 65$ and $\mu_{1/r} >= 5$).The above modification of the expression of the normalised axial force allows for the contribution of the steel section. The original EC8 expression for RC columns is:
$$v_d = N_{sd}/(A_c f_{cd}).$$

Similar modifications are included in the AISC and NEHRP codes. In both of these codes, the ACI rule for minimum areas of confining reinforcement is adopted*, but with the required area being reduced by the factor $(1 - A_a f_y/N_{sd})$. The new EC8 proposal above is a more conservative modification of the existing RC rule, as it assumes that axial load is shared equally between the steel and concrete.

* for 'special systems' (AISC) and 'performance categories D and E' (NEHRP).

4. In addition to 3 above, the spacing of the confining hoops, s, in the critical regions shall not exceed

$$s \leq \min (b_o/3, 150mm, 7d_{bl})$$

in which b_o is the minimum dimension of the concrete core and d_{bL} is the diameter of the longitudinal bars.

5. The diameter of the hoops, d_{bw} , shall be at least

$$d_{bw} \geq \max (0,35 \, d_{bl,max} \, [f_{ydL}/f_{ydw}]^{0.5}, 6mm)$$

C. Rules 3-5 ensure protection of the concrete at the top and bottom of each column length. This will be very unconservative in braced frames where no moment transfer between beams and columns takes place. The values given are the same as those specified for RC columns in DC'M' frames.

The minimum hoop spacing in DC'H' frames is min ($b_o/4$, 100mm, $5d_{bl}$), while for DC'L' frames it is min ($b_o/2$, 200mm, $9d_{bl}$).

For DC'H' frames, the coefficient of 0,35 in the expression for minimum d_{bw} size is modified to 0,4. There is no similar expression for DC'L' frames. The 6mm minimum applies for all ductility classes. The rules for minimum hoop spacing adopted in the US guidance for composite columns, is similar to the EC8 RC column detailing rules, and the values given in the AISC Provisions are identical to those given in the NEHRP Recommendations. The minimum spacing specified for columns of 'special frames' is similar to that proposed above, namely s $\leq$ min ($b_c/4$, 102mm, $24d_{bw}$) in the critical regions and s $\leq$ min ($6d_{bL}$, 152mm) throughout the column. (b_c is the least dimension of the column cross-section). The requirements for 'intermediate frames' (AISC) and frames in the middle Seismic Performance Category (NEHRP) are similar to those for EC8 DC'L'.

In this context, the most useful test data available relates to a series of 8 two-thirds scale tests conducted by Ricles and Paboojian [12]. These indicated that a maximum hoop spacing of s_{max} = $6.9d_{bl}$ should be used if a rotation ductility of at least 6.0 is required.

The minimum diameter of hoop reinforcement required by both the NEHRP and AISC is either $b_c/50$ or 10mm, whichever is the smaller.

6. Multiple hoop patterns (as described in Figure 2.13 for RC columns – Section 2 of this Chapter 1.3.) must be used in the critical regions of dissipative members. The distance restrained by hoop bends or cross-ties should not exceed 150mm in the critical regions of dissipative members or 250mm in the critical regions of non-dissipative members.

C. This is the EC8 rule for RC columns in DC'H'. Multiple hoops are not required for DC'M' or DC'L' where the minimum distance above is increased to 200mm and 250mm respectively.

The NEHRP rules for composite columns also adopts the hoop detailing requirements for RC columns in special moment frames, although the commentary to this code suggests that this is a conservative approach taken in the absence of adequate research. The commentary also presents an example of a suitable 'closed hoop detail' for encased columns. The DC'H' rule is retained because in the test series described above [12], specimens with only single hoop pattern reinforcement displayed lower rotation ductility capacities ($\mu < 6.0$).

7. In the lower two stories, hoops according to 4 - 6 shall be provided beyond the critical regions for an additional length equal to half the length of the critical regions.

8. The minimum cross-sectional dimension of the column shall not be less than 250 mm.

C. These are the EC8 rules for RC columns in DC'M'.
C. In EC4, the interaction of the steel and concrete parts of columns can be achieved through the use of bond alone. Allowable design bond stresses are specified for this purpose. The same approach is employed in Japanese design practice, but not in the ACI and AISC codes of the US [13], in which all shear transfer between the steel and concrete is based entirely on direct bearing. However, the AISC code does allow bond stress alone to be employed in fully-encased beams.
These EC4 design bond stresses may not be reliable in cyclic response conditions. Wherever the bond stress is inadequate, it is necessary to provide shear connectors. In braced frames, a small number of such connectors may be required close to the beam-column connection. For filled columns, Japanese practice achieves shear transfer through the connection detail in the interior of the column.
In non-dissipative columns, it may be valid to assume that composite action can be achieved through bond alone. However, to allow for cyclic response effects, the allowable design values of the bond stress should be lower than in EC4. Tests by Roeder on embedded steel sections [13] suggest that only one-third of the static bond capacity should be allowed. Such a code provision may allow a column without shear connectors to be designed as composite whenever the seismic action effects are not too high.

9. For earthquake-resistant design, the design values given in Eurocode 4 (Cl. 4.8.2.6 to 4.8.2.8) on the design shear strengths due to bond and friction. must be reduced by a

factor of [0.3]. Where insufficient shear transfer is achieved through bond and friction, shear connectors shall be provided to ensure full composite action.

10. Wherever a composite column is subjected to predominately axial forces, sufficient shear transfer must be provided to ensure that the steel and concrete parts share the loads applied to the column at connections to beams and bracing members.

   C. In the columns of braced frames, axial forces are more significant than bending moments and the response will normally be elastic. Here, it must be assured that the axial force is shared between the concrete and steel parts. There will be a high bond demand wherever the column axial force increases. This will occur at floor levels where bracing members and beams are connected to the column [14].

11. Wherever a composite column is subjected to predominately flexure, and the normalised axial load $v_d \leq 0.3$, the requirements of 9 are satisfied if shear connectors are provided along the axis of the structural steel member from the point of inflection to the point of maximum bending moment designed for a force equal to $f_y(A_{st} - A_{sc})$ where f_y is the yield strength of the structural steel and A_{st} and A_{sc} are the cross-section areas of structural steel on the tension and compression sides of the neutral axis, respectively. These shear connectors should be attached to the outside face of the embedded shape.

   C. For fully-encased composite columns under vertical loading, sufficient transfer of forces is normally provided through bond and friction. It is not certain that this approach remains valid under cyclic loading conditions. The proposal in 9 allows this practice to be applied to non-dissipative members in braced frames. The proposal in 11 relates to the behaviour of columns in moment resisting frames and is taken from the NEHRP Recommended provisions and should be conservative. Haajier [15] has shown that the global response of moment-resisting frames with filled columns, as displayed in push-over analyses, does not vary significantly with the stiffness or strength of the shear connection achieved through bond and friction. That is, the same response is displayed with and without composite action. However, the provisions of the NEHRP Recommended Provisions stipulate that shear connectors must be provided in composite columns subjected primarily to flexure. A conservative approach, in the absence of better information, would be to adopt this provision in EC8.

12. The following provisions concerning anchorages and splices in the design of RC columns also apply to the reinforcement details of partially encased composite beams: 2.6.2.1(1), 2.6.2.1(2) with $k_b = 5$, 2.6.2.1(3) and 2.6.3(1)-(5).

   C. These are the relevant EC8 rules for RC columns (Section 2 Chapter 1.3.). The NEHRP stipulates the ACI rules for longitudinal reinforcement.

6.6. Partially-encased columns

6.6.1. General.

1. The design of partially-encased composite columns may consider the resistance of the steel section alone or the composite resistance of the steel section and concrete encasement.

2. The design of partially-encased columns in which only the steel section is assumed to contribute to member resistance may be carried out according to the provisions of Chapter 3 of this part and the capacity design provisions of 5.3.

    C. Clause 5.3 should ensure that the design shear force is determined on the basis of the moment of resistance of the full composite section of beams.

3. When the concrete encasement is assumed to contribute to the resistance of the member, the design rules given in 6.6.2 apply. These rules ensure full shear interaction between the concrete and steel parts and protect the dissipative zones against premature inelastic failure.

4. In moment-resisting frames, the following inequality must apply for all composite columns:

$$N_{sd} / N_{pl,Rd} < 0,30$$

C. This is a conservative rule which could be relaxed with more test data. However, it should not be too restrictive as moment capacity will decrease significantly above this level of axial load. A cross-reference to this clause should be given in the rules for moment-resisting frames (7.1).

6.6.2. Rules for composite behaviour.

1. Critical regions at top and bottom ends of a column extend for a distance l_{cr}, equal to:
    $l_{cr} = l_{cl}/4$ in dissipative zones, or
    $l_{cr} = l_{cl}/8$ in non-dissipative zones.

C. The value for the column base is a likely upper bound based on parametric studies. The smaller value for other regions reflects the fact that these will not be intended dissipative zones. The lengths of these critical regions are shorter than those for fully-encased columns.

2. In critical regions, transverse reinforcement shall be provided to prevent premature failure of the concrete encasement.

3. The requirement of 2 can be assumed to exist if one of the two transverse reinforcement details (a) and (b) shown below are provided throughout the length of the column.

    C. Fig (a) showing hoops welded to web
    Fig (b) showing straight bars welded to flanges

4. The diameter of the transverse reinforcement, d_{bw}, shall be at least 6mm.
    In addition, whenever transverse links are employed to prevent local flange buckling as described in 6 and 7 below, d_{bw} should not be less than

$$d_{bw} \geq [(b_f T_f/4)(f_{ydf}/f_{ydw})]^{0.5}$$

    in which b_f and T_f are the width and thickness of the flange outstand and f_{ydf} and f_{ydw} are the design strengths of the flange and reinforcement.

    C. These rules are intended to reflect minimum conditions in test specimens, and may be conservative. The latter requirement seeks to prevent premature yielding of the reinforcement when acting as a tie by ensuring that the yield resistance of the reinforcement is at least 3% of that of the flange.

5. The longitudinal spacing of the transverse reinforcement, s, should not be less than

    $s \leq \min [0.3d, 50mm]$ in the critical regions or
    $s \leq \min [0.5d, 150mm]$ elsewhere

    where d is the depth of the steel section between its flanges.

    C. These values are presented for discussion. They reflect the minimum conditions in test specimens. Tests at Imperial College (152 UC) used a spacing of 40mm in the critical region and 80mm elsewhere. Tests at Liege, Milan and Darmstadt (HEB300 and HEA260) used a spacing of 150mm throughout.

6. In the dissipative zones, the presence of the concrete encasement and, when provided, straight bars as shown in 3. above help to prevent local flange and web buckling. This can be taken into account by allowing higher flange width-to-thickness ratios (b_f/T_f) to be used for a particular design behaviour factor. When straight bars are provided in section type (b), the maximum allowable b_f/T_f value depends on the bar spacing, s.

7.   Table of b_f/T_f ratios for section types (a) and (b).

q	Section Type [as per 6.6.2.3]			
	(a)	(b)		
		s/b = 1.6	s/b = 1.0	s/b = 0.5
6	5.5 ε	5.5 ε	8.0 ε	12.5 ε
4	8.0 ε	8.0 ε	9.5 ε	14.5 ε
2	9.5 ε	9.5 ε	13.0 ε	20.0 ε

$\varepsilon = (f_y/235)^{0.5}$

Values other than those shown in the above table may be obtained by linear interpolation.

C. The rules in 6 , 7 and 8 are related. In 6, the absolute maximum link spacing for concrete protection is specified. In 7, the relaxation of the b/T ratios from the steel part is allowed. In 8 allowable b/T ratios are specified for given spacings and q-values or section classes. These values are obtained from a parametric study which extended test results. They correspond to the b_f/T_f ratios required to provide a rotation ductility capacity of 6, 4 or 2 given an allowable axial load of 30% of section capacity.

8.   In dissipative members, the shear resistance of the column shall be determined on the basis of the structural steel section alone.

9.   In non-dissipative members, the shear resistance of the column shall be determined according to the rules given in Eurocode 4.

C. As with fully-encased columns, the rule in 8 is a conservative proposal in the absence of better research. It is possible to ensure that the concrete encasement can contribute to the shear resistance of the member, but there will be additional detailing requirements (ref Bouwkamp proposal). While this may be worthwhile in the case of beams (esp. in eccentrically braced frames), the shear force in column members is not expected to be as high. Hence the conservative approach of 8 should not be too restrictive. The contribution of tie reinforcement to the shear resistance has not been investigated in tests.

10. For earthquake-resistant design, the design values given in Eurocode 4 (Cl. 4.8.2.6 to 4.8.2.8) on the design shear strengths due to bond and friction. must be reduced by a factor of [0.3]. Where insufficient shear transfer is achieved through bond and friction, shear connectors shall be provided to ensure full composite action.

11. Transverse reinforcement welded to the steel section as described in 6.6.2.3 may be employed as shear connectors.

12. Wherever a composite column is subjected to predominately axial forces, sufficient shear transfer must be voided to ensure that the steel and concrete parts share the loads applied to the column at connections to beams and bracing members.

13. Wherever a composite column is subjected to predominately flexure, and the normalised axial load $v_d \leq 0.3$, the requirements of 10 are satisfied if shear connectors are provided along the axis of the structural steel member from the point of inflection to the point of maximum bending moment designed for a force equal to $f_y(A_{st} - A_{sc})$ where f_y is the yield strength of the structural steel and A_{st} and A_{sc} are the cross-section areas of structural steel on the tension and compression sides of the neutral axis, respectively. These shear connectors should be attached to the outside face of the embedded shape.

C. The provisions of 12 and 13 are the same as those proposed for fully-encased columns. Most test data on beams and beam-column refers to the case where mechanical connection using welded reinforcement is present. In most cases, full shear connection was achieved. In some tests at Liege, the reinforcement spacing was too great and full shear connection was not achieved. It is possible that sufficient shear transfer could be achieved through bond alone, but the cases where this applies are not known for cyclic inelastic response.

14. The following provisions concerning anchorages and splices in the design of RC columns also apply to the reinforcement details of partially encased composite beams: 2.6.2.1(1), 2.6.2.1(2) with $k_b = 5$, 2.6.2.1(3) and 2.6.3(1)-(5).

C. These are the relevant EC8 rules for RC columns (Section 2 Chapter 1.3.). The NEHRP stipulates the ACI rules for longitudinal reinforcement.

6.7.   Filled Composite Columns

6.7.1. General.

1. The design of fully-encased composite columns may consider the resistance of the steel section alone or the composite resistance of the steel section and concrete infill.

2. The design of filled composite columns in which only the steel section is assumed to contribute to member resistance may be carried out according to the provisions of Chapter 3 of this part and the capacity design provisions of 5.3.

    C. Clause 5.3 should ensure that the design shear force is determined on the basis of the moment of resistance of the full composite section.

3. In moment-resisting frames, the following inequality must apply for all composite columns: $N_{sd} / N_{pl,Rd} < 0,30$

4. When the concrete infill is assumed to contribute to the resistance of the member, the design rules hereunder apply. These rules ensure full shear interaction between the concrete and steel parts and protect the dissipative zones against premature inelastic failure.

6.7.2. Rules for composite behaviour.

1. In the dissipative zones, the presence of the concrete fill helps to prevent local buckling. This can be taken into account by allowing column depth to-thickness ratios (D/t) to be used for a particular design behaviour factor.

2. Table of D/t ratios for cross-section types.

This Table still has to be developed

C. The local detailing of a filled column is more straightforward than that for partially-or full-encased columns, and depends only on the D/t ratio. Rectangular cross-sections are less ductile than circular ones, so lower D/t ratios should be required for the same design q. The table to be completed should reflect this.

3. In dissipative members, the shear resistance of the column shall be determined on the basis of the structural steel section alone.

4. In non-dissipative members, the shear resistance of the column shall be determined according to the rules given in Eurocode 4.

5. For earthquake-resistant design, the design values given in Eurocode 4 (Cl. 4.8.2.6 to 4.8.2.8) on the design shear strengths due to bond and friction. must be reduced by a factor of [0.3]. Where insufficient shear transfer is achieved through bond and friction, shear connectors shall be provided to ensure full composite action.

6. Wherever a composite column is subjected to predominately axial forces, sufficient shear transfer must be provided to ensure that the steel and concrete parts share the loads applied to the column at connections to beams and bracing members.

7. Wherever a composite column is subjected to predominately flexure, and the normalised axial load $v_d \leq 0.3$, the requirements of 5 are satisfied if shear connectors are provided along the axis of the structural steel member from the point of inflection to the point of

maximum bending moment designed for a force equal to $f_y(A_{st} - A_{sc})$ where f_y is the yield strength of the structural steel and A_{st} and A_{sc} are the cross-section areas of structural steel on the tension and compression sides of the neutral axis, respectively. These shear connectors should be attached to the outside face of the embedded shape.

C. Haajier [15] has shown that the global response of moment-resisting frames with filled columns, as displayed in push-over analyses, does not vary significantly with the stiffness or strength of the shear connection achieved through bond and friction. That is, the same response is displayed with and without composite action.

However, the provisions of the NEHRP Recommended Provisions stipulate that shear connectors must be provided in composite columns subjected primarily to flexure. A conservative approach, in the absence of better information, would be to adopt this provision in EC8.

6.8.   Partially-encased beams

6.8.1. General.

1.   The design of partially-encased composite beams may consider the resistance of the steel section alone or the composite resistance of the steel section and concrete encasement.

2.   The design of partially-encased beams in which only the steel section is assumed to contribute to member resistance may be carried out according to the provisions of Chapter 3 of this part and the capacity design provisions of 5.2. and 5.3.

C. Clause 5.3.4 should ensure that the design shear force is determined on the basis of the moment of resistance of the full composite section.

3. In partially-encased beams, N_{sd} should be such that: $N_{sd} / N_{pl,Rd} < 0,15$

    C. This rule is similar to the one for steel structures. It is given here for homogeneity in EC8 rules. In practice, this condition is almost always fulfilled.

4. When the concrete encasement is assumed to contribute to the resistance of the member, the design rules given in 6.8.2 hereunder apply. These rules ensure full shear interaction between the concrete and steel parts and protect the dissipative zones against premature inelastic failure.

6.8.2. Rules for composite behaviour.

1. Critical regions exist wherever a beam experiences yielding under the design seismic load combination. At beam ends, these critical regions extend for a distance l_{cr}, while at mid-span these critical regions extend for a distance $2l_{cr}$, in which $l_{cr} = 1,5d$, where d is the depth of the beam.

    C. This is the EC8 rule for RC beams in DC'M'. This rule is not present in EC4, but is needed in EC8 because ductility is needed in seismic design. As this ductility is needed over a length, it is necessary to define this length l_{cr}. The value of l_{cr} is taken as 2.0d in DC'H' and 1.0d in DC'L'. An alternative approach would be to relate the length of the plastic hinge to the ultimate and yield moments of the cross-section, i.e.

    $$l_{cr} = [(M_u - M_y)/M_u]l_{cf}$$

    in which l_{cf} is the distance from the location of maximum bending moment to the adjacent point of contraflexure. This is closer to the definition used for partially-encased columns.

2   In critical regions, transverse reinforcement shall be provided to prevent premature failure of the concrete encasement.

3   The requirement of 2. can be assumed to exist if one of the two transverse reinforcement details (a) and (b) shown below are provided throughout the length of the beam.

    C. *Fig (a) showing hoops welded to web*
        Fig (b) showing straight bars welded to flanges

4   The diameter of the transverse reinforcement, d_{bw}, shall be at least 6mm.
    In addition, whenever transverse links are employed to prevent local flange buckling as described in 6 and 7 below, d_{bw} should not be less than

$$d_{bw} \geq [(b_f t_f/4)(f_{ydf}/f_{ydw})]^{0.5}$$

in which b_f and t_f are the width and thickness of the flange outstand and f_{ydf} and f_{ydw} are the design strengths of the flange and reinforcement.

C. *These rules are intended to reflect minimum conditions in test specimens, and may be conservative. The latter requirement seeks to prevent premature yielding of the reinforcement when acting as a tie by ensuring that the yield resistance of the reinforcement is at least 3% of that of the flange.*

5   The longitudinal spacing of the transverse reinforcement, s, should not be less than
   $s \leq min [0.3d, 50mm]$ in the critical regions or
   $s \leq min [0.5d, 150mm]$ elsewhere.
   d is the depth of the steel section between its flanges.

C. *These values are presented for discussion. They reflect the minimum conditions in test specimens. Tests at Imperial College (152 UC) used a spacing of 40mm in the critical region and 80mm elsewhere. Tests at Liege, Milan and Darmstadt (HEB300 and HEA260) used a spacing of 150mm throughout.*

6   In the dissipative zones, the presence of the concrete encasement and, when provided, straight bars as shown in (3) above help to prevent local flange and web buckling. This can be taken into account by allowing higher flange width-to-thickness ratios (b_f/T_f) to be used for a particular design behaviour factor. When straight bars are provided in section type (b), the maximum allowable b_f/T_f value depends on the bar spacing, s.

7   Table of b_f/T_f ratios for section types (a) and (b).

q	Section Type [as per 6.6.8.2(3)]			
	(a)	(b)		
		s/b = 1.6		
6	8.5 ε	8.5 ε	s/b = 1.0	s/b = 0.5
4	10.5 ε	10.5 ε	10.0 ε	17.0 ε
2	13.5 ε	13.5 ε	12.5 ε	20.0 ε

$$\varepsilon = (f_y/235)^{0.5}$$

   Values other than those shown in the above table may be obtained by linear interpolation.

C. *The above rules are the same as for partially-encased columns.*
The rules in (6) ,(7) and (8) are related. In (6) the absolute maximum link spacing for concrete protection is specified. In (7), the relaxation of the b/T ratios from the steel part is allowed. In (8) allowable b/T ratios are specified for given spacings and q-

values or section classes. These values are obtained from a parametric study which extended test results. They correspond to the b_f/T_f ratios required to provide a rotation ductility capacity of 6, 4 or 2 given an allowable axial load of 15% of section capacity

8   In dissipative members, the shear resistance of the beam shall be determined on the basis of the structural steel section alone, unless special details are provided to mobilise the shear resistance of the concrete encasement.

C. In dissipative zones and under dynamic loading, the bond between steel and concrete is soon destroyed and may thus not be relied on to adopt the same evaluation as EC4, which refers to static loading.

9   In non-dissipative members, the shear resistance of the beam shall be determined according to the rules given in Eurocode 4.

C. As with columns, the rule 9 is a conservative proposal in the absence of better experimental evidence.

10   Transverse reinforcement welded to the steel section as described in 6.8.2(3 may be employed as shear connectors.

11   The following provisions concerning anchorages and splices in the design of RC columns also apply to the reinforcement details of partially encased composite beams: 2.6.2.1(1), 2.6.2.1(2) with $k_b = 5$, 2.6.2.1(3) and 2.6.3(1)-(5).

C. These are the relevant EC8 rules for RC columns (Section 2 of Chapter 1.3. of Eurocode 8)

7. DESIGN AND DETAILING RULES FOR MOMENT FRAMES

7.1 Specific criteria.

1.   The moment resisting frames shall be designed so that plastic hinges form in the beams and not in the columns. This requirement is waived at the base of the frame, at the top floor of multistorey buildings and for one storey buildings.

2.   The moment resisting frames with limited sway correspond to the ones classified "braced" or "non sway" in Eurocode 4 ; they are such that second order (P–Δ) effects need not be considered, because the condition given in 3.1. is satisfied at every storey.
     $$V_{tot} . h$$

C. This condition is defined in 4.2.2. of Part 2 of this Eurocode 8 ; it is similar to the condition in 4.9.4.2. of Eurocode 4. When SLS (Serviceability Limit State) requirements on drift are strong, this criteria is practically always fulfilled

3. The composite beams shall be designed such that they have adequate ductility and that the integrity of concrete is maintained.

4. The beam to column joints shall have adequate overstrength to allow the plastic hinges to be formed in the beams or in the connections of the beams (see 5.4.).

5. The required hinge formation pattern should be achieved by observing the rules given in 7.3

7.2 Analysis

1. The analysis of the structure is made on the basis of the section properties defined in 4.

2. In beams, two different flexural stiffnesses are considered : EI_1 for the part of the spans submitted to positive (sagging) bending moment (uncracked section) and EI_2 for the part of the span submitted to a negative (hogging) bending moment (cracked section).

3. The analysis can alternatively be made considering for beams an equivalent moment of inertia (second moment of area) I_{eq} constant on the whole span of beams :
$I_{eq} = 0.6\,I_1 + 0.4\,I_2$.
C1. This value of the equivalent moment of inertia is a proposal which needs further validation. No such explicit value has been proposed until now in Eurocode 4 for the analysis of moment frames submitted to horizontal loading. Under vertical loading the rule is I_2 over 15 % the span on each side of each internal support and I_1 elsewhere, which correspond to $E_{eq} = 0,7\,I_1 + 0,3\,I_2$.
In fact, there is not one single answer to the definition of I_{eq}, because it is a function of the ratio between vertical and horizontal loadings and of the ratio between I_1 and I_2.
If vertical loading is zero, which correspond to the smallest extension of the positive bending zone and is thus on the safe side, it can be established that the I_{eq} giving the same drift as the real distribution of I_1 and I_2 is :

$$I_{eq} = \frac{I_1^2 + I_2^2}{I_1 + I_2}$$

For $I_1 = 2\,I_2$ this formula give $I_{eq} = 1,66\,I_2$,
while $I_{eq} = 0,6\,I_1 + 0,4\,I_2 = 1,60\,I_2$. These two values match well.

C2. In the FEMA 267 (1997) and the AISC seismic provisions (1997), the elastic analysis is based on the properties of the bare steel frame, that is $I_{eq} = I_l$. The resulting calculation of building stiffness and period underestimates the actual properties substantially. Although this approach result in unconservative estimates of design force levels (period T being overstimated, pseudo acceleration β (T) are underestimated), it produces conservative estimates of interstorey drift demands. It is considered in the U.S. that since the design of most moment-resisting frames are controlled by considerations of drift, this approach is preferable to methods that would over-estimate building stiffness because many of the elements that provide additional stiffness may be subject to rapid degradation under severe cyclic lateral deformation, so that the bare frame stiffness would be a reasonable estimate of the effective stiffness under long duration ground shaking response.

This option can be justified by a temporary lack of experimental data. It is an approximation, even when there is a degradation of stiffness under cyclic deformations. This approximation may be acceptable for high beams (beam depth above 700 mm) for which I_l is not too much different of I_2 ; for beams depth under 500 mm, the analysis made considering a bare frame is unrelated to the real behaviour of the structure, especially when measures are taken to achieve ductility in the connection zone.

C3. It has been estimated by the drafting panel that three facts support the option of considering composite structures : the experimental basis allow this definition ; the application field, which does not include sway frames of high flexibility (see 6.7.1), provide a safety margin covering potential discrepancies between real and design elastic flexibility of structures. Earthquake durations expected in most parts of Europe are not such that section   degradation is likely to take place.Furthermore, it does not seem a good option to underestimate the stiffness of sections to take into account potential degradation of dissipative zones ; this fatigue aspect of composite sections should rather be defined as such, by the development of a fatigue ULS which should be considered in the design as is at present the deformation ULS.

4.   In columns, $(EI)_c = E_a I_a + 0.5\, E_{cm}\, I_c + E_s\, I_s$

7.3.   Detailing rules

1. Composite T beam design should comply with 6.2.

2. Beams should be verified as having sufficient safety against lateral or lateral torsional buckling failure according to clause 4.6.2. of Eurocode 4, assuming the formation of a negative plastic moment at one end of the beam.

3. For plastic hinges in the beams, it should be verified that the full plastic moment resistance and rotation capacity is not decreased by compression and shear forces. To this end, the following inequalities should be verified at the location where the formation of hinges is expected.

$$\frac{M_{Sd}}{M_{pl,Rd}} \leq 1,0 \qquad \frac{N_{Sd}}{N_{pl,Rd}} \leq 0,15$$

$$\frac{V_{G,Sd} + V_{M,Sd}}{V_{pl,Rd}} \leq 0,5$$

where

N_{Sd}, M_{Sd} — design action effects (resulting from the structural analysis),

$N_{pl,Rd}$, $M_{pl,Rd}$, $V_{pl,Rd}$ — design resistances according to clause 5.4. of Part 1-1 of Eurocode 4,

$V_{G,Sd}$ — shear force due to the non-seismic actions,

$V_{M,Sd}$ — shear force due to the application of the resisting moments $M_{Rd,A}$ and $M_{Rd,B}$ with opposite signs at the extremities A and B of the beam.

4. For the verification of columns the most unfavourable combination of the axial force N and the bending moments M_x and M_y shall be assumed.

5. The sum of the design moment resistance in the cross sections of the columns adjacent to the beams shall not be less than the sum of the upper bound resisting moments $M_{u,Rd}$ of the beams connected to the column, as defined in 6.3.

6. The transfer of the forces from the beams to the columns should comply with the design rules given in clause 4.10 of ENV 1994-1-1 (or Section 8 of pr EN 1994-1-1).

7. The resistance verifications of the columns should be made according to 4.8. of Part 1-1 of Eurocode 4.

8. The column shear force V_{Sd} (resulting from the structural analysis) should be limited to

$$\frac{V_{Sd}}{V_{pl,Rd}} \leq 0,5$$

9. In framed web panels of beam/column connections (see Figure 7.1.) the following assessment is permitted :

$$\frac{V_{wp,Sd}}{V_{wp,Rd}} \leq 1,0$$

where

$V_{wp,Sd}$ design shear force in the web panel due to the action effects,

$V_{wp,Rd}$ Shear resistance of the web panel according to annex J of Part 1-1 of Eurocode 3 (steel panel) or to Annex J. Paragraph 3.5.2. of Eurocode 4 (composite steel concrete panel).

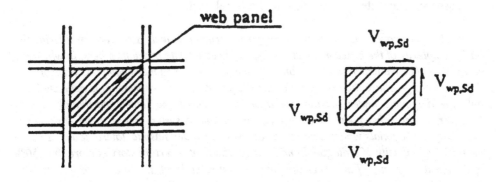

Figure 7.1. : Web panel framed by flanges or stiffeners

7.4   Condition for disregarding the composite character of beams with slab.

The plastic resistance of a composite section can be computed considering only the steel section if the slab is totally disconnected from the steel frame in a circular zone around a column of diameter $2b_{eff}$, b_{eff} being the greater of the effective width of the beams connected to that column. Total disconnection means no contact between slab and any vertical side of steel element (columns, shear connectors, connecting plates, corrugated flange, omega steeldeck nailed to flange of steel section, ...).

8.   DESIGN   AND   DETAILING   RULES   FOR   COMPOSITE CONCENTRICALLY BRACED   FRAMES

8.1   Specific criteria

1 Composite concentrically braced frames consist of concentrically connected members and shall be designed so that the dissipative action will occur primarily through tension yielding and buckling of the braces. All other truss members should not yield or buckle

and connections should not fail. Columns and beams shall be either structural steel or composite structural steel; braces shall be structural steel.

C. In principle, braces could also be of composite structural steel. However, considering the basic complexity of the brace member design, using composite brace members, in which the concrete is effective under compression only, the design is even more complex. Basically, in the initial design phase the resistance of the diagonals under compression is neglected and only the diagonals under tension are assumed to resist the lateral earthquake loads. However, in the subsequent capacity analysis the entire system needs to be considered, including the diagonals which experience compression under seismic lateral loads. In case the buckling capacity of a diagonal under compression at a certain story is more than 50% of the tensile capacity of the other diagonal at that same level, the lateral story resistance provided by the two diagonals at the moment of first buckling is larger than the design lateral load which was used initially to design the tensile-loaded diagonals. Hence, at that moment during an earthquake, the columns will experience larger axial forces than first used in the initial design phase. Of course, subsequently, the lateral force capacity will reduce to the corresponding capacity of the tension diagonal only and the columns will not be overloaded any longer. (Assuming that at each level the buckling resistance of the diagonals is 60% of the tensile capacity of the corresponding diagonals under tension, the columns would experience a load of 120%, or an overload of 20%).

In order to limit such possible overloads, it is essential to develop a design in which the compression capacity of the diagonals is low. The clause 3.5.5.2 (1) of ENV 1998 which limits for diagonal members the non-dimensional slenderness to $\lambda \leq [1,5]$ does not intend to reach that goal. It intends to avoid the possibility of shocks due to sudden elastic re-tensioning of the diagonal when the reversal of forces takes place and the non-dimensionless slenderness limit is there to "prevent elastic buckling" of the diagonal. The limit on λ corresponds to a design resistance in compression N_{Rd} which is about 35% of the tensile capacity, but N_{Rd} is a lower characteristic value, so that the real resistance in compression may well be 60% of the tensile capacity or more; the overloading of the columns with respect to their design resistance can then be critical. As, in order to design a dissipative system, the diagonals should buckle to allow the tensile diagonals to yield it seems more effective to put both an upper and a lower limit on λ
The proposal is [1,5]< λ <[1,8]

Under these conditions, introducing a composite brace would increase the difficulties to design a slender diagonal member which would buckle early.

2   The diagonal elements of bracings should be placed in such a way that the structure exhibits similar load deflection characteristic at each floor and in every braced direction under load reversals. To this end, the following rule should be met storey by storey:

$$(A+ - A-)/(A+ + A-) \leq \quad 0,05$$

where A^+ and A^- are the areas of the horizontal projections of the cross-sections of the tension diagonals, when horizontal seismic actions have a positive or negative direction respectively (see fig. 3.3).

8. 2   Analysis

1. Under gravity load conditions, only beams and columns shall be considered to resist such loads, without considering the bracing members.

2. Under seismic action, the entire system shall be considered with only the tension diagonals assumed to react the corresponding given seismic action.

8.3.   Diagonal members

1. Diagonal members shall meet the following requirement: $[1,5] < \overline{\lambda} < [1,8]$

2. The tension force Nsd should be limited to the yield resistance $N_{pl,Rd}$ of the gross cross-section.

3. In V-bracings the compression diagonals should be designed for the compression resistance computed according to clause 5 of Part 1-1 of Eurocode 3.

4. The connections of the diagonals to any member should fulfil the overstrength condition

    $R_d \geq 1,20 . N_{pl,Rd}$ where $N_{pl,Rd}$ is the axial resistance of the diagonal in tension .

8.4.   Beams and Columns

1. Beams and columns with axial forces should meet the following minimum resistance requirement:
    $$N_{Rd} (M) \geq \qquad 1,20 . (N_{Sd,G} + \alpha. N_{Sd,E})$$

    where
    $N_{Rd}(M)$ buckling resistance of the beam or the column according to Eurocode 3 or 4,

    $N_{Sd,G}$ axial force in the beam or in the column due to the non-seismic actions included in the combination of actions for the seismic design situation,

$N_{Sd,E}$ axial force in the beam or in the column due to the design seismic action multiplied by the importance factor,

α minimum value of $\alpha_i = N_{pl,Rdi}/N_{Sdi}$ over all diagonals of the braced frame system, where

$N_{pl,Rdi}$ design resistance of diagonal i,

N_{Sdi} design value of the axial force in the same diagonal i in the seismic design situation.

C. At present, it is proposed for the steel part of EC8 that α_i should not differ from the minimum value by more than 20 %. This rule will probably be implemented in this Section as well.

2.   In V-bracings the beams should be designed to resist all non-seismic actions without considering the intermediate support given by the diagonals.

3.   The connections of the frames to the foundations should fulfill the overstrength condition as given in 5.5.

9.   DESIGN AND DETAILING RULES FOR COMPOSITE ECCENTRICALLY BRACED FRAMES

9.1   Specific criteria

1.   Composite eccentrically braced frames consist of braced systems for which each brace intersects a beam either at an eccentricity from the intersection of the centreline of the beam and the column or at an eccentricity from the intersection of the centreline of the beam and an adjacent brace. The beam portions between the points of intersection of either a brace and the adjacent column or two adjacent braces are called links.

2.   Composite eccentrically braced frames shall be designed so that the dissipative action will occur only through yielding in shear of the links. All other members should remain elastic and failure of connections should be prevented.

3.   Columns, beams and braces can be structural steel or composite.

C. The above scope leaves in principle the door open to all design options for what concerns all type of members. These can be made of steel only or composite. However, there are still uncertainties, in both options.

For instance, assuming that the capacity of a composite link can be assessed correctly, the braces and columns can be designed correctly to withstand the seismic action forces. However, cyclic non linear seismic exposure, which for eccentric bracings correspond at the ultimate stage to plastic rotations more important than those met in moment frames, generates necessarily a   concrete deterioration which leads to a link with reduced resistance, reflecting basically the steel section capacity only, or less.

Should the composite action be underestimated, e.g. neglecting the concrete presence, the underestimated link capacity leads to an under-design of both braces and columns. In that condition, the larger than considered link capacity leads to an overload of the brace and column members and their failure.

Another problem arise when the evaluation of effective width of collaborating slab is needed, when for instance composite links active in bending would be designed; there is clearly a lack of experimental background on this topic.

The gap in knowledge is even greater for what concerns "disconnection" of the slab in order to allow a reference to pure steel as being the only resisting part of sections in bending: it is not obvious that a local disconnection allows a reference to steel resistance only; on the contrary, in the context of moment frames, there has been experimental evidence that too local disconnections had a limited effect and that a composite resistance was still operative; membrane forces in the slab might realise force transfers which are not really mastered at present.

From these considerations, it can be concluded that links working in bending in beam elements with slab always raise an evaluation problem. On the contrary, vertical steel links correspond to a totally mastered situation. Beam links made of an encased steel section with possible slab and working in shear correspond also to an adequately mastered situation, because the slab contribution in the shear resistance of the links is negligible.

4. The braces, columns and beam segments outside the link segments shall be designed to remain elastic under the maximum forces that can be generated by the fully yielded and/or cyclically strain-hardened beam link.

9.2.   Analysis.

The analysis of composite EBF is similar to the one of pure steel

9.3   Links

1   Links may be made of steel sections, possibly composite slabs. They may not be encased.

6.   The links shall be designed to yield in shear and to provide a reliable plastic resistance under cyclic conditions corresponding to a structure global relative drift of 3,5%

9.4.   Columns, beams and braces.

1.   Composite columns can be designed as stipulated by EC 4. Where a link is adjacent to a fully encased composite column, transverse reinforcement meeting the requirements of clause 6.5. or 6.6. shall be provided above and below the link connection.

2.   Structural steel braces shall meet the requirements for steel eccentrically braced frames as stipulated in clause 3.5.6.3. Composite braces shall meet the requirements for composite compression members of 3.3 to 3.5. In case of a composite brace under tension, only the cross-section of the structural steel section should be considered in the evaluation of the resistance of the brace.

3.   Connections should meet the requirements of steel eccentrically braced frames as in clause 3.5.6.

10. DESIGN AND DETAILING RULES FOR STRUCTURAL SYSTEMS MADE OF REINFORCED CONCRETE SHEAR WALLS COMPOSITE WITH STRUCTURAL STEEL ELEMENTS.

10.1. Specific criteria.

1.   The provisions in this Section apply to three types of structural systems using reinforced concrete walls Figure 10.1. In Type 1, reinforced concrete walls serve as infill panels in steel or composite frames. Typical sections at the wall-to-column interface for such cases are shown in Figure 10.2. In Type 2, encased steel sections are used as vertical reinforcement in reinforced concrete shear walls. In Type 3, steel or composite beams are used to couple two or more reinforced concrete walls. Examples of coupling beam-to-wall connections are shown in Figure 10.3.

C. When properly designed, these systems have shear strength and stiffness comparable to those of pure reinforced concrete shear wall systems.

The structural steel sections in the boundary members however increase the flexural resistance of the wall and delay plastic flexural hinging (tall walls).

Like for reinforced concrete structures, two levels of ductility and two values of the behaviour factor are defined, depending on the level of requirements in the detailing rules.

2. Structural systems Types 1 and 2 are designed to dissipate energy in the concrete shear walls.

3. Type 3 is designed to dissipate energy in the concrete shear walls and in the coupling beams.

Figure 10.1. Composite Structural Systems with shear walls

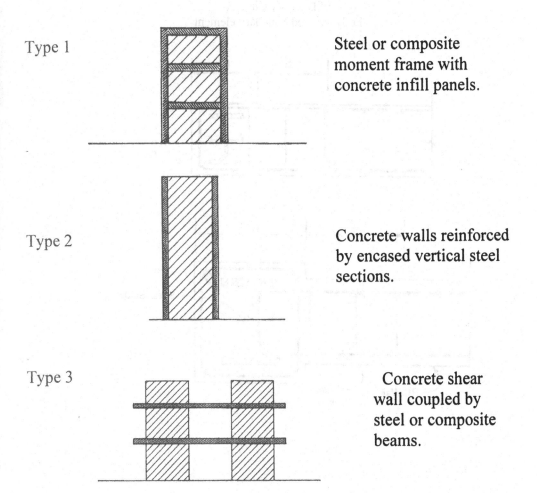

Type 1   Steel or composite moment frame with concrete infill panels.

Type 2   Concrete walls reinforced by encased vertical steel sections.

Type 3   Concrete shear wall coupled by steel or composite beams.

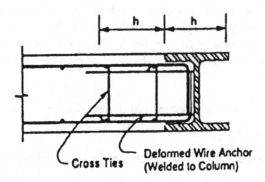

Partially encased steel boundary element

Figure 10.2. – Detail of partially and fully encased composite boundary elements in Structures
of Ductility Class H

Fully encased boundary element

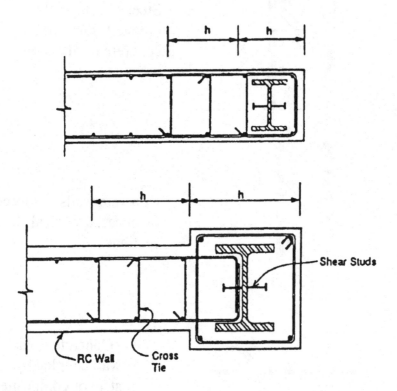

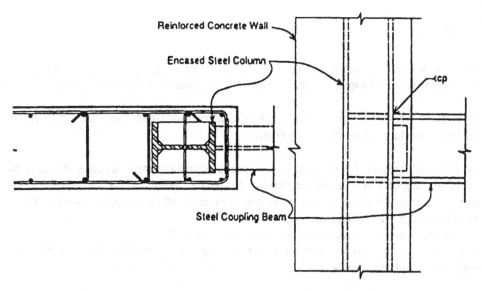

Steel coupling beam to reinforced concrete wall with composite boudary member

Figure 10.3 Example of coupling beam to wall connections
in structures of ductility class H

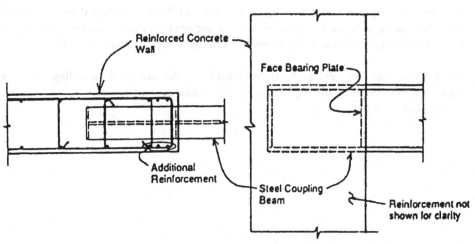

Steel coupling beam to reinforced concrete wall

10.2   Analysis.

1. The analysis of the structure is made on the basis of the section properties defined in 4.2. for composite beams and in Section 2 of part 1.3. of Eurocode 8 for concrete walls.

2. Two levels M and H of ductility can be considered, for which two different values of the behaviour factor q are defined in Figure 3.1

3. In structural system Type 1, when unencased structural steel sections function as boundary members in reinforced concrete infill panels, design will be made considering that the earthquake action effects in the boundary elements are axial forces only. These axial forces are determined assuming that :
    -the shear forces are carried by the reinforced concrete wall
    -the entire gravity and overturning forces are carried by the boundary members in conjunction with the shear wall.

4. In structural system Type 1, when encased structural steel sections function as boundary members in reinforced concrete infill panels, the analysis shall be based upon a transformed concrete section using elastic material properties.

    C. This requirement is based on the assumption that the boundary elements may have to carry all compressive forces at the critical section of the wall when maximum horizontal forces are acting.

5. In composite coupling beams, two different flexural stiffnesses are considered : EI_1 for the part of the spans submitted to positive bending moment (uncracked section) and EI_2 for the part of the span submitted to a negative bending moment (cracked section).

6. The analysis can alternatively be made considering for composite coupling beams an equivalent moment of inertia I_{eq} constant on the whole span of beams :
    $I_{eq} = 0.6\, I_1 + 0.4\, I_2$.

10.3. Detailing rules for Ductility class M.

1. Unencased steel sections used as boundary members of reinforced concrete panels shall belong to a class of cross section related to the behaviour factor of the structure as indicated in 6.1.3.

2. The reinforced concrete of infill panels shall meet the requirements of Chapter 2 of this Section 1.3.

3. Fully encased structural steel sections used as boundary members in reinforced concrete panels shall be designed to meet the requirements of 6.2.3.

   C. This is a logical consequence of 3 above.

4. Partially encased structural steel sections used as boundary members in reinforced concrete panels shall be designed to meet the requirements of 6.6. and 6.8.

5. Headed shear studs or welded reinforcement anchors shall be provided to transfer vertical shear forces between the structural steel of the boundary elements and the reinforced concrete.

6. If the wall elements are interrupted by steel beams at floor levels, shear connectors are needed at the wall-to-beam upper and lower interfaces.

   C. Tests on concrete infill walls have shown that if shear connectors are not present, story shear forces are carried primarily through diagonal compressions struts in the wall panel (Chrysostomou, 1991). This behaviour often includes high forces in localised areas of the walls, beams, columns, and connections. The shear stud requirements will improve performance by providing a more uniform transfer of forces between the infill panels and the boundary members.

7. Coupling beams shall have an embedment length into the reinforced concrete wall that is sufficient to develop the maximum possible combination of moment and shear that can be generated by the nominal bending and shear strength of the coupling beam. The embedment length shall be considered to begin inside the first layer of confining reinforcement in the wall boundary member Figure 10.3.

   C. Two examples of connections between steel coupling beams to concrete walls are shown in Figure 10.3. The requirements for coupling beams and their connections are based largely on recent tests of unencased steel coupling beams (Harries, et al., 1993 ; Shahrooz et al., 1993). These test data and analyses show that properly detailed coupling beams can be designed to yield at the face of the concrete wall and provide stable hysteretic behaviour under reversed cyclic loads. Under high seismic loads, the coupling beams are likely to undergo large inelastic deformations through either flexural and/or shear yielding.

8. The design of the connection of the steel part of coupling beams in order to assure the transfer of loads between the coupling beam and the wall shall meet the requirements of Chapter 3 of the Section 1.3.

9. Vertical wall reinforcement with design axial strength equal to the nominal shear strength of the coupling beam shall be placed over the embedment length of the beam with two-thirds of the steel located over the first half of the embedment length. This wall reinforcement shall extend a distance of at least one anchorage length above and below the flanges of the coupling beam. It is permitted to use vertical reinforcement placed for other purposes, such as for vertical boundary members, as part of the required vertical reinforcement.

10.4. Additional detailing rules for Ductility Class H.

1. Transverse reinforcements for confinement of the composite boundary members, either partially or fully encased, shall extend to a distance of 2 h. into the concrete wall ; h is the depth of the boundary element in the plane of the wall. Figure 10.2.

2. Face bearing plates shall be provided on both sides of the coupling beams at the face of the reinforced concrete wall. These stiffeners will be full depth stiffeners provided on both sides of the coupling beam ; their combined width will not be less than ($b_f - 2\,t_w$) ; their thickness will not be less than 0,75 t_w or 8 mm. (b_f and t_w are the coupling beam flange width and web thickness respectively. Figure 10.3.

3. The requirements to links in Eccentric Braced Frames apply to coupling beams.

    C. In particular, the coupling rotation can be assumed as 0,08 radian, unless a smaller value is justified by a rational analysis of the inelastic deformations.

4. The vertical wall reinforcement as specified in 10.3.9 shall be confined by transverse reinforcement that meets the requirements for frame members or boundary members ; these transverse reinforcements shall be anchored within the confined core of the boundary element.

11. DESIGN AND DETAILING RULES FOR COMPOSITE STEEL PLATES SHEAR WALLS.

11.1. Specific criteria.

1. The provisions in this Section apply to structural walls consisting of steel plates with reinforced concrete encasement on one or both sides of the plate and structural steel or composite boundary members.

2. Composite steel plate shear walls are designed to yield in shear in the steel plate.

3. The steel plate shall be adequately stiffened by encasement and attachment to the reinforced concrete, so that the shear resistance of the steel plate is :

$$V_{Rd} = A_{pl} \times f_{yd} / 3^{0.5}$$

where f_{yd} is the design yield strength of the plate
       A_{pl} is the horizontal area of the plate

C. Steel plate reinforced composite shear walls can be used most effectively where story shear forces are large and the required thickness of conventionally reinforced shear walls is excessive. The provisions limit the shear strength of the wall to the yield strength of the plate because there is insufficient basis from which to develop design rules for combining the yield strength of the steel plate and the reinforced concrete panel. Moreover, since the shear strength of the steel plate usually is much greater than that of the reinforced concrete encasement, neglecting the contribution of the concrete does not have a significant practical impact.

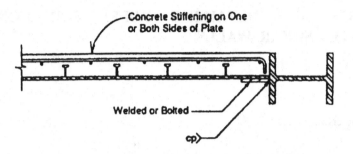

*Concrete stiffened steel shear wall with steel
boundary member.*

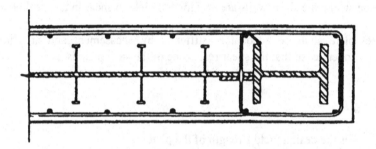

*Concrete stiffened steel shear wall with
composite (encased) boundary member.*

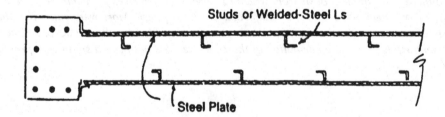

*Concrete filled composite shear wall
with two steel plates.*

Figure 11.1. Possible sections of steel shear walls

11.2.   Analysis

1.   The analysis of the structure is made on the basis of the section properties defined in 6.6.

11.3. Detailing rules

1.   It will be checked that :

$$V_{Sd} < V_{Rd}$$

where V_{Rd} is computed according to 11.1.3.

C. See Commentary in 11.1.3.

2.   The connections between the plate and the boundary members (columns and beams) as well as the connections between the plate and the concrete encasement will be designed such that full yield strength of the plate can be developed.

   C. This rule means that local as well as overall plate buckling has to be avoided. For what concerns local buckling, this can for instance be achieved by limiting studs spacing in such a way that h/t values of unstiffened parts of the steel plate remain under limiting values of h/t. It is recommended that overall buckling of the composite panel be checked using elastic buckling theory considering the section stiffness of the composite wall.

3.   The steel plate shall be continuously connected on all edges to structural steel framing and boundary members with welds and/or slip-critical high-strength bolts to develop the nominal shear strength of the plate.

4.   Structural steel and composite boundary members shall be designed to meet the requirements in Section 6.10.

5.   Boundary members shall be provided around openings in the steel plate as required by analysis.

REFERENCES

AISC – Seismic provisions for structural steel buildings.

Richard Yen, J.Y., Lin Y., and Lai, M.T., "Composite beams subjected to static and fatigue loads"

J. of Struct. Eng. June 1997, 765-771.

Bursi, O.S. and Zandonini, R., "Quasi-static Cyclic and Pseudo-dynamic Tests on Composite Substructures with Softening Behaviour" selected paper in Stability and Ductility of Steel Structures, Usami T. and Itoh Y. ed. Elsevier, 1998, 119-130.

Bursi, O.S. and Gramola, G., "Behaviour of headed stud shear connectors under low-cycle high amplitude displacements", Revised for Materials and Structures, Vol.32, May 1999, pp.290-297.

Plumier, A., Problems and options in the seismic design of composite frames, SPEC internal report, 1996.

Ricles, J.M. and Paboojian, SD., 'Seismic performance of steel-encased composite columns', J. Struct. Engng, 120, 8, 1994 (2474-2494).

Doneux, C. Research on energy dissipation capacity of composite steel/concrete structures. Tests on 3 composite joints at Technical University of Darmstadt – Report II – April 1999.

Paulay, T. and Priestley, M.J.N. Seismic design of Reinforced Concrete and Masonry buildings ed. John Wiley & sons, inc. 1992.

Sanchez, L. and Plumier, A. Particularities raised by the evaluation of load reduction factors for the seismic design of composite steel concrete structures paper accepted for SDSS 1999 (Timisoara).

. Design guide for partially restrained composite connections, J. of Struct. Eng. ASCE, October 1998, 1099-1114.

Astaneh-Asl, A. Seismic design of bolted steel moment-resisting frames, in Steel Tips, Structural Steel Education Council Moraga, Calif., (1995) 82 pp.

Ricles, JM and Paboojian, SD., 'Seismic performance of steel-encased composite columns', J. Struct. Engng, 120, 8, 1994 (2474-2494).

CHAPTER 7

DESIGN OF CONNECTIONS

L. Calado
Instituto Superior Tècnico, Lisbon, Portugal

1 Introduction

The use of steel connections is inherent of every structural steel building, whether it is of one story or one hundred stories. Therefore, the beam-to-column connection, because of its importance to all construction, is significant both economically and structurally. Savings in connection costs as well as improved connection quality has an impact on all types of buildings. Because of the repetitive nature of the connections, even minor material or labour savings in one connection is compounded and expanded throughout the entire building. It is important, then, for a design engineer to understand the behaviour of the connection, not only from the point-of-view of the connection as a structural element, but also from the point-of-view of the connection as a part of the complete structural system.

Following the Northridge (1994) and Hyogoken-Nanbu (1995) earthquakes, the confidence of structural engineers in welded moment resisting connections was strongly compromised due to the extensive brittle damage detected in several frames. As a consequence of these observations, a great deal of theoretical and experimental research activity is presently being developed in USA, Japan and Europe on the cyclic behaviour of beam-to-column connections.

Since ecently bolted connections, in particular top and seat with web angles connections, have not been considered appropriate for structures in seismic areas, due to their partial strength and semirigidity characteristics. However, as pointed out by several researchers amongst which Astaneh (1995), Leon (1997), Elnashai et al. (1998) and De Matteis et al. (2000), the dynamic behaviour of semirigid frames can be particularly favourable due to the period elongation, related to the connection flexibility, to the damping increase and to highly dissipative friction mechanism deriving from a proper "slip capacity design". Both these effects act as a sort of self-isolation of the frame structure, thus leading to remarkable reduction of the seismic actions.

It is worth to emphasise that in the context of the European Inco-Copernicus Joint Research Project "Reliability of Moment Resistant Connections of Steel Building Frames in Seismic Areas" (RECOS) and also of the SAC Steel Project, which started immediately after the Northridge earthquake to address the specific problem of beam-to-column connections, a great interest in bolted configurations as alternative to the standard welded connections (Roeder, 1998), was developed.

In this chapter a comparative assessment of the cyclic behaviour of welded and bolted beam-to-column connections is discussed aiming to define the effect of the column size and of the panel zone design. An analytical model to simulate the cyclic behaviour of bolted connection is discussed and a design methodology based on damage accumulation is presented.

2 Experimental Behaviour of Beam-to-Column Connections

In this chapter reference is essentially made to full-scale experimental tests recently carried out by Calado et al. (1999a, 1999b, 2000) at Material and Structures Test Laboratory of the Instituto Superior Técnico, Lisbon. Two alternative connection solutions, namely fully welded connections (WW) and top and seat with web angle (TSW) designed for the same beam-to-column joints were experimentally investigated. The test program was planned with the aim of assessing the comparative behaviour of bolted and welded connections, and of defining the effect of the column size and of the panel zone design on the behaviour of the two types of connection, with variation of the applied loading history.

2.1 Experimental Program

The specimens. Two series of full-scale specimens were designed and tested namely a WW specimen series (BCC5, BCC6 and BCC8) and a TSW specimen series (BCC9, BCC7 and BCC10). The specimens of the two series were T-shaped beam-to-column subassemblages, consisting approximately of a 1000 mm long beam and a 1800 mm long column. Due to the characteristics of the test set-up the column was the horizontal element while the beam was the vertical one. The material used for the columns, beams, and angles was steel S235 JR. For each series the cross section of the beam was the same (IPE300), while the column cross section was varied, being respectively HEB160 for the BCC5 (WW) and BCC9 (TSW) specimens, HEB200 for the BCC6 (WW) and BCC7 (TSW) specimens, and HEB240 for the BCC8 (WW) and BCC10 (TSW) specimens. In both series, the continuity of the connection through the column was ensured by 12 mm thick plate stiffeners, fillet welded to the column web and flanges.

WW specimens. In the WW specimens, the beam flanges were connected to the column flange by means of complete joint penetration (CJP) groove welds, while fillet welds were applied between both sides of the beam web and the column flange, Figure 1.

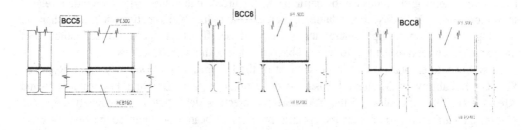

Figure 1. Detail of the fully welded connections.

In Table 1 the mean values of the coupon test performed for each section type and for both web and flange elements are presented.

Table 1. Material characteristics.

Member Section	Section element	f_y (MPa)	ε_y (%)	f_u (MPa)	ε_u (%)
HEB160	web	395.56	0.56	490.08	22.35
	flange	323.13	0.43	460.22	27.71
HEB200	web	401.62	0.49	489.35	22.30
	flange	312.56	0.38	434.91	36.75
HEB240	web	309.00	0.55	469.40	28.40
	flange	300.50	0.51	457.40	34.00
IPE 300 (HEB160)	web	305.54	0.35	412.60	32.65
	flange	274.78	0.38	404.55	36.05
IPE 300 (HEB200)	web	304.62	0.40	418.03	34.45
	flange	278.62	0.45	398.76	37.10
IPE 300 (HEB240)	web	299.50	0.56	450.55	37.70
	flange	292.00	0.44	445.15	35.80

In Table 2, the adopted nomenclature is reported for each analysed specimen. In particular, the size of the column, the distance h relative to the point of application of loading actuator to the external upper flange of the column and the displacement history type are reported.

Table 2. Adopted nomenclature for tested specimens.

Displacement History	Column Section (Actuator position h (mm))		
	HEB160	HEB200	HEB240
$d = \pm 37.5$ mm	BCC5D ($h = 862$)	BCC6D ($h = 862$)	BCC8B ($h = 760$)
$d = \pm 50.0$ mm	BCC5A ($h = 862$)	BCC6A ($h = 862$)	BCC8A ($h = 765$)
$d = \pm 75.0$ mm	BCC5B ($h = 862$)	BCC6B ($h = 862$)	BCC8C ($h = 760$)
ECCS	BCC5C ($h = 862$)	BCC6C ($h = 862$)	BCC8D ($h = 763$)

The flexural strengths of the beam (M_{pb}), column (M_{pc}) and panel zone ($M_{p,PZ}$) were computed on the basis of the nominal and actual yield stress and are reported in Table 3.

Table 3. Moment capacities (in kNm) of the WW specimens.

Specimen	Values	M_{pb}	M_{pc}	$M_{p,PZ}$
BCC5	Nominal	147.6	83.2	91.1
	Actual	173	114	149.8
BCC6	Nominal	147.6	151.1	132.4
	Actual	175	201	220.7
BCC8	Nominal	147.6	247.5	182.9
	Actual	183	316	239.7

From the simple comparison among the nominal plastic moments reported in Table 3, it can be observed that in the three WW specimens the weakest component of the joint configuration was respectively: the column for the BCC5 specimen, the panel zone for the BCC6 specimen and the beam for the BCC8 specimen.

TSW specimens. In the BCC9, BCC7 and BCC10 (TSW) specimens, 120x120x10 angles were adopted. Two rows of bolts were placed on each leg of the flange angles, while on the legs of the web angles there was only one row of three bolts. The bolts were M16 grade 8.8 (yield stress f_{yb}=640 MPa, ultimate stress f_{ub}=800 MPa, A_s=157 mm^2), preloaded according to the Eurocode 3 (1994) provisions, i.e. at $F_{P,CD}$= 0.7 $f_{ub} A_s$ = 87.9 kN.

In Table 4 the mean values of the coupon test performed for each section type, for both web and flange elements and angles are presented.

Table 4. Material characteristics.

Member Section	Section element	f_y (MPa)	ε_y (%)	f_u (MPa)	ε_u (%)
HEB160	web	348.63	0.53	490.31	31.5
	flange	303.43	0.84	453.12	44.5
HEB200	web	302.31	0.55	434.39	38.2
	flange	282.29	0.75	433.27	43.3
HEB240	web	290.28	0.52	447.91	39.6
	flange	304.74	0.81	454.40	39.1
IPE 300	web	315.67	0.66	451.26	41.1
	flange	304.71	0.68	452.63	42.7
L 120x120x10		252.23	0.52	420.14	44.5

In Table 5, the adopted nomenclature and the distance h relative to the loading actuator and displacement history type are reported.

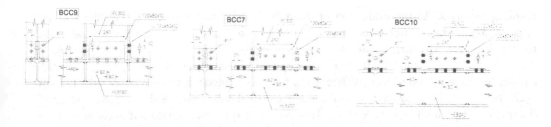

Figure 2. Detail of the top and seat with web angle connections.

Table 5. Adopted nomenclature for tested specimens.

Displacement History	Column Section (Actuator position h (mm))		
	HEB160	**HEB200**	**HEB240**
$d = \pm 37.5$ mm	BCC9C ($h = 793$)	BCC7D ($h = 780$)	BCC10B ($h = 765$)
$d = \pm 50.0$ mm	BCC9A ($h = 792$)	BCC7A ($h = 780$)	BCC10A ($h = 765$)
$d = \pm 75.0$ mm	BCC9B ($h = 790$)	BCC7B ($h = 780$)	BCC10D ($h = 765$)
ECCS	BCC9D ($h = 793$)	BCC7C ($h = 770$)	BCC10C ($h = 760$)

It is well known that two major phenomena characterise the behaviour of the TSW connection: the slippage of bolts and the yielding of the tension angle. For the TSW specimens herein described the bending moment corresponding to bolt slippage (M_{slip}) and angle yielding ($M_{y,angle}$) are reported in Table 6, together with the beam (M_{pb}) and column (M_{pc}) moment capacities. From the comparison between the bending moments corresponding to bolt slippage and angle yielding, it derives that the specimens are "slip critical" connections, since slippage of top and seat angle bolts occurs at a load level higher than the one corresponding to yielding of the tension angle.

Table 6. Moment capacities (in kNm) of the TSW specimen.

Specimen	Values	M_{slip}	$M_{y,angle}$	M_{pb}	M_{pc}
BCC9	Nominal	32 - 47.5	23.3	147.6	83.2
	Actual		28.1	198.2	123.4
BCC7	Nominal	32 - 47.5	23.3	147.6	151.1
	Actual		28.1	198.2	194.4
BCC10	Nominal	32 - 47.5	23.3	147.6	247.5
	Actual		28.1	198.2	305.7

Loading histories. As the evaluation of ultimate capacity under severe earthquake was the principal aim of the research, each set of specimen was tested up to failure under several cyclic rotation histories. The different rotation histories applied to the specimens can be grouped in the following sets: 1) monotonic, 2) cyclic stepwise increasing amplitude, 3) cyclic constant amplitude. The adoption of these different sets for each connection allowed to consider the low cycle fatigue loading together with the more classical monotonic and the more "seismic" (increasing amplitude) type of loading. This latter test type was carried out according to the basic loading history suggested by the ECCS Guidelines (1986). In this loading history the cycles were symmetric in peak deformation d, the history was divided into steps, with three cycles performed in each step and peak deformation given by an increasing predetermined value of the normalised beam tip deflection, d/d_y. In Figure 3 the ECCS loading history is schematically reported, providing both the d/d_y and d/H values for the single load steps, where:

d applied beam tip displacement;
d_y theoretical yield displacement;
H distance between beam tip and centreline of the column.

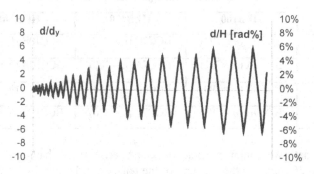

Figure 3. Increasing amplitude loading history.

Experimental set-up and specimen instrumentation. The test set-up, Figure 4, consisted mainly of a foundation, a supporting girder, a reaction r.c. wall, a power jackscrew and a lateral frame. The power jackscrew (capacity 1000 kN, stroke ± 200mm) was attached to a specific frame, pre-stressed against the reaction wall and designed to accommodate the screw backward movement. The specimen was connected to the supporting girder through two steel elements. The supporting girder was fastened to the reaction wall and to the foundation by means of pre-stressed bars.

An automatic testing technique was developed to allow computerised control of the power jackscrew, of the displacement and of all the transducers used to monitor the specimens during the testing process. Specimens were instrumented with electrical displacement transducers (LVDTs), which recorded the displacement histories at several points in order to obtain a careful documentation of the various phenomena occurring during the tests, Figure 5.

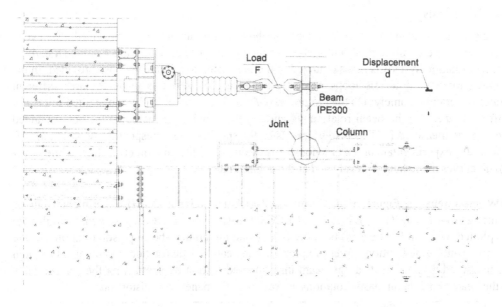

Figure 4. Test set-up.

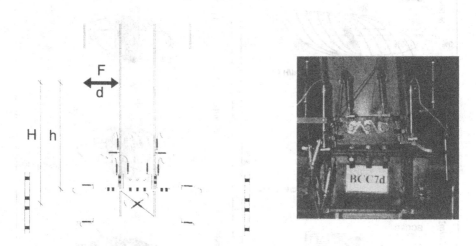

Figure 5. Typical set-up instrumentation.

Similar arrangement of LVDTs were adopted for the three WW and the three TSW specimen series.

2.2 Cyclic Tests

Premise. The experimental results obtained in the test program are herein provided. The cyclic behaviour and the failure modes observed for the six sets of specimens are described, and the moment rotation hysteresis loops obtained in the increasing amplitude tests are provided. In the moment rotation hysteresis loops hereafter presented, reference is made to three different values of rotation, namely: (1) the "*unprocessed*" total rotation given by the applied interstory drift angle d/H; (2) the beam rotation Φ_b and (3) the panel zone rotation Φ_{PZ}, both obtained through the measured LVTDs displacements of the specimens. Correspondingly, in the M-d/H and M-Φ_{PZ} experimental curves the moment is evaluated at the column centreline, while in the M_b-Φ_b curves the moment is evaluated at the column face.

WW specimens. In Figure 6 (a) the moment-global rotation (M-d/H) experimental curves resulting from the BCC5C, BCC6C and BCC8D tests (cyclic increasing stepwise amplitude) are plotted, while in Figure 6 (b) both the corresponding moment-beam plastic rotation and the moment-panel zone rotation curves are plotted. The beam plastic rotation was obtained through the measured displacements at the beam instrumented section by subtracting the contributions of the beam and column elastic rotations as well as of the panel zone distortion.

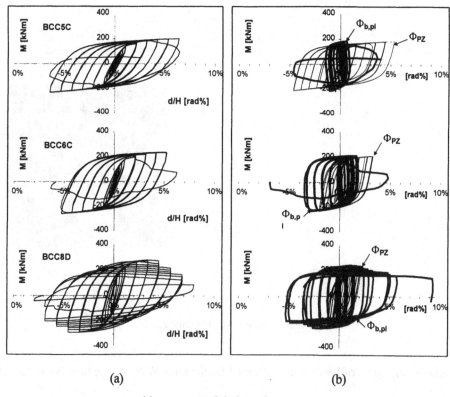

(a)   (b)

(a)   moment-global rotation curves
(b)   moment-beam plastic rotation and moment-panel rotation curves
Figure 6. WW specimens.

BCC5	N	Nf	Failure modes
A	20	16	Crack on the beam flange close to the weld, propagated in the beam web.
B	9	5	Fracture of the beam flange near the weld.
C	24	15	Crack on the beam flange close to the weld, propagated in the beam web.
D	27	22	Fracture of the beam flange.

BCC6	N	Nf	Failure modes
A	19	14	Plastic hinge. Crack on the beam flange in the buckled zone at 15 cm from the column face.
B	15	8	Plastic hinge. Crack on the beam flange in the buckled zone at 10 cm from the column face.
C	21	12	Less evident plastic hinge. Crack in the beam flange along to the weld line.
D	22	18	Less evident plastic hinge. Crack in the beam flange developed close to the weld.

BCC8	N	Nf	Failure modes
A	16	12	Plastic hinge. Crack in the beam flange at the buckled zone at 10 cm from the column face.
B	20	16	Plastic hinge. Crack on the beam flange along the weld line.
C	6	2	Plastic hinge. Crack in the beam flange at the buckled zone.
D	21	15	Plastic hinge. Crack in the beam flange, developed in the proximity of the weld.

Figure 7. WW specimens after the tests.

In Figure 7 some photos of WW specimens at the end of the test are provided together with a table summarising the imposed cycles (N), the number of cycles of the conventional failure (Nf) and the failure modes.

BCC5. As can be derived from the curves reported in Figure 6 (a) and (b), and as demonstrated also throughout the experimental program, the cyclic behaviour of the specimen BCC5 is characterised by a great regularity and stability of the hysteresis loops up to failure, with no deterioration of stiffness and strength properties. In the very last cycle the specimen has collapsed with a sudden and sharp reduction of strength, due to fracture initiated in the beam flange and propagated also in the web. During the tests, significant distortion of the joint panel zone was observed, while not remarkable plastic deformation in the beam occurred.

BCC6. Throughout the test program, two different kinds of cyclic behaviour were observed for the BCC6 specimens. In some cases the behaviour of the specimens was close to the behaviour observed for the BCC5 type, with almost no deterioration of the mechanical properties up to the last cycle, during which the collapse occurred. For the other tests a gradual reduction of the peak moment at increasing number of cycles was evident. In these cases, starting from the very first plastic cycles, local buckling of the beam flanges occurred, and a well-defined plastic hinge was formed in the beam. The contribution of the panel zone deformation was not as significant as in the BCC5 specimen type.

BCC8. The hysteresis loops obtained from the tests on the BCC8 specimens show a gradual reduction of the peak moment starting from the second cycle, where the maximum value of the applied moment was usually registered. This deterioration of the flexural strength of the connection was related to occurrence and spreading of local buckling in the beam flanges and web. A well-defined plastic hinge in the beam was formed in all the tested specimens. In the specimens BCC8 the panel zone deformation was not remarkable, and the plastic deformation took place mainly in the beam.

TSW specimens. In Figure 8 (a) the moment-global rotation (M-d/H) experimental curves resulting from the BCC9D, BCC7C and BCC10C tests (cyclic increasing stepwise amplitude) are shown, while in Figure 8 (b) both the corresponding moment-beam rotation and the moment-panel zone rotation curves are plotted. In Figure 9 some photos of TSW specimens at the end of the test are provided together with a table summarising the imposed cycles (N), the number of cycles of the conventional failure (Nf) and the failure modes.

As can be derived from the curves reported in Figure 8, the shape of hysteresis loops of the three TSW specimens is very similar. The cyclic behaviour, the phenomena observed during the tests and the collapse modes were the same for the three specimen series, thus the following unique paragraph is devoted to describe the above issues for the three specimens.

BCC9 / BCC7 / BCC10. The cyclic behaviour of the TSW connections was characterised by bolt slippage and yielding and spreading of plastic deformation in the top and bottom angles, cyclically subjected to tension. Plastic ovalization of the bolt holes was also observed mainly in the leg of the angle adjacent to column flange. The experimental curves, typical of this type of connection, showed pinched hysteresis loops, with a large slip plateau (very low slope of the experimental curve) and subsequent sudden stiffening. In fact, when the specimen position was at $d = 0$, due to the concomitant effects of bolt slippage, hole ovalization and the plastic deformation of the angle legs adjacent to the column flange, the beam was completely separated from the column (gap open).

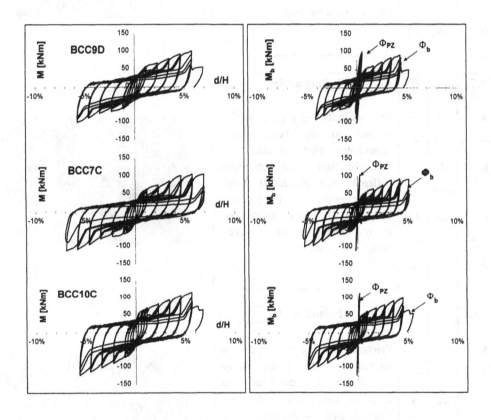

(a)   (b)

(a) moment-global rotation curves
(b) moment-beam plastic rotation and moment-panel rotation curves
Figure 8. TSW specimens.

BCC9	N	Nf	Failure modes
A	12	5	Fracture in the leg angle in tension located on the beam flange immediately after the fillet. Bolt holes of flange and web cleats suffered large ovalization. Large slippage emphasising the pinching behaviour. The connecting elements always represent the weakest component of the joint.
B	7	3	
C	20	6	
D	18	12	

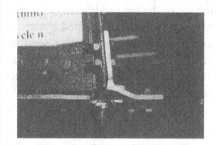

BCC7	N	Nf	Failure modes
A	12	6	Fracture in the leg angle in tension located on the beam flange immediately after the fillet. Bolt holes of flange and web cleats suffered large ovalization. Large slippage emphasising the pinching behaviour. The connecting elements always represent the weakest component of the joint.
B	7	2	
C	20	13	
D	17	6	

BCC10	N	Nf	Failure modes
A	12	5	Fracture in the leg angle in tension located on the beam flange immediately after the fillet. Bolt holes of flange and web cleats suffered large ovalization. Large slippage emphasising the pinching behaviour. The connecting elements always represent the weakest component of the joint.
B	21	4	
C	18	12	
D	6	2	

Figure 9. TSW specimens after the tests.

At large applied displacements, which imposed large rotations to the connection, the contact of the compression angle and the beam web to the column flange (gap closure) gave rise to sudden stiffening of the connection, which is evident in the experimental curves. No significant rotation of the column and distortion of the panel zone was observed throughout the experimental tests carried out on the three specimens. At each step on the test, slight deterioration of the joint resistance in the three applied cycles could be observed in the experimental curves, mainly due to yielding and spreading of plastic deformation in the top and bottom angles, cyclically subjected to tension.

In all the test carried out on the three specimen series, the collapse of the connection occurred due to fracture in the leg angle located on the beam flange, immediately after the fillet. Negligible scatters could be observed in the moment capacity of the three connection series, as it was expected, since the inelastic behaviour of the connection was governed by the angle. Also the maximum values of global rotation experienced by the specimens were the same for the BCC9 and BCC10 series, and slightly larger for the BCC7 one.

2.3 Monotonic Tests

The moment rotation curves obtained from the monotonic tests carried out on the six specimens are presented in Figures 10 (a) and (b), which respectively report the results of the WW and TSW specimens. In these curves the moment is evaluated at column centreline and the rotation is given by the total interstory drift angle (d/H). In each Figure also the moment panel zone rotation is reported. From the experimental results on the two series of specimens the effect both of the connection typology (TSW and WW) and of the column cross section (HEB160, HEB200, and HEB240) can be derived.

By comparing the two series of experimental curves it must be noticed that the three WW specimens showed significant differences in the initial stiffness, maximum strength and deformation capacity, thus confirming the strong effect of the column cross section size already observed in the cyclic tests.

On the contrary, the three TSW specimens presented quite close experimental responses. This difference between the behaviour of WW and TSW specimens was mainly related to the design of the specimens, since in the TSW connections the weakest component was the same in the three specimens (the angle in tension), thus the beam, column and panel zone strength ratios did not affect the response of the specimens. Slight scatters could be observed in the initial stiffness, due to the different column and panel zone deformability, but the non-linear portion of the curve and the maximum bending moment were very similar.

As already evidenced, the behaviour of the WW connections was affected by the column dimensions since the three combinations of beam and column framing in the joint gave rise to panel zone strength values respectively: smaller than, approximately equal to and larger than the plastic moment of the beam, for the BCC5, BCC6 and BCC8 specimens. These observations are confirmed by analysing the main test data provided in Table 7.

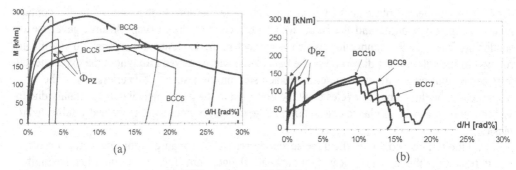

(a)   (b)

(a) WW specimens; (b) TSW specimens
Figure 10. Monotonic experimental curves.

Table 7. Main experimental data from the monotonic tests on the WW and TSW specimens.

		M (kNm)	M_b (kNm)	d/H (%)	Φ_{PZ} (%)	$\Phi_{b,pl} - \Phi_{PZ}$ (%)
WW	BCC5	214	196	21.4	16.5	0.66
	BCC6	231	207	10.5	4.5	4.9
	BCC8	292	251	8.3	3.3	5.3
		M (kNm)	M_b (kNm)	d/H (%)	Φ_{PZ} (%)	Φ_{slip} (%)
TSW	BCC9	135	122.5	10.8	2.4	5.2
	BCC7	144	127	10.5	1.15	6.8
	BCC10	146	126	9.4	0.19	4.8

It can be noticed that while in the WW specimens the panel zone distortion Φ_{PZ} significantly contributed to the specimen global rotation d/H (at the maximum value of the bending moment registered in the relevant test), though at different extent in the three specimens, a completely different order of magnitude of this contribution was registered in the TSW connections. For the TSW specimens, instead, the rotations due to bolt slippage Φ_{slip} computed on the basis of the LVDTs measured displacements constituted a major contribution to the total rotation d/H.

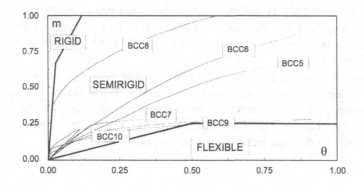

Figure 11. Classification of WW and TSW according to Eurocode 3 (1997).

Figure 11 shows that both the TSW and the WW specimens can be classified, according to the Eurocode 3 (1997), as semirigid joints, even though the TSW specimens are very close to the boundary of the flexible joint behaviour, while all the WW connections are able to sustain bending moment larger than the beam plastic moment capacity (as can be derived by comparing the values of Table 7). Thus, even though WW specimens can be classified as full strength connections, they experience the maximum bending moment at large deformation levels.

2.4 Classification of the Specimens According to other Proposals

The Eurocode 3 (1997) classification system for joints has been recently criticised for several reason. One of the main drawbacks of the approach is that it considers the stiffness and the strength criteria separately, thus possibly leading to some inconsistencies when both the serviceability and the ultimate state are considered; concerning this point Nethercot et al. (1998) propose an unified classification system in which connection stiffness and strength are considered simultaneously. Furthermore Hasan et al. (1998) discuss the weakness of the idea of adopting the stiffness of the beam as the only parameter for defining the initial stiffness of the connection and emphasise the need of defining a non-linear classification system, with no reliance on the beam stiffness. However, the definition of a unique approach, fully exhaustive, is not a simple task.

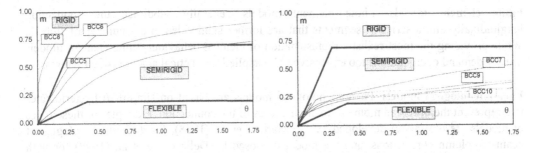

Figure 12. Classification of WW specimens and TSW specimens according to Bjorhovde et al. (1990).

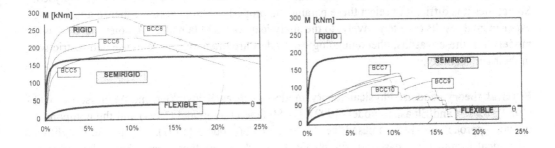

Figure 13. Classification of WW specimens and TSW specimens according to Hasan et al. (1998).

In order to show the differences arising from the application of the different methods, Figures 12 and 13 present the application of the approach proposed by Bjorhovde et al. (1990), quite close to the Eurocode 3 (1997) one, and the new non-linear system proposed by Hasan et al. (1998) for classifying the TSW and WW large scale connections described in this chapter. From Figure 12 it derives that the TSW connections can be defined as semirigid according to both the two proposals as well as to the Eurocode 3 (1997). On the contrary, for the WW connections (Figure 13) the two systems provide results quite different from the Eurocode 3 ones, which defines the three connections as semirigid (see Figure 11). Both systems classify the BCC8 specimen as fully rigid, while the BCC6 and BCC5 specimens behave as semirigid joints at serviceability limit state and as rigid joints at the ultimate limit state.

3   Numerical Modelling of Beam-to-Column Connections

3.1   Introduction

Several analytical models have been developed in the last years to represent the cyclic behaviour of beam-to-column connections. These models can be divided into three different general types: the finite element, the phenomenological and the physical theory models.

Finite element models. Finite element models generally subdivide the connection longitudinally into a series of segments that are further subdivided into a number of fibbers. While providing the most realistic representation of element behaviour, finite element models usually demand computations too expensive to be applied to practical analyses of structures.

Phenomenological models. Phenomenological models are based on simplified hysteric rules that represent the observed moment–rotation curves of the connection. Examples of these types of models are the model developed by De Martino et al. (1988), for different typologies of beam-to-column connections, and the model proposed by Della Corte et al. (1999) for both welded and bolted connections. This last model includes in its formulation pinching phenomena, stiffness degradation and strength deterioration. Currently, phenomenological modelling is the most common approach used in parametric studies. However, users of such models have to specify numerous empirical input parameters for each connection analysed. Sometimes it is difficult to select these parameters properly without access either to appropriate experimental results or, alternatively, to the analytical results obtained using other more refined models. For these reasons, phenomenological models are often expensive to use and restrictive in their applicability.

Physical theory models. Physical theory models such as Krawinkler and Mohasseb model (1987), Madas and Elnashai model (1992), De Stefano et al. model (1994) and the component method approach in which is based the Annex J of Eurocode 3 (1997), incorporate simplified theoretical formulations based on the physical considerations that permit the cyclic inelastic behaviour to be computed. Unlike the prior empirical information on cyclic inelastic behaviour required for phenomenological models, the input parameters for physical theory models are based on material properties and common geometric or derived engineering properties of the connection (e.g., the cross-sectional area, cross-sectional moment of inertia, effective member length, plastic section modulus, etc). However, the geometric representation of the connection is considerably simpler than that used for a finite element model. Thus, physical theory models attempt to combine the realism of finite element approaches with the computational simplicity of phenomenological modelling.

When fully developed and verified, physical theory models can be valuable in selecting input parameters for phenomenological models as well as in analysing large-scale structures. Phenomena like deterioration of mechanical features arising from plastic deformation of the steel, inelastic behaviour of bolts and angles must be correctly simulated in order to describe the whole cyclic response of the joint up to failure and to perform sophisticated cyclic design analyses of steel frame structures. This is essential when joints constitute the weakest element class of the structure and collapse is based upon local criteria.

3.2   Stress-Strain Relationship for Steel

Introduction. The model proposed by Guiffré and Pinto (1970) for the stress-strain relationship of the steel under cyclic loading can be represented by the following equation (1):

$$\sigma_s^* = \frac{\varepsilon_s^*}{\sqrt[R]{1+\left|\varepsilon_s^*\right|^R}} \tag{1}$$

In this equation ε_s^* and σ_s^* represent respectively the relative strain and relative stress. For the 1st half-cycle they can be obtained by the following equations (2):

$$\varepsilon_s^* = \frac{\varepsilon_s}{\varepsilon_{sy}} \qquad\qquad \sigma_s^* = \frac{\sigma_s}{\sigma_{sy}} \tag{2}$$

For the following half-cycles these parameters can be obtained by the equations (3):

$$\varepsilon_s^* = \frac{\varepsilon_s - \varepsilon_{sr}}{2\varepsilon_{sy}} \qquad\qquad \sigma_s^* = \frac{\sigma_s - \sigma_{sr}}{2\sigma_{sy}} \tag{3}$$

In these equations the variables have the following meaning:

ε_s strain

ε_{sr} strain at the last loading inversion

ε_{sy} yield strain

σ_s stress

σ_{sr} stress at the last loading inversion

σ_{sy} yield stress

In this model the stress-strain relationship is elasto-plastic and has a transition curve that is a function of the parameter R. The larger the value of the parameter R the similar the curve is to the elastic perfectly plastic behaviour. The model proposed by Guiffré and Pinto (1970) has two asymptotic lines that are respectively $\sigma = \pm\sigma_{sy}$. Unloadings are obtained with an initial stiffness equal to the modulus of elasticity (4):

$$E = \frac{\sigma_{sy}}{\varepsilon_{sy}} \tag{4}$$

In this model the parameter R takes into account the Bauschinger effect and defines the shape of the transition curve between the elastic and plastic zones and is a function of the maximum plastic excursion (ξ_{max}) obtained until the instant. Thus, the parameter R depends on the loading history and can be assessed by the following equation (5):

$$R = R_o - \frac{A_1 \xi_{max}}{A_2 + \xi_{max}} \tag{5}$$

In this equation R_0, A_1 and A_2 are constants. The maximum plastic excursion can be obtained by the following equation:

$$\xi_{max} = \max_i \left\{ \left| \varepsilon_{sr}^{(i)} - \varepsilon_{sr}^{(i-1)} + \frac{\sigma_{sr}^{(i-1)} - \sigma_{sr}^{(i)}}{E} \right| \right\} \tag{6}$$

where $\left(\varepsilon_{sr}^{(i)}, \sigma_{sr}^{(i)} \right)$ represent the i inversion of the load.

The hardening was later included in the model (Menegotto and Pinto, 1973):

$$\sigma_s^* = b \varepsilon_s^* + \frac{(1-b)\varepsilon_s^*}{\sqrt[R]{1 + |\varepsilon_s^*|^R}} \tag{7}$$

where b is the ratio between the kinematic stiffness hardening and the modulus of elasticity:

$$b = \frac{E_{kin}}{E} \tag{8}$$

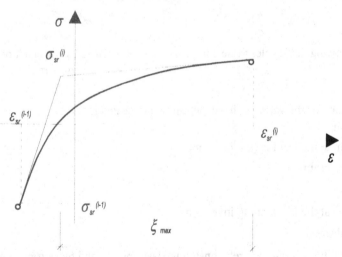

Figure 14. Plastic excursion.

This stress-strain relationship has as asymptotes the lines with slope E_{kin} and which include points $(\pm\varepsilon_{sy},\pm\sigma_{sy})$. The plastic excursion (Figure 14) can be obtained from the equation (9):

$$\xi_{max} = \max_i \left\{ \frac{1}{1-b} \times \left| \varepsilon_{sr}^{(i)} - \varepsilon_{sr}^{(i-1)} + \frac{\sigma_{sr}^{(i-1)} - \sigma_{sr}^{(i)}}{E} \right| \right\} \tag{9}$$

The model developed by Menegotto and Pinto (1973) model has some limitations. It does not consider the partial unloading and it imposes the equality between the positive and negative yield stresses.

Model for σ–ε relationship with unloading. The model developed by Brito (1999) is based on Menegotto and Pinto (1973) model, which was reformulated to take into account the effects of unloading. In this reformulation new expressions are proposed to assess the relationship between the relative stress (σ_s^*) and the relative strain (ε_s^*):

$$\sigma_s^* = b\,\varepsilon_s^* + \frac{(1-b)\varepsilon_s^*}{\sqrt[R]{1+\left|\varepsilon_s^*\right|^R}} \tag{7}$$

The relative stress and relative strain can be obtained (Brito, 1999) by the following equations (10):

$$\sigma_s^* = \frac{\hat{\sigma}_s - \hat{\sigma}_{sr}}{k_\sigma \sigma_{sy}} \qquad\qquad \varepsilon_s^* = \frac{\varepsilon_s - \varepsilon_{sr}}{k_\varepsilon \varepsilon_{sy}} \tag{10}$$

where

$\hat{\sigma}_s$ stress without damage accumulation

$\hat{\sigma}_{sr}$ stress at the last inversion without damage accumulation

The parameters k_σ and k_ε related with the amplitude of the stress and strain are obtained by imposing that the stress-strain curve is asymptotic to the kinematic hardening curve, as is shown in Figure 15:

$$\hat{\sigma}_s \to \sigma_{yo}^+ + E_{kin}\varepsilon_s \tag{11}$$

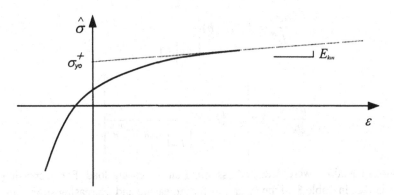

Figure 15. Asymptotic behaviour of the stress-strain relationship.

From the equation proposed by Menegotto and Pinto (1973) it is possible to obtain the following equation (12):

$$\sigma_s^* \rightarrow b\,\varepsilon_s^* + (1-b) \tag{12}$$

Replacing the relative stress and the relative strain (10) in the equation (12), it can be re-written as follows:

$$\hat{\sigma}_s \rightarrow \left[\frac{k_\sigma \sigma_{sy}}{k_\varepsilon \varepsilon_{sy}} b\right]\varepsilon_s + \left[k_\sigma \sigma_{sy}(1-b) - \frac{k_\sigma \sigma_{sy}}{k_\varepsilon \varepsilon_{sy}} b\varepsilon_{sr} + \hat{\sigma}_{sr}\right] \tag{13}$$

Comparing (11) with (13) the following equations can be obtained (14-15):

$$\frac{k_\sigma}{k_\varepsilon} = \frac{E_{kin}}{Eb} = 1 \tag{14}$$

$$k_\sigma = \frac{\sigma_{yo}^+ - \hat{\sigma}_{sr} + E_{kin}\varepsilon_{sr}}{(1-b)\sigma_{sy}} = 1 - \frac{\hat{\sigma}_{sr}}{(1-b)\sigma_{sy}} + \frac{b}{1-b}\frac{\varepsilon_{sr}}{\varepsilon_{sy}} \tag{15}$$

The implementation of this stress-strain relationship needs the assessment that the stress is a function of the strain but also the evaluation of the tangent stiffness $\dfrac{d\sigma}{d\varepsilon}$.

The main expressions to assess this stress-strain relationship which takes into account the partial unloading are (Brito, 1999):

$$\frac{d\hat{\sigma}_s}{d\varepsilon_s} = \frac{d\hat{\sigma}_s}{d\sigma_s^*} \times \frac{d\sigma_s^*}{d\varepsilon_s^*} \times \frac{d\varepsilon_s^*}{d\varepsilon_s} = \frac{\dfrac{d\sigma_s^*}{d\varepsilon_s^*}}{\dfrac{d\sigma_s^*}{d\hat{\sigma}_s}} \times \frac{d\varepsilon_s^*}{d\varepsilon_s} \tag{16}$$

$$\frac{d\sigma_s^*}{d\hat{\sigma}_s} = \frac{1}{k_\sigma \sigma_{sy}} \tag{17}$$

$$\frac{d\varepsilon_s^*}{d\varepsilon_s} = \frac{1}{k_\varepsilon \varepsilon_{sy}} \tag{18}$$

$$\frac{d\sigma_s^*}{d\varepsilon_s^*} = b + (1-b)\frac{1}{\left[1 + \left|\varepsilon_s^*\right|^R\right]^{1+\frac{1}{R}}} \tag{19}$$

$$\frac{d\hat{\sigma}_s}{d\varepsilon_s} = E \times \left[b + (1-b)\frac{1}{\left[1 + \left|\varepsilon_s^*\right|^R\right]^{1+\frac{1}{R}}}\right] \tag{20}$$

The previous equations were deducted assuming an increasing load. For decreasing load the deduction is similar. In Table 8 all the equations for increasing and decreasing strains are written.

Table 8. Equation for the stress-strain relationship considering the partial unloading.

Eqn	ε_s increasing	ε_s decreasing
10	$\hat{\sigma}_s^* = \dfrac{\hat{\sigma}_s - \hat{\sigma}_{sr}}{k_\sigma \sigma_{sy}}$ $\quad \varepsilon_s^* = \dfrac{\varepsilon_s - \varepsilon_{sr}}{k_\varepsilon \varepsilon_{sy}}$	$\hat{\sigma}_s^* = \dfrac{\hat{\sigma}_s - \hat{\sigma}_{sr}}{k_\sigma \sigma_{sy}}$ $\quad \varepsilon_s^* = \dfrac{\varepsilon_s - \varepsilon_{sr}}{k_\varepsilon \varepsilon_{sy}}$
11	$\hat{\sigma}_s \rightarrow \sigma_{yo}^+ + E_{kin}\,\varepsilon_s$	$\hat{\sigma}_s \rightarrow \sigma_{yo}^- + E_{km}\,\varepsilon_s$
12	$\sigma_s^* \rightarrow b\varepsilon_s^* + (1-b)$	$\sigma_s^* \rightarrow b\varepsilon_s^* - (1-b)$
13	$\hat{\sigma}_s \rightarrow \dfrac{k_\sigma}{k_\varepsilon} E_{kin}\varepsilon_s + \left[k_\sigma \sigma_{sy}(1-b) - \dfrac{k_\sigma}{k_\varepsilon} E_{kin}\varepsilon_{sr} + \hat{\sigma}_{sr} \right]$ $\hat{\sigma}_s \rightarrow \dfrac{k_\sigma}{k_\varepsilon} E_{kin}\varepsilon_s + \left[k_\sigma \sigma_{sy}(1-b) - \dfrac{k_\sigma}{k_\varepsilon} E_{kin}\varepsilon_{sr} + \hat{\sigma}_{sr} \right]$	
14	$\dfrac{k_\sigma}{k_\varepsilon} = 1$	$\dfrac{k_\sigma}{k_\varepsilon} = 1$
15	$k_\sigma = 1 - \dfrac{\hat{\sigma}_{sr}}{(1-b)\sigma_{sy}} + \dfrac{b}{1-b}\dfrac{\varepsilon_{sr}}{\varepsilon_{sy}}$	$k_\sigma = 1 + \dfrac{\hat{\sigma}_{sr}}{(1-b)\sigma_{sy}} - \dfrac{b}{1-b}\dfrac{\varepsilon_{sr}}{\varepsilon_{sy}}$

Model for $\sigma-\varepsilon$ relationship with damage accumulation. To take into account the damage accumulation due to the cyclic loading it is necessary to include into the stress-strain relationship a damage index that should consider the deterioration in the previous complete semi-cycles and the deterioration in the actual semi-cycle, according to the following equation (21) (Brito, 1999):

$$I_d^T = I_d^c + \Delta I_d^c \tag{21}$$

where

I_d^T total damage index;

I_d^c damage index related with the complete semi-cycles;

ΔI_d^c damage index related to the actual semi-cycle.

The damage index related with the complete semi-cycles is constant during each semi-cycle while the damage index related to the actual semi-cycle is function of the actual strain and can be assessed through the following equation (Brito, 1999) (22):

$$\Delta I_d^c = \frac{1}{N_j} = \frac{\left(\dfrac{|\varepsilon_s - \varepsilon_{sr}|}{\varepsilon_y} \right)^m}{K} \tag{22}$$

Thus the value of stress taking into account the damage accumulation can be obtained by the following equation (23):

$$\sigma_s = \hat{\sigma}_s \times (1 - f(I_d)) \tag{23}$$

where:

σ_s stress with damage accumulation;

$\hat{\sigma}_s$ stress without damage accumulation;

f function for the decrease of the stress.

The function for the decrease of the stress (f) can be obtained through experimental tests (Proença, 1996) and is shown in Figure 16.

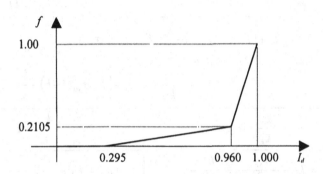

Figure 16. Function for the decrease of the stress.

Analytically the function f can be represented by the following equations (24):

$$f(I_d) = \begin{cases} 0 & \Leftarrow & I_d \leq 0.295 \\ 0.31681 I_d - 0.0936 & \Leftarrow & I_d \in]0.295;0.960] \\ 19.7332 I_d - 18.7332 & \Leftarrow & I_d > 0.960 \end{cases} \qquad (24)$$

In this model the assessment of the stress is carried out in two steps. A first step where the stress is evaluated without the consideration of the damage accumulation ($\hat{\sigma}_s$) and a second one where the stress is corrected to take into account the damage of the actual semi-cycle. In this methodology the correction of the stress is made at the end of the actual semi-cycle. The new equation for the tangent stiffness can be obtained by the following equation (25):

$$\frac{d\sigma_s}{d\varepsilon_s} = \frac{d\hat{\sigma}_s}{d\varepsilon_s} \times \left[1 - f(I_d)\right] - \frac{df}{d\varepsilon_s} \hat{\sigma}_s \qquad (25)$$

To assess the value of equation (25), $\dfrac{df}{d\varepsilon_s}$ may be obtained by the following equation:

$$\frac{df}{d\varepsilon_s} = \frac{df}{dI_d^T} \times \frac{dI_d^T}{d\varepsilon_s} = \frac{df}{dI_d^T} \times \frac{d\Delta I_d^p}{d\varepsilon_s} \qquad (26)$$

with

$$\frac{df}{dI_d^T} = \begin{cases} 0 & \Leftarrow & I_d^T < 0.295 \\ 0.3168 & \Leftarrow & I_d^T \in [0.295;0.960[\\ 19.332 & \Leftarrow & I_d^T \geq 0.960 \end{cases} \tag{27}$$

and

$$\frac{d\Delta I_d^P}{d\varepsilon_s} = \frac{m}{K \varepsilon_y} \left(\frac{|\varepsilon_s - \varepsilon_{sr}|}{\varepsilon_y} \right)^{m-1} \times SGN(\varepsilon_s - \varepsilon_{sr}) \tag{28}$$

The *SGN* function takes into account the sign of the argument.

Accuracy of the σ–ε model developed by Brito (1999). To analyse the accuracy and the validity of the model developed by Brito (1999) for the stress-strain relationship of the steel which takes into account the unloading and the damage accumulation, an experimental result, Figure 17, is compared with a numerical one, Figure 18.

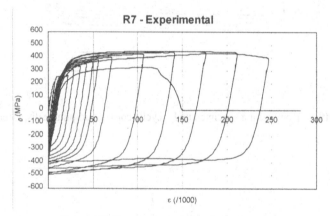

Figure 17. Tension test of a steel specimen.
(adapted from Proença, 1996)

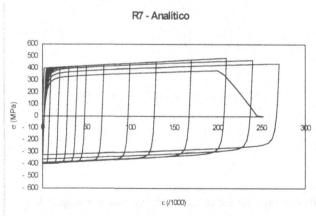

Figure 18. Numerical simulation of the tension test.

Comparison between experimental data and numerical simulations allows to conclude that the proposed numerical model is able to simulate the cyclic behaviour of the steel, the Bauschinger effect and the damage accumulation until the failure.

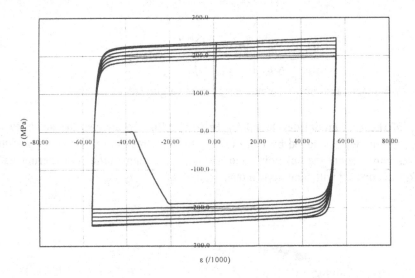

Figure 19. σ–ε curve for a specimen under cyclic constant amplitude displacement.

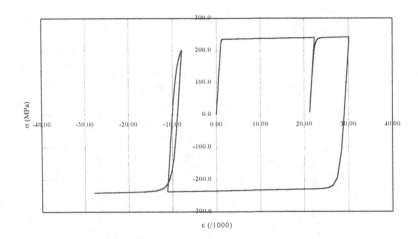

Figure 20. σ–ε curve for a specimen under cyclic random amplitude displacement.

In Figures 19 and 20 some numerical simulations are presented in order to show the ability of the proposed model to represent the damage accumulation and the partial unloading.

3.3 Numerical Model for Bolts in Shear

To adequately model the cyclic behaviour of bolted connections it is fundamental to simulate the behaviour of bolts in shear, Figure 21.

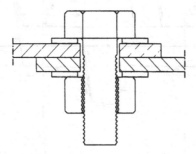

Figure 21. Bolt in shear.

The model developed by Brito (1999) is based on an elasto-plastic behaviour with kinematic hardening and takes into account the slip between the connected elements. In this model the ovalization of the hole and the changes in the preloading force of the bolt are not considered. The damage accumulation is considered in the stress-strain relationship of the material.

Under monotonic loading the behaviour of a bolt in shear can be represented by the following force-displacement curve, Figure 22:

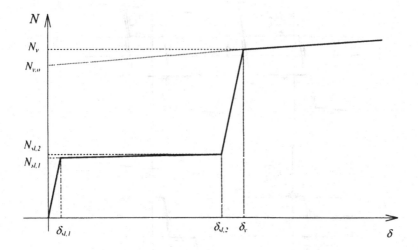

Figure 22. Behaviour of a bolt in shear under a monotonic load.

In Figure 22 four different zones can be defined: elastic behaviour before the slip; slip; elastic behaviour after the slip; plastic behaviour with hardening.

Under cyclic loading several possibilities can be considered for the force-displacement curve of the bolt, Figure 23.

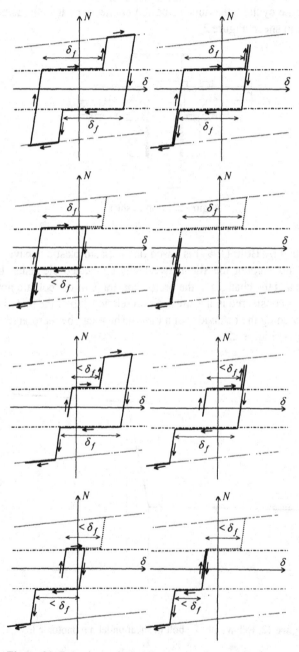

Figure 23. Behaviour of a bolt in shear under a cyclic loading.

As the damage accumulation is considered in the material, the lines that define the kinematic hardening are constant during all the cyclic behaviour. To completely define the behaviour of the bolt in shear under cyclic loading it is necessary to define the following parameters, (Brito, 1999) Figure 24:

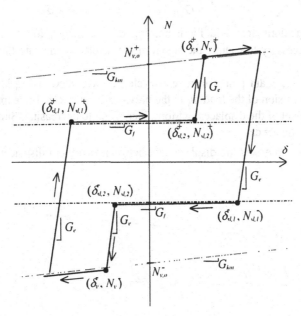

Figure 24. Parameters used to define the complete cycle.

where:

$N^+_{sl,1}\ N^-_{sl,1}$ forces to start the slip;

$N^+_{sl,2}\ N^-_{sl,2}$ forces at the end of the slip;

$N^+_y\ N^-_y$ yield forces;

$N^+_{y,o}\ N^-_{y,o}$ yield forces for zero displacement;

$\delta^+_{sl,1}\ \delta^-_{sl,1}$ displacement at the beginning of the slip;

$\delta^+_{sl,2}\ \delta^-_{sl,2}$ displacement at the end of the slip;

$\delta^+_y\ \delta^-_y$ yield displacement.

In order to define the N-δ curve it is necessary to know, at the beginning of each semi-cycle, the main points of the behaviour of the bolt, namely the points related with the beginning and the end of the slip and the point associated with the yield. Through the knowledge of these points it is possible to define the force N as a function of the displacement:

$$N(\delta) = \begin{cases} N_{sl,1} + G_e \times (\delta - \delta_{sl,1}) & \Leftarrow & \delta < \delta_{sl,1} \\ N_{sl,1} + G_f \times (\delta - \delta_{sl,1}) & \Leftarrow & \delta \in [\delta_{sl,1}; \delta_{sl,2}[\\ N_{sl,2} + G_e \times (\delta - \delta_{sl,2}) & \Leftarrow & \delta \in [\delta_{sl,2}; \delta_y[\\ N_{y,0} + G_{kin} \times \delta & \Leftarrow & \delta \geq \delta_y \end{cases} \quad (29)$$

The previous equations are valid for increasing displacements. In the case of decreasing displacement it is necessary to change the signal of the displacements and the forces except for $\delta_f(\delta, N)$.

In the following the main points in a semi-cycle are presented taking into account all the possibilities of the inversion of the loading in the previous semi-cycle. The expressions presented are valid for the reversal of the displacement form increasing to decreasing. Similar equations can be obtained for the opposite case.

It is thus necessary to assess the displacement between two yield displacements ($\delta_y^+ - \delta_y^-$), Figure 25:

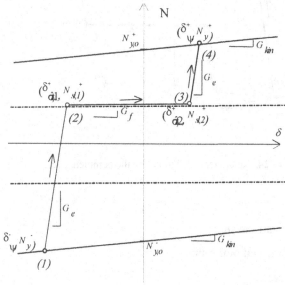

Figure 25. Assessment of the displacement between two yield displacements.

From this figure it is possible to obtain the following equations (30-34):

Point 1 $\quad N_y^- = N_{y,0}^- + G_{kin} \times \delta_y^-$ $\qquad\qquad\qquad\qquad\qquad$ (30)

Point 2 $\quad \delta_{sl,1}^+ = \delta_y^- + \dfrac{N_{sl,1}^+ - N_y^-}{G_e}$ $\qquad\qquad\qquad\qquad$ (31)

Point 3 $\quad \delta_{sl,2}^+ = \delta_{sl,1}^+ + \delta_f \qquad\qquad N_{sl,2}^+ = N_{sl,1}^+ + G_f \times \delta_f$ $\qquad$ (32)

Point 4 $\quad N_y^+ = N_{y,0}^+ + G_{kin}\delta_y^+ \qquad N_y^+ = N_{sl,2}^+ + G_e \times (\delta_y^+ - \delta_{sl,2}^+)$ $\qquad$ (33)

$$\delta_y^+ = \frac{N_{y,o}^+ - N_{y,o}^- + (G_e - G_f) \times \delta_f}{G_e - G_{kin}} + \delta_y^- \tag{34}$$

The distance δ between the yield points (1) and (4) has a constant value and can be assessed by the following equation (35):

$$\delta_y^+ - \delta_y^- = \frac{N_{y,o}^+ - N_{y,o}^- + (G_e - G_f) \times \delta_f}{G_e - G_{kin}} \tag{35}$$

There are two different cases for the assessment of the main points which are the attainment or not of the yield in the previous semi-cycle, as shown in Figure 26:

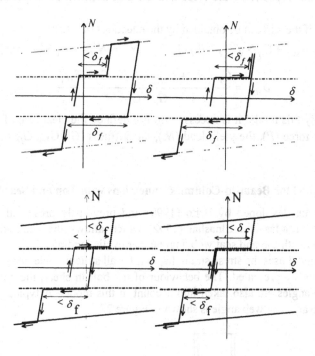

Figure 26. New variable after the inversion of the displacement.

In the case that the yield is not reached in the previous semi-cycle, the negative yield point can be assessed from the following equations (36-37):

$$\delta_y^- = \delta_y^+ - \frac{N_{y,o}^+ - N_{y,o}^- + (G_e - G_f) \times \delta_f}{G_e - G_{kin}} \tag{36}$$

$$N_y^- = N_{y,o}^- + G_{kin} \times \delta_y^- \tag{37}$$

For the case that the yield is reached in the previous semi-cycle the negative yield point can be obtained through the following equations (38-39):

$$\delta_y^- = \delta_r - \frac{N_{y,o}^+ - N_{y,o}^- + (G_e - G_f) \times \delta_f}{G_e - G_{kin}} \tag{38}$$

$$N_y^- = N_{y,o}^- + G_{kin} \times \delta_y^- \tag{39}$$

The other points are obtained in a similar way, the point at the beginning of the slip being assessed by the imposition of a load equal to N_{sl}^- :

$$N_r + G_e \times (\delta_{sl}^- - \delta_r) = N_{sl,l}^- \Rightarrow \delta_{sl}^- = \delta_r + \frac{N_{sl,l}^- - N_r}{G_e} \tag{40}$$

where N_r and δ_r are respectively the force and the displacement corresponding to the actual inversion.

The final point of the slip can be obtained by the equations (41-42):

$$N_{sl,2}^- = N_{sl,l}^- + G_f \times (\delta_{sl,2}^- - \delta_{sl,l}^-) \qquad N_y^- = N_{sl,2}^- + G_e \times (\delta_y^- - \delta_{sl,2}^-) \tag{41}$$

$$\delta_{sl,2}^- = \frac{G_e \times \delta_y^- + N_{sl,l}^- - N_y^- - G_f \times \delta_{sl,l}^-}{G_e - G_f} \tag{42}$$

In order to apply this model it is necessary to pre-defined the clearance of the bolt (δ_f), the applied preloading force (P), the yield load (N_y), the stiffness (G_e, G_{kin}, G_f) and the slip factor (μ).

3.4 Numerical Model for Beam-to-Column Connections with Top and Seat Web Angles

The numerical model developed by Brito (1999) and previously presented is based on the model proposed by Madas and Elnashai (1992) which allows the simulation of beam-to-column connections with top and seat web angle under cyclic loading.

The proposed model uses the stress-strain diagram for all structural elements with the damage accumulation previously presented. The behaviour of the bolt in shear, the panel zone and the deformation of the angles are also taken into account in this model. A typical beam-to-column connection with top and seat web angle is shown in Figure 27.

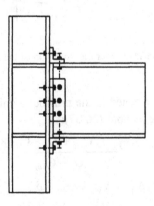

Figure 27. Connection type considered in the Brito's model (1999).

In this model all structural elements are simulated by finite elements of beam with geometric and mechanical properties analogous to the real elements. In Figure 28 the numerical model for a bolted beam-to-column connection is shown.

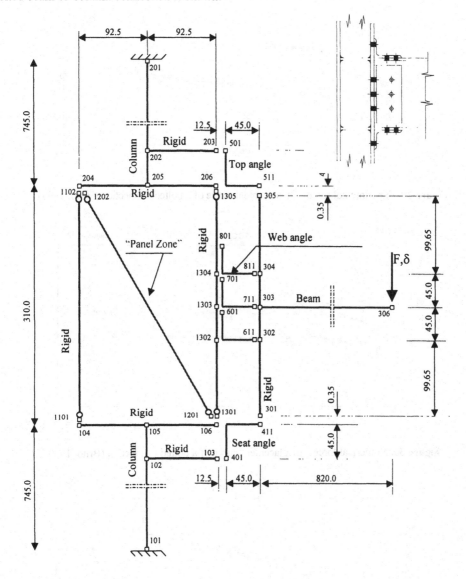

Figure 28. Numerical model for beam-to-column connection with top and seat web angle.

The accuracy of the numerical model proposed by Brito (1999) can be analysed by the comparison of the force-displacement curves and accumulated energy presented in Figures 29 to 31.

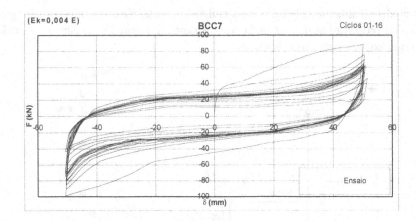

Figure 29. Experimental force – displacement curve of a bolted connection (Brito, 1999).

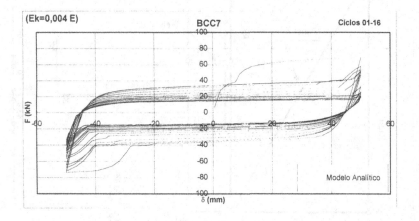

Figure 30. Numerical force – displacement curve of a bolted connection (Brito, 1999)

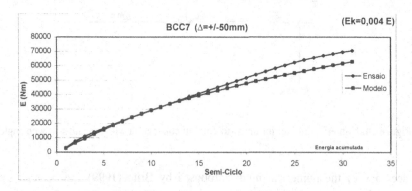

Figure 31. Accumulated energy of a bolted connection (Brito, 1999).

Comparison between numerical and experimental results allows to conclude that the numerical model proposed by Brito (1999) is able to simulate with a good level of accuracy the behaviour of beam-to-column connection with top and seat web angle until the failure.

The force-displacement curve and the accumulated energy are appropriately modeled allowing to conclude the suitableness of this numerical model to simulate the damage accumulation of the material and the non-linear behaviour of the bolts. One of the advantages of this numerical model is that it does not need experimental tests for its calibration. It needs as input data the geometric and mechanical characteristics of the connection.

4   Fatigue Resistance Design

4.1   Introduction

The success of an accurate prediction of fatigue failures (Peeker, 1997) depends on the selection of a good fatigue failure prediction function, which can be defined as the function that relates a parameter *(S)*, representative of the imposed cyclic actions to the fatigue endurance *(N)* of the structural detail.

Most common approaches to the fatigue behaviour modelling can be classified into three categories, depending on the fatigue failure prediction function adopted:
  - the *S-N* line approach, which assumes *S* to be the nominal stress range $\Delta\sigma_0$;
  - the local strain approach, which considers the local non-linear strain range $\Delta\varepsilon$;
  - the fracture mechanic approach, which adopts the stress intensity factor range.

The *S-N* line approach. The fatigue failure prediction function, used by the *S-N* curve approach, can be expressed by the following equation (43):

$$N.S^m = K \tag{43}$$

where *N* is the number of cycles to failure at the constant stress (strain) range *S*. The non-dimensional constant *m* and the dimensional parameter *K* depend on both the typology and the mechanical properties of the considered steel component. In the *Log-Log* domain equation (43) can be re-written as:

$$Log(N) = Log(K) - mLog(S) \tag{44}$$

Equation (44) represents a straight line with a slope equal to *-1/m* called fatigue resistance line, which identify the safe and unsafe regions (Figure 32).

Once identified the pertinent *S-N* line, for an undamaged component, if the point of co-ordinates *Log (n_{tot})*, *Log (S)* representative of a generic loading event (where n_{tot} is the total number of cycles sustained by the component in the loading history), falls in the safe region, failure is not to be expected for the component under that particular loading event.

Because of its simplicity, the *S-N* curve approach has been introduced into many fatigue design codes. The fatigue resistance curves adopted in design standards are built using a statistical analysis of constant amplitude fatigue tests data.

If variable amplitude loads are used, the direct assessment of the fatigue resistance is not possible and reference should be made to a cycle counting (Matsuishi and Endo, 1968) method *and to a suitable damage accumulation rule.* Usually the linear damage accumulation rule proposed by Miner (1945) is adopted for calculation of an effective value, S_{eq}, which is used instead of S as argument in the fatigue failure prediction function.

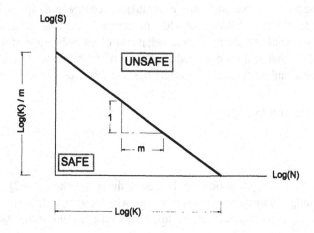

Figure 32. Fatigue resistance line in the *Log (S)-Log (N)* scale.

The main advantage of the *S-N* curve approach is its simplicity. As this approach is able to interpret correctly the phase of stable crack propagation, it is commonly adopted in civil engineering for the assessment of the fatigue strength of welded details where, due to the presence of fabrication imperfections, the crack initiation phase is practically absent.

The local strain approach. This fatigue model considers S to be the strain range, $\Delta\varepsilon$, which can be expressed as the sum of the elastic strain amplitude $(\Delta\varepsilon_{el})$ and the plastic strain amplitude $(\Delta\varepsilon_{pl})$ i.e. as:

$$\Delta\varepsilon = \Delta\varepsilon_{el} + \Delta\varepsilon_{pl} \tag{45}$$

The elastic and plastic contributions of the strain range can be expressed as:

$$\frac{\Delta\varepsilon_{el}}{2} = \left(\frac{\sigma_f - \sigma_m}{E}\right).(2.N)^b \tag{46}$$

and

$$\frac{\Delta\varepsilon_{pl}}{2} = \varepsilon_f.(2.N)^c \tag{47}$$

It should be remarked that the coefficient 2 before the term N was introduced in order to keep into account loading histories with cycles of variable amplitude. In this case, reference should be made to half-cycles, between two subsequent reversal points. Equation (47) is known as the Coffin-Manson equation (Coffin, 1954; Mason, 1954). In addition to the mean stress value σ_m, the strain relationship (equation (45)) contains five material dependent constants: E, σ_f, b, ε_f, c.

The determination of these constants is standardised (ANSI/ASTM E466-76, 1977; Landgraf et al., 1969) and their values are available in different handbooks for various materials.

Although the high precision in the fatigue failure prediction, usually the local strain approach can only be used for the assessment of the crack initiation period. Calculation of $\Delta\varepsilon$ at the crack initiator is relatively simple and if $\Delta\varepsilon$ is known, equation (45) allows the valuation of the crack initiation period. On the other hand, calculation of $\Delta\varepsilon$ at the fatigue crack tip is more complex. Due to crack growth, the cyclic strain field at the crack tip changes at each cycle, and consequently calculation of the stable crack growth period using equation (45) is impossible.

The fracture mechanic approach. This approach, based on the fracture mechanic, adopts the concept of the stress intensity factor, allowing to analyse the behaviour of a great variety of fatigue cracks. However, as it is based on crack propagation laws, this approach can be adopted only for the prediction of the stable crack growth period; furthermore, it requires the determination of the stress intensity factor, which might be a relatively complex procedure.

It is commonly accepted that a modern methodology to assess the low-cycle fatigue endurance of civil engineering structures should adopt parameters related to the global structural behaviour, such as displacements, rotations, bending moments, etc. Taking this into account, the S-N curve approach may be adopted, considering as S parameters related with the global structural ductility (e.g., interstorey drifts or joint rotations).

In order to apply equation (43), both parameters S and the number of cycles to failure (N) should be defined. The number of cycles to failure N can be identified on the basis of the failure criterion, while S can be defined with reference to the definitions given in the literature by various authors.

4.2   Definition of the Strain Range (S)

Some of the most relevant proposals available in literature for the definition of the strain range S are shortly presented.

Krawinkler and Zohrei proposal. The concept to connect the parameter (S) of the fatigue failure prediction functions with the global structural ductility was originally proposed by Krawinkler and Zohrei (1983). They proposed a relationship between the fatigue endurance and the plastic portion of the generalised displacement component, which can be expressed as:

$$N\left(\Delta\delta_{pl}\right)^m = K^{-1} \tag{48}$$

where $\Delta\delta_{pl}$ represents the plastic portion of the deformation range.

Ballio and Castiglioni proposal. Ballio and Castiglioni (1995) proposed a unified approach for the design of steel structures under low and/or high-cycle fatigue, which is based on global displacement parameters instead of local deformation parameters. The fundamental hypothesis is the validity of the following equation (49):

$$\frac{\Delta\varepsilon}{\varepsilon_y} = \frac{\Delta v}{v_y} = \frac{\Delta\phi}{\phi_y} \tag{49}$$

where ε represents the strain, v the displacement, ϕ the rotation (or the curvature), Δ the range of variation in a cycle and the subscript y identifies the yielding of the material (ε_y) as well as the conventional yielding with reference to the generalised displacement (v_y, ϕ_y) assumed as the test control parameter.

For an ideal elastic material, the relationship between the strains ε and the load F (or bending moment M) causing the displacement v (or rotation ϕ) can be written as:

$$E.\varepsilon = \sigma(F) \tag{50}$$

and, at yield:

$$E.\varepsilon_y = \sigma(F_y) \tag{51}$$

With reference to the previous equations (49)-(51), the following parameters can be defined:

$$\Delta\sigma^* = E.\Delta\varepsilon = \frac{E.\Delta v}{v_y} \varepsilon_y = \frac{\Delta v}{v_y}\sigma(F_y) = \frac{\Delta\phi}{\phi_y}\sigma(F_y) \tag{52}$$

The term $\Delta\sigma^*$ represents the effective stress range, associated with the real strain range $\Delta\varepsilon$ in an ideal member made of an indefinitely linear elastic material. In the case of high-cycle fatigue (i.e. under cycles in the elastic range), $\Delta\sigma^*$ coincides with the actual stress range $\Delta\sigma$. In general, $\Delta\sigma^*$ can be expressed in terms of the generalised displacement component δ as follows:

$$\Delta\sigma^* = \frac{\Delta\delta}{\delta_y}\sigma(F_y) \tag{53}$$

Using the S-N curves and assuming the effective stress range $\Delta\sigma^*$ as S, equation (43) can be re-written as:

$$N\left(\frac{\Delta\delta}{\delta_y}\sigma(F_y)\right)^m = K \tag{54}$$

Feldmann, Sedlacek, Weynand and Kuck proposal. On the basis of an extensive finite element analysis, Feldmann (1994), Sedlacek et al. (1995) and Kuck (1994) investigated the behaviour of beam-to-column joints under constant amplitude cyclic loading. It appeared that the relationship between the plastic strain, ε_{pl}, at the hot spot (strain at the relevant place where first crack occurs) and the plastic rotation (ϕ_{pl}) is linear, i.e.:

$$\frac{\varepsilon_{2,pl}}{\varepsilon_{1,pl}} = \frac{\phi_{2,pl}}{\phi_{1,pl}} \tag{55}$$

where subscripts 1 and 2 refer to two different loading steps.

This linear relationship, when plotted in a Log-Log scale with the number of cycles on the vertical axis and the plastic rotations on the horizontal one, plots as a "Wöhler" line, the slope of which (m) is the exponent of the Manson-Coffin equation:

$$Log(\phi_{pl}) = Log(\phi_{0,pl}) - \frac{1}{m}Log(N) \tag{56}$$

where $\phi_{0,pl}$ represents the maximum theoretical plastic rotation under static loading. In general, equation (56) can be re-written in terms of the plastic portion δ_{pl} of the generalised displacement component δ as:

$$Log(\delta_{pl}) = Log(\delta_{0,pl}) - \frac{1}{m} Log(N) \tag{57}$$

In accordance with the procedure proposed by Ballio and Castiglioni (1995) and with reference to equation (55), it is possible to define and effective plastic stress range $\Delta\sigma_{pl}^*$ which can be expressed in terms of the generalised displacement component δ_{pl} as follows:

$$\Delta\sigma_{pl}^* = \frac{\Delta\delta_{pl}}{\delta_y} \sigma(F_y) \tag{58}$$

Considering the usual S-N curves and assuming S as the effective stress range $(\Delta\sigma_{pl}^*)$, equation (43) can be re-written as:

$$N\left(\frac{\Delta\delta_{pl}}{\delta_y} \sigma(F_y)\right)^m = K \tag{59}$$

Bernuzzi, Castiglioni and Calado proposal. Based on the re-elaboration of an extensive experimental data (Castiglioni and Calado, 1996) of beam-to-column joints as well as of beams and beam-columns, they proposed to use the total interstorey drift (i.e., the total displacement range) Δv as parameter S, and consequently equation (43) can be re-written, in the form:

$$N.\Delta\delta^m = K \tag{60}$$

4.3 Definition of the Fatigue Endurance (N)

For the prediction of the low-cycle fatigue endurance of steel structural connections, it seems convenient to adopt a failure criterion based on parameters (i.e. stiffness, strength or dissipated energy) associated with the response of the component. Three failure criteria are reviewed in the following.

Calado, Azevedo and Castiglioni criterion $(N_{\alpha=0.5})$. These authors (Calado and Azevedo, 1989; Castiglioni and Calado, 1996) developed a failure criterion of general validity for structural steel components under both constant and variable amplitude loading histories that can be written in the following form:

$$\frac{\eta_l}{\eta_0} \le \alpha \tag{61}$$

In this equation η_l represents the ratio between the absorbed energy of the considered component at the last cycle before collapse and the energy that might be absorbed in the same cycle if it had an elastic-perfectly plastic behaviour while η_0 represents the same ratio but with reference to the first cycle in plastic range. The value of α, which depends on several factors (such as type of the joint and the steel grade of the component) should be determined by fitting the experimental results. As it is particularly interesting to identify α a priori, in order to define a unified failure criterion, a value of $\alpha=0.5$ is recommended for a satisfactory and conservative appraisal of the fatigue life. In the case of variable amplitude loading histories the same criterion remains valid but should be applied on semi-cycles, which can be defined in plastic range as the part of the hysteresis loop under positive or negative loads or as two subsequent load reversal points.

Bernuzzi, Calado and Castiglioni criterion ($N_{\Delta Wr}$). In this criterion, valid only for constant amplitude loading, it is assumed that the drop of the hysteretic energy dissipation is the main parameter for the definition of the low-cycle fatigue endurance. In particular, focusing attention only on the cycles performed at the displacement range Δv_i, the relative energy drop, ΔW_r, can be defined as:

$$\Delta W_r = \left(\frac{W_{init} - W_i}{W_{init}} \right) \tag{62}$$

where W_{init} represents the dissipated energy in the first cycle at the assumed displacement range while W_i is the energy associated with the ith cycle at the same displacement range. Failure is assumed to occur when the energy drop is evident, i.e. the generic point of co-ordinates n_k, ΔW_{rk}, is not on the straight line.

Calado, Castiglioni and Bernuzzi criterion (Δv_{TH}). A re-analysis of the test data on beam-to-column connections (Calado et al., 1998) seems to indicate that, with reference to the force-displacement relationship, if the ductility range is lower than a threshold value (Δv_{Th}), corresponding to the displacement related to the maximum strength in a monotonic test, a brittle failure mode is to be expected, while if Δv is greater than Δv_{Th} a ductile failure mode is attained. For cycles with Δv in the range of Δv_{Th} a mixed failure mode was observed. Despite the small number of joint types analysed in this study, this seems to be a general consideration, independently on the size or the type of components (Agatino et al., 1997, Barbaglia, 1998).

Hence, a simplified criterion to predict the type of failure expected in a steel component for a given ductility range can be proposed based on the previous considerations. It is assumed that the main parameters governing the failure mode are: the ductility range $\Delta v/v_y$; the beam web slenderness ratio $\lambda_w = d/t_w$ of the depth of the profile (d) to the web thickness (t_w); the beam flange slenderness ratio $\lambda_f = c/t_f$ of the half width of the flange (c) to its thickness (t_f); the weld quality and the severity of the detail, globally accounted for, by the introduction of the numerical coefficient ξ.

For strong column weak beam frame structures the following equation can be proposed for the assessment of the threshold value of the ductility range cycle amplitude (Δv_{Th}) associated with transition from one failure mode to the other:

$$\Delta v_{Th} = \frac{\gamma . v_y}{\xi . \lambda_f . \lambda_w} \tag{63}$$

In equation (63) ξ is a parameter related to the weld quality as well as to the severity of the detail, that might range from $\xi = 1$ for good quality (or not) welds, to $\xi = 0.5$ for poor quality welds.

For the non-dimensional coefficient γ, based on the available database, a value of $2000 \pm 15\%$ is proposed, independently on the component under consideration. A value of $\gamma = 1700$ gives a threshold value such that lower ductility ranges are generally associated with a "brittle" failure mode, while a value of $\gamma = 2300$ gives a threshold value such that larger ductility ranges are expected to be associated with a "ductile" failure mode. The range of variation ± 15% around the value $\gamma = 2000$ accounts for possible "mixed" failure modes.

Since ξ is related to the weld quality as well as to the typology of the welded connection, (i.e. full penetration, simple bevel, double bevel, presence or absence of the backing bar, etc.) it was assumed, for the connections under consideration, to appraise it from the test results. For each test, the experimental value of Δv_{Th}, corresponding to the displacement related to the maximum strength in a monotonic test, was derived. As previously noticed, in the tests where Δv is lower than Δv_{Th}, the maximum strength was observed at reversal points and these values were not included for the evaluation of Δv_{Th}.

In this research it is proposed that the value of α_f (Calado, Azevedo and Castiglioni criterion) should be related with $\Delta v/\Delta v_{Th}$ in order to take into account the failure modes. The value of Δv_{Th} was appraised for all the available tests giving particular attention on the value of ξ that is chosen depending on the different failure mode of the specimen and on the weld quality. In Figure 33, the value of α_u (i.e. the actual value of the ratio E/E_0, experimentally determined at complete failure of the specimen) is plotted versus, $\Delta v/\Delta v_{Th}$. A conservative and satisfactory appraisal of the fatigue life can be obtained assuming a value of α_f determined by fitting the experimental data in the failure criterium so that all the performed tests plot below the $\alpha_f(\Delta v/\Delta v_{Th})$ line.

$$\alpha_f = 1 - 0.235*(\Delta v/\Delta v_{Th}) \qquad\qquad \Delta v/\Delta v_{Th} < 0.85$$
$$\alpha_f = 1.65 - (\Delta v/\Delta v_{Th}) \qquad\qquad 0.85 < \Delta v/\Delta v_{Th} < 1.15 \qquad\qquad (64)$$
$$\alpha_f = 0.5 \qquad\qquad \Delta v/\Delta v_{Th} > 1.15$$

In Figure 33 it can be seen that, in the range of $0.85 < \Delta v/\Delta v_{Th} < 1.15$, where a mixed failure mode is to be expected, a strong dependence of α_f on the cycle amplitude Δv was accounted for.

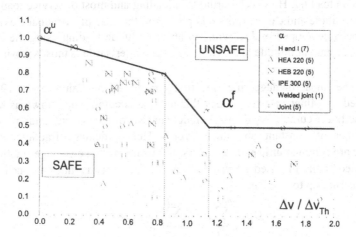

Figure 33. Failure criterion based on Δv of threshold.
(adapted from Calado et al. (1998))

For values of $\Delta v/\Delta v_{Th}$ greater than 1.15, a ductile failure mode is to be expected, and a constant value of $\alpha_f = 0.5$ is assumed, in agreement with the Calado and Castiglioni proposal. For

$\Delta v/\Delta v_{Th}$ lower than 0.85 a brittle failure mode is to be expected, and α_j is assumed linearly decreasing when increasing the cycle amplitude Δv.

4.4 Definition of *m*

The parameter *m* identifies the slope of the line interpreting, in a *Log-Log* scale, the relationship between the number of cycles to failure *(N)* and the stress (strain) range *(S)*.

Focusing the attention on *m*, some discrepancies can be found in the literature among the proposals of various authors. In particular it can be noticed that Ballio and Castiglioni suggested to adopt a value of *m* = 3; on the contrary, Krawinkler and Zohrei proposed an exponent *m* = 2. The Feldmann approach tried to re-write the Coffin-Mason equation in terms of global deformation parameters, proposing a value *m* = 2 in agreement with Krawinkler and Zohrei. Both Feldmann and Krawinkler adopted the plastic range of the assumed parameters, while Ballio adopted the total cyclic excursion (i.e. elastic and plastic).

Hence, it seems that the value of the exponent *m* depends on the definition of *S*. In order to assess the exponent *m*, the data obtained in tests with cycles of constant amplitude should be plotted in terms of number of cycles to failure *(N)* and stress (strain) range *(S)* in a *Log-Log* diagram and fitted by a straight line. The slope *m* of the best fitting line can be considered the exponent to be adopted in equation (43).

4.5 Variable Amplitude Loading

Up to this point the discussion of the models for the assessment of the *S-N* lines has dealt with constant amplitude loading. However, earthquake loading and most of service loading histories have variable amplitude and can be quite complex as in the case of earthquake events. Several methods have been developed to deal with variable amplitude loading using the baseline data generated from constant amplitude tests. However, the Miner's rule is most known.

Miner's rule. The linear damage rule was firstly proposed by Palmgren in 1924 and was further developed by Miner in 1945. Today this method is commonly known as Miner's rule. The basis for the linear cumulative damage model consists of converting random cycles into an equivalent number of constant amplitude cycles. Hence, damage fractions due to each individual cycle are summed until failure occurs. In this method, the damage fraction D_i for the ith cycle is defined as the life used up to the ith event. Failure is assumed to occur when these damage fractions sum up to or exceed 1,

$$D_T = \sum_{i=1}^{n_{TOT}} D_i = \sum_{i=1}^{n_{TOT}} \left(\frac{1}{N_{fi}} \right) \geq 1 \tag{65}$$

where $D_i = (1/N_{fi})$ is the damage fraction for the ith cycle and (N_{fi}) is the fatigue life at the stress σ_i.

Often, in a sequence of variable amplitude loadings, there is a repeated number of cycles of the same amplitude. Under these circumstances, it is convenient to sum the cycles at the same amplitude. Consider a situation of variable amplitude loadings as illustrated in Figure 34. A certain stress σ_1 is applied for a number of cycles N_1, where the number of cycles to failure is N_{f1}. The fraction of the life used is then N_1/N_{f1}. Now let consider another stress amplitude σ_2 in which

N_2 cycles were imposed and in which N_{f2} corresponds to attend the failure. The Miner's rule states that fatigue failure is expected when such life fractions sum to unit, that is, when 100% of the life is exhausted:

$$D_T = \frac{N_1}{N_{f1}} + \frac{N_2}{N_{f2}} + \frac{N_3}{N_{f3}} + \ldots = \sum_{i=1}^{n_{TOT}} \left(\frac{N_i}{N_{fi}} \right) \geq 1 \tag{66}$$

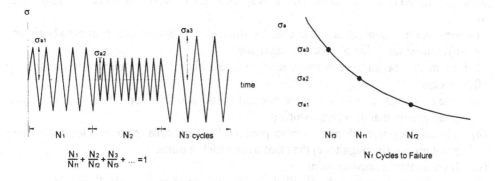

$$\frac{N_1}{N_{f1}} + \frac{N_2}{N_{f2}} + \frac{N_3}{N_{f3}} + \ldots = 1$$

Figure 34. Use of the Miner's rule for the life prediction for variable amplitude loading. (adapted from Dowling, 1993)

Considerable test data have been generated in an attempt to verify Miner's rule. In most cases these tests use a two-step history. This involves testing at an initial stress level σ_l for a certain number of cycles. The amplitude is then changed to a second level, σ_2, until failure occurs. If $\sigma_l > \sigma_2$ it is called a high-low test, and if $\sigma_l < \sigma_2$, a low-high test. The results of Miner's original tests showed that the cycle ratio corresponding to failure ranged from 0.61 to 1.45 (Sines and Waisman, 1959). Other researchers have shown variations as large as 0.18 to 23.0. Most results tend to fall between 0.5 and 2.0. In most cases the average value is closed to Miner's proposed value of 1. There is a general trend that for high-low tests the values are less than 1, and for low-high tests the values are greater than 1. In other words, Miner's rule is nonconservative for high-low tests. One problem with two-level step tests is that they do not relate to many service load histories. Most load histories do not follow any step arrangement and instead are made up of random distribution of loads of various amplitudes. Tests using random histories with several stress levels show very good correlation with Miner's rule.

The difficulty in applying this method to non completely reversal amplitude loading or random loading arise from the fact that the loading histories in seismic events or loading such as wind or traffic often do not have well-defined cycles. To overcome the irregularities of real load histories, several cyclic counting methods have been developed. One of the most widely used methods is the well-known "rainflow counting method".

Rainflow counting method. The original rainflow method of cycle counting derived its name from an analogy used by Matsuishi and Endo (1968) in their early work on this subject. Since that time "rainflow counting" has become a generic term that describes any cycle counting method (Bannantine et al., 1990), which attempts to identify closed hysteris loops in the stress-strain response of material subjected to cyclic loading.

The first step in implementing this procedure is to draw the strain-time history so that the time axis is oriented vertically, with increasing time downward. One could now image that strain history forms a number of "pagoda roofs". Cycles are then defined by the manner in which rain is allowed to "drip" of "fall" down the roofs. A number of rules are imposed on the dripping rain so as to identify closed hysteresis loops. The rules specifying the manner in which rain falls are as follows:

1) To eliminate the counting of half cycles, the strain-time history is drawn so as to begin and to end at the strain value of greatest magnitude.

2) A flow of rain began at each strain reversal in the history and is allowed to continue to flow unless:

    (a) The rain began at a local maximum point (peak) and falls opposite a local maximum point greater than that from which it came.

    (b) The rain began at a local minimum point (valley) and falls opposite a local minimum point greater (in magnitude) than that from which it came.

    (c) It encounters a previous rainflow.

The foregoing procedure can be clarified through the use of an example. Figure 35 shows a stress history and the resulting flow of rain.

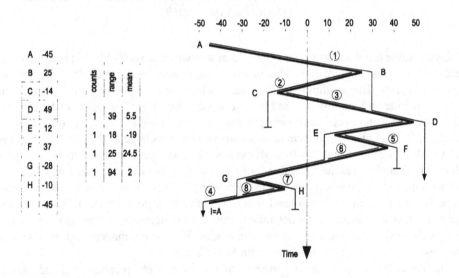

Figure 35. Example of rainflow cycle counting.
(adapted from Bannantine et al., 1990)

The following discussion describes in detail the manner in which each rainflow path was determined.

- Rain flows from point A over points B and D and continues to the end of the history since none of the conditions for stopping rainflow are satisfied.
- Rain flows from point B over C and stops opposite point D, since both B and D are local maximums and the magnitude of D is greater than B (rule 2a).
- Rain flows from point C and must stop upon meeting the rain flow from point A (rule 2c).
- Rain flows from point D over points E and G and continues to the end of the history since none of the conditions for stopping the rainflow are satisfied.
- Rain flows from point E over point F and stops opposite point G, since both E and G are local minimum and the magnitude of G is greater than E (rule 2b)
- Rain flows from point F and must stop upon meeting the flow from point D (rule 2c).
- Rain flows from point G over point H and stops opposite point A, since both G and A are local minimum's and the magnitude of A is greater than G (rule 2b)
- Rain flows from point H and must stop upon meeting the rainflow from point D (rule 2c).

Having completed the above, we are now able to combine events to form completed cycles. In Figure 35 is also shown a table with results of this example.

Equivalent strain range (S_{eq}). In case of variable amplitude loading, it appears convenient to make reference to an equivalent strain range, S_{eq}, in order to use equation (43). Using the Miner's rule the equivalent strain range, S_{eq}, can be defined as follows:

$$S_{eq} = \left(\frac{\sum_{i=1}^{L} N_i \cdot S_i^m}{N_{TOT}} \right)^{\frac{1}{m}} \tag{67}$$

where N_{TOT} is the total number of cycles in the loading history, N_i is the number of cycles at the same range S_i, and L is the number of different ranges S_i identified in the loading history, for instance with the rainflow counting. The parameter S_i can correspond to one of the definitions previously described and the parameter m should be in accordance with the definition used for S_i as previously mentioned.

4.6 Assessment of Design Fatigue Resistance Lines

According to the limit state design method, the parameter governing the design should be defined on the basis of a statistical analysis, allowing to make reference to a value associated with a given probability of failure P_f (or of survival, $1- P_f$). Such a probability should be defined with reference to suitable levels of safety and reliability of the structure and its components. Hence, the S-N lines should be defined with reference to a given probability of failure P_f. In this section, the procedure proposed in the JWG XIII-XV - Fatigue Recommendations of the International Welding Institute (1994) is briefly summarised herein.

The scope of the procedure is the definition of the value $Log(K_k)$ such that the line with a slope $(-1/m)$, and intersecting the x-axis at $Log(K)$, is associated with a probability of 5% that test data plot below it. In the following, reference is made to the set of values $Log(K)$ representing the intersections with the N axis of the S-N lines, each one having a constant slope $(-1/m)$, and passing for every single test data point.

The basic formula to determine the design value of the random variable $X=Log$ (K) can be assumed as:

$$X_d = \mu - \phi^{-1}(\alpha) \cdot \sigma \qquad (68)$$

where X_d is the design value, μ is the mean value of X, σ is the standard deviation of X, ϕ is the normal law of probability and α is the probability of survival.

It is possible to show that μ can be considered distributed according to the t-Student model and the standard deviation σ^2 follows the chi-square law (χ^2). As σ and μ are unknown values, their estimated values from tests, $\bar{\mu}$ e $\bar{\sigma}$ respectively, should be associated with β, confidence level. By assuming

$$\bar{\mu} = \frac{\sum_{i=1}^{n} x_i}{n} \qquad (69)$$

as the estimated mean value of X_i, and

$$\bar{\sigma} = \sqrt{\frac{n\sum (x_i)^2 - (\sum x_i)^2}{n(n-1)}} \qquad (70)$$

as the estimated standard deviation, the design value of X can be expressed as:

$$X_d = \bar{\mu} - \bar{\sigma} \cdot \left[\frac{t(\beta, n-1)}{\sqrt{n}} + \phi^{-1}(\alpha) \cdot \sqrt{\frac{n-1}{\chi^2(\frac{1-\beta}{2}, n-1)}} \right] \qquad (71)$$

According to Eurocode 3 (1994) the confidence range should be 75% of 95% probability of survival on the Log (N) axis. Hence, α e β should be assumed respectively equal to 0.95 and 0.75. Naming γ as:

$$\gamma = \left[\frac{t(\beta, n-1)}{\sqrt{n}} + \phi^{-1}(\alpha) \cdot \sqrt{\frac{n-1}{\chi^2(\frac{1-\beta}{2}, n-1)}} \right] \qquad (72)$$

and in the case of limited amount of data the value of γ can be obtained directly from Table 9 on the basis of the number data (n).

In the same table, assuming a number of degrees of freedom equal to $n-1$, the value of function t for $\alpha = 0.75$ and function χ for $\beta = 0.125$ are reported together with the value of function ϕ in correspondence with $P_f = 0.95$. Equation (72) can hence be re-written as:

$$X_d = \bar{\mu} - \bar{\sigma} \cdot \gamma \qquad (73)$$

The test data were analysed following the proposed procedure in order to evaluate K_d (i.e. the design value of the constant K) which defines the design low-cycle fatigue of the considered component.

Table 9. Coefficients for statistical evaluation.

n	$t\,(0.75,n\text{-}1)$	$\chi^2\,(0.125,n\text{-}1)$	$\phi^{-1}\,(0.95)$	γ
2	2.4142	0.0247	1.6449	12.1632
3	1.6036	0.2471	1.6449	5.4271
4	1.4226	0.6924	1.6449	4.1353
5	1.3444	1.2188	1.6449	3.5811
6	1.3009	1.8082	1.6449	3.2663
7	1.2733	2.4411	1.6449	3.0600
8	1.2543	3.1063	1.6449	2.9127
9	1.2403	3.7965	1.6449	2.8011
10	1.2297	4.5070	1.6449	2.7132
11	1.2213	5.2341	1.6449	2.6418
12	1.2145	5.9754	1.6449	2.5823
13	1.2089	6.7288	1.6449	2.5319
14	1.2041	7.4929	1.6449	2.4884
15	1.2001	8.2661	1.6449	2.4505

4.7 S-N lines for Beam-to-Column Connections

To ilustrate the validity and accuracy of the methodology proposed by Ballio and Castiglioni (1995) it was applied to the tests performed in the Laboratory of Structures and Strength of Material of Instituto Superior Técnico, Lisbon on fully welded (WW) and top and seat with web angle (TSW) connections. In this re-elaboration the fatigue endurance *(N)* was assessed according the Calado, Azevedo and Castiglioni *($N_{a=0.5}$)* proposal.

For each test the relevant parameters of each specimen *(Δv, v_y, f_y)* are presented in Table 10 as well as the total number of imposed cycles *(N)* and the fatigue endurance, i.e. the number of cycles of the conventional failure N_f. The value of S_{eq} and the slope m of the S-N lines obtained with the best fitting are also presented. From Table 10, Figure 36 and 37 it can be observed that the slope of the S-N line when assessed with the Ballio and Castiglioni method varies from 1.23 to 3.46. The discrepancy of the slope values can be imputed to the small number of tests for each typology and the vicinity of the results. According to Eurocode 3 (1994) to assess a fatigue line the number of data points to be considered in the analysis must be not lower that 10. On the other hand the vicinity of the results is not the most suitable to assess the slope of the line. In effect, the numbers of cycles to failure vary from 2 to 22.

Re-elaboration of other experimental tests performed by Ballio et al. (1997) on beams, beam-to-column connections and cruciform welded joint has demonstrated that the slope of the S-N line is approximately equal 3.

Table 10. Relevante parameters for the assessment of the S-N lines.
WW and TSW connections

Type of column	Load case	Nf	Δv (mm)	v_y (mm)	f_y (Mpa)	S_{eq} (MPa)	m tests	Log K tests	Log K (m=3)	Log K_d statistic
WW	A	16	100.0	8.78	274.8	3129			11.69	
HEB160	B	5	150.0	8.54	274.8	4826	2.17	8.67	11.75	11.31
BCC5	C	15	ECCS	8.73	274.8	2741			11.49	
	D	22	75.0	8.68	274.8	2374			11.47	
WW	A	14	100.0	8.33	278.6	3344			11.72	
HEB200	B	8	150.0	8.26	278.6	5059	1.23	5.41	12.02	10.97
BCC6	C	12	ECCS	8.22	278.6	2529			11.29	
	D	18	75.0	8.41	278.6	2484			11.44	
WW	A	12	100.0	6.87	292.0	4250			11.96	
HEB240	B	16	75.0	6.94	292.0	3155	3.46	13.49	11.70	11.55
BCC8	C	2	150.0	6.85	292.0	6394			11.72	
	D	15	ECCS	6.74	292.0	3765			11.90	
TSW	A	5	100.0	5.07	252.2	4974			11.79	
HEB160	B	3	150.0	5.22	252.2	7247	1.93	7.83	12.06	11.25
BCC9	C	6	75.0	5.26	252.2	3596			11.45	
	D	12	ECCS	5.31	252.2	3533			11.72	
TSW	A	6	100.0	4.12	252.2	6121			12.14	
HEB200	B	2	150.0	4.10	252.2	9226	2.82	11.41	12.20	11.72
BCC7	C	13	ECCS	4.06	252.2	4884			12.18	
	D	6	75.0	3.95	252.2	4788			11.82	
TSW	A	5	100.0	3.80	252.2	6636			12.16	
HEB240	B	4	75.0	3.72	252.2	5084	2.92	11.82	11.72	11.57
BCC10	C	12	ECCS	3.65	252.2	5147			12.21	
	D	2	150.0	3.72	252.2	10169			12.32	

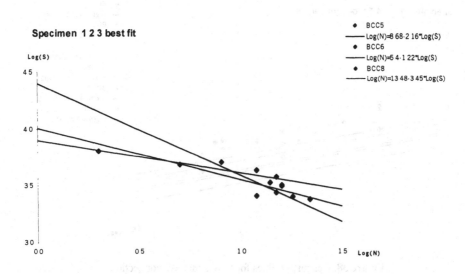

Figure 36. Evaluation of the *S-N* line for WW connections.

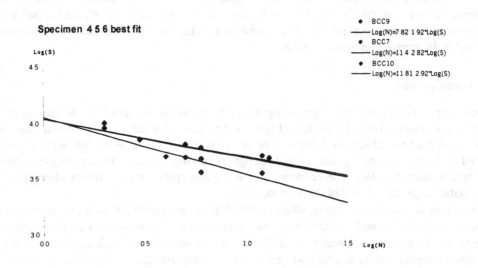

Figure 37. Evaluation of the *S-N* line for TSW connections.

To assess design fatigue lines ($m = 3$ and K_d) with an acceptable level of probability the methodology proposed by the JWG XIII-XV - Fatigue Recommendations of the International Welding Institute (1994) was used with a $\gamma = 2$, due to the small number of tests for each type of connection (Table 10).

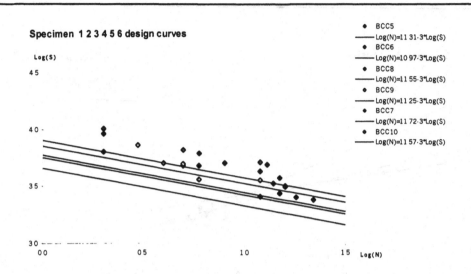

Figure 38. Design *S-N* lines for WW and TSW connections.

It can be observed (Figure 38) that the *S-N* lines for these types of connections are not so different. However as already stated by Leon (1997) and Elnashi et al. (1998) the top and seat with web angle connections exhibit a better behaviour under cyclic loading due to its flexibility and highly dissipative friction mechanism.

5   Conclusions

In this chapter the major aspects governing the cyclic behaviour of bolted (TSW) and welded (WW) connections have been evidenced against experimental results. It was shown that the panel zone does not affect the behaviour of the TSW connections, which instead is mainly related to the tension angle geometry and strength properties. On the contrary the panel zone has demonstrated to affect at large extent all the response parameters (stiffness, strength and deformation capacity) of welded connections, WW.

A numerical model for the stress-strain relationship of the steel which takes into account the damage accumulation and the unloading was presented and its accuracy was discussed. A physical theory model developed by Brito (1999) for top and seat with web angle connections was also presented and its accuracy was compared with experimental results showing that it is able to simulate until failure the non-linear behaviour of the connection.

There were also presented methods for the assessment of the fatigue resistance design of beam-to-column connections under cyclic loading based on the *S-N* line approach. Possible definitions of the fatigue endurance *(N)* and procedures for the assessment of the stress (strain) range *(S)* were discussed, including its influence on the slope *m*.

The results presented in this chapter have shown that bolted connections under cyclic loading exhibit a favourable behaviour under cyclic loading due to its flexibility and highly dissipative mechanism.

References

Agatino, M. R., Bernuzzi, C., Castiglioni, C. A., and Calado, L., (1997), Ductility and strength of structural steel joints under low-cycle fatigue, *Proceedings of the XVI Congresso C.T.A. - Collegio dei Tecnici dell'Acciaio*, Ancona, Italy.

ANSI/ASTM E466-76, (1977), Standard recommended practice for constant axial fatigue tests on metallic materials, *Annual Book of ASTM Standards*, Part 10, American Society for Testing and Materials (ASTM), Philadelphia, U.S.A.

Astaneh-Asl, A. (1995). Seismic design of bolted steel moment-resisting frames. *Steel Tips*, Structural Steel Educational Council, Moraga, California, U.S.A.

Ballio, G., and Castiglioni, C. A., (1995), A unified approach for the design of steel structures under low and high cycle fatigue, *Journal of Constructional Steel Research*, Vol. 34.

Ballio, G., Calado, L., and Castiglioni, C. A., (1997), Low cycle fatigue behaviour of structural steel members and connections, *Fatigue and Fracture of Engineering Materials and Structures*, Vol. 20, N° 8.

Bannantine, J., Comer, J., and Handrock, J., (1990), *Fundamentals of metal fatigue analysis*, Prentice-Hall.

Barbaglia P., (1998), *Low-cycle fatigue behaviour of welded steel beam-to-column connections*, Laurea Thesis, Politecnico di Milano, Italy.

Bjorhovde, R., Colson, A., and Brozzetti, J., (1990) Classification system for beam-to-column connections. *Journal of Structural Engineering*, ASCE, Vol. 116.

Brito, A., (1999), *Semi-rigid steel connections: cyclic behaviour analysis*, MSc Thesis, Institute Superior Técnico, Lisbon, Portugal. (in Portuguese)

Calado, L., and Azevedo, J., (1989), A model for predicting failure of structural steel elements, *Journal of Constructional Steel Research*, Vol. 14.

Calado, L., Castiglioni, C. A., Barbaglia, P., and Bernuzzi, C., (1998), Procedure for the assessment of low-cycle fatigue resistance for steel connections, *Proceedings of the International Conference on Control of the Semi-Rigid Behaviour of Civil Engineering Structural Connections*, Liege, Belgium.

Calado, L., De Matteis, G., Landolfo, R., and Mazzolani, F. M., (1999a), Cyclic behaviour of steel beam-to-column connections: interpretation of experimental results, *Proceedings of the 6th International Colloquium on Stability and Ductility of Steel Structures - SDSS'99*, Timisoara, Romania.

Calado, L., Mele, E., and De Luca, A., (1999b). Experimental investigation on the cyclic behaviour of welded beam-to-column connections. *Proceedings 2nd European Conference on Steel Structures*, Praha, Czech Republic.

Calado, L., and Mele, E., (2000), Experimental behaviour of steel beam-to-column joints: fully welded vs bolted connections, *Proceedings of the 12th World Conference on Earthquake Engineering*, Auckland, New Zealand.

Castiglioni, C. A., and Calado, L., (1996), Seismic damage assessment and failure criteria for steel members and connections, *Proceedings of the International Conference on Advances in Steel Structures*, Hong Kong.

Chen, W., and Lui, E., (1991), *Stability design of steel frames*, CRC Press, Boca Raton.

Coffin, L. F., (1954), A study on the effect of cyclic thermal stresses on a ductile metals, Transaction of the *American Society of Mechanical Engineers*, ASME, Vol. 76.

Della Corte, G., De Matteis, G., and Landolfo, R. (1999), A mathematical model interpreting the cyclic behaviour of steel beam-to-column joints. *Proceedings of the XVII Congresso C.T.A. - Collegio dei Tecnici dell'Acciaio*, Napoli, Italy.

De Martino, A., Faella, C., and Mazzolani, F. M., (1988), Simulation of beam-to-column joint behavior under cyclic loads. *Costruzioni Metalliche, N° 6*.

De Matteis, G., Landolfo, R., and Calado, L., (2000), Cyclic behaviour of semi-rigid angle connections: a comparative study of tests and modelling, *Proceedings of the 3rd International Conference on the Behaviour of Steel Structures in Seismic Areas - STESSA 2000*, Montreal, Canada.

De Stefano, M., De Luca, A., and Astaneh-Asl, A., (1994), Modelling of cyclic moment-rotation response of double-angle connections, *Journal of Structural Engineering*, ASCE, Vol. 120, N° 1.

Dowling, N., (1993), *Mechanical behaviour of materials – engineering methods for deformation, fracture and fatigue*, Prentice-Hall International Editions.

ECCS (1986). Seismic design. Recommended testing procedure for assessing the behaviour of structural steel elements under cyclic loads. Technical Committee 1 - Structural Safety and Loadings, *TWG1.3 – Report N°. 45.*

Elnashai, A.S., Elghazouli, A.Y., and Denesh-Ashtiani, F.A., (1998) Response of semirigid steel frames to cyclic and earthquake loads. *Journal of Structural Engineering, ASCE*, Vol.124, N°.8.

Eurocode 3 - Design of Steel Structures - Part 1: General Rules and Rules for Buildings, (1994), Commission of the European Communities.

Eurocode 3, Part 1.1. Rev. Annex J. Joints in building frames, (1997), CEN TC250/SC3-PT9, Ed. Approved Draft: January.

Feldman, M., (1994), *Zur Rotationskapazitat vin I-profilen Statisch und Dynamisch Belasteter Trager*, (in German), PhD Thesis, Institute of Steel Construction, RWTH Aachen, Germany.

Giuffré, A. and Pinto, P., (1970), Il comportamento del cemento armato per sollecitazioni cicliche di forte intensità, *Giornale del Genio Civile*, Maggio, Italy.

Hasan, R., Kishi, N., and Chen, W. F., (1998) A new non-linear connection classification system. *Journal of Constructional Steel Research*, Vol. 47.

International Welding Institute, IIW, JWG XIII-XV, (1994), Fatigue recommendations, Doc. XIII-1539-94/XV-845-94, September.

Krawinkler, H. and Mohasseb, S., (1987), Effects of panel zone deformations on seismic response, *Journal of Constructional Steel Research*, Vol. 8.

Krawinkler, H. and Zohrei, M., (1983), Cumulative damage model in steel structures subjected to earthquake ground motion, *Computers and Structures*, Vol. 16, N° 1-4.

Kuck, V. J., (1994), *The application of the dynamic plastic hinge method for the assessment of limit states of steel structures subjected to seismic loading*, (in German), Ph.D. Thesis, Institute of Steel Construction, RWTH Aachen, Germany.

Landgraf, R. W., Morrow, J. D., and Endo, T., (1969), Determination of cyclic stress-strain curve, *Journal of Materials*, Vol. 4, N° 1, American Society for Testing and Materials (ASTM), Philadelphia, U.S.A.

Leon, R. T., (1997), Seismic Performance of Bolted and Riveted Connections. In *Program to Reduce the Earthquake Hazards of Steel Moment Frames Structures*, FEMA-288, Report No. SAC-95-09, Sacramento, California, U.S.A.

Madas, P. J., and Elnashai, A. S., (1992), A component-based model for beam-column connections, *Proceedings of the 10ᵗʰ World Conference of Earthquake Engineering*, Madrid, Spain.

Manson, S. S., (1954), Behaviour of materials under conditions of thermal stress, *National Advisory Commission on Aeronautics*: Report 1170, NACA, Cleveland, Lewis Flight Propulsion Laboratory.

Matsuishi, M., and Endo, T., (1968), Fatigue of metals subjected to varying stress, *Proceedings of the Japan Society of Mechanical Engineering*, March, Fukuoka, Japan.

Menegotto, M., Pinto, P., (1973), Method of analysis for cyclically loaded R. C. plane frames including changes in geometry and non-elastic behaviour of elements under combined normal force and bending, Preliminary Report, *Proceedings of the IABSE Symposium- International Association for Bridge and Structural Engineering*, Vol. 13, Lisbon, Portugal.

Miner, M. A., (1945), Cumulative damage in fatigue, Trans. ASME, *Journal of Applied Mechanics*, Vol. 67, pp. A159-A164.

Nethercot, D. A., Li, T. Q., and Ahmed, B., (1998) Unified classification system for beam-to-column connections. *Journal of Constructional Steel Research*, Vol.45.

Palmgren, A., (1924), Durability of ball bearing, *ZVDI*, Vol. 68, n° 14, pp. 339-341 (in German)

Peeker, E., (1997), *Extended numerical modelling of fatigue behaviour*, Ph.D. Thesis, Department de Génie Civil, Ecole Polytecnique Federale de Lausanne, Switzerland.

Proença, J. M., (1996), *Seismic behaviour of precast structures*, Ph.D. Thesis (in Portuguese), Instituto Superior Técnico, Lisbon, Portugal.

Roeder, C. W., (1998). Design models for moment resisting steel constructions. *Proceedings of the Structural Engineering World Wide*, San Francisco, U.S.A.

Sedlacek, G., Feldmann, M., and Weynand K., (1995), Safety consideration of Annex J of Eurocode 3, *Proceedings of 3rd International Workshop on Connections in Steel Structures*, Trento, Italy.

Sines, G., and Waisman, J. L., (1959), *Metal fatigue*, McGraw-Hill, New York.